21-12-70

"ASPECT" GEOGRAPHIES

A GEOGRAPHY OF MANUFACTURING

"ASPECT" GEOGRAPHIES

A GEOGRAPHY OF MANUFACTURING

H. R. JARRETT
B.A., M.Sc.(Econ.), Ph.D., F.R.G.S.
Formerly
Senior Lecturer in Geography
in the
University of Newcastle, New South Wales

MACDONALD & EVANS LTD
8 John Street, London W.C.1
1969

First published September 1969

©
MACDONALD AND EVANS LTD
1969

S.B.N: 7121 0707 X

This book is copyright and may not be reproduced in whole *or in part* (except for purposes of review) without the express permission of the publishers in writing.

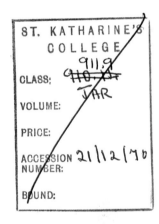

Printed in Great Britain by Richard Clay (*The Chaucer Press*), Ltd.,
Bungay, Suffolk

Introduction to the Series

THE study of modern geography grew out of the medieval cosmography, a random collection of knowledge which included astronomy, astrology, geometry, political history, earthlore, etc. As a result of the scientific discoveries and developments of the seventeenth and eighteenth centuries many of the component parts of the old cosmography hived off and grew into distinctive disciplines in their own right as, for example, physiography, geology, geodesy and anthropology. The residual matter which was left behind formed the geography of the eighteenth and nineteenth centuries, a study which, apart from its mathematical side, was encyclopaedic in character and which was purely factual and descriptive.

Darwinian ideas stimulated a more scientific approach to learning, and geography, along with other subjects, was influenced by the new modes of thought. These had an increasing impact on geography, which during the present century has increasingly sought for causes and effects and has become more analytical. In its modern development geography has had to turn to many of its former offshoots—by now robust disciplines in themselves—and borrow from them: geography does not attempt to usurp their functions but it does use their material to illuminate itself. Largely for this reason geography is a wide-ranging discipline with mathematical, physical, human and historical aspects: this width is at once a source of strength and weakness, but it does make geography a fascinating study and it helps to justify Sir Halford Mackinder's contention that geography is at once an art, a science and a philosophy.

Naturally the modern geographer, with increasing knowledge at his disposal and a more mature outlook, has had to specialise, and these days the academic geographer tends to be, for example, a geomorphologist or climatologist or economic geographer or urban geographer. This is an inevitable development since no one person could possibly master the vast wealth of material or ideas encompassed in modern geography.

This modern specialisation has tended to emphasise the importance of systematic geography at the expense of regional geography, although it should be recognised that each approach to geography is incomplete without the other. The general trend, both in the universities and in the school examinations, is towards systematic studies.

This series has been designed to meet some of the needs of students pursuing systematic studies. The main aim has been to provide introductory texts which can be used by sixth formers and first-year university students. The intention has been to produce readable books which will provide sound introductions to various aspects of geography, books which will introduce the students to new ideas and concepts as well as more detailed factual information. While one must employ precise scientific terms, the writers have eschewed jargon for jargon's sake; moreover, they have aimed at lucid exposition. While, these days, there is no shortage of specialised books on most branches of geographical knowledge, there is, we believe, room for texts of a more introductory nature.

The aim of the series is to include studies of many aspects of geography embracing the geography of agriculture, the geography of manufacturing industry, biogeography, land use and reclamation, food and population, the geography of settlement and historical geography. Other new titles will be added from time to time as seems desirable.

H. ROBINSON
Geographical Editor

Author's Preface

It is hoped that this book will provide some guidance and help to those students who are embarking for the first time on the study of Industrial Geography as a discipline in its own right, and also to those general readers who wish to gain some understanding of a subject which is of the greatest importance in our time. We live in a technological age—that is easily enough said—and it surely behoves us to make some attempt to understand at least some of the forces which have brought that age into being and which help to sustain it—that is less easily accomplished. The present book may make some small contribution towards this end.

The interested reader will have little difficulty in understanding the scope of the term "industry" as it is used in this book. It refers, as the title of the book suggests, to "manufacturing industry," the various processes with the help of which raw materials are shaped and fashioned into acceptable articles of commerce. Agriculture is not included in the survey though admittedly farmers as a class are industrious enough, neither is the so-called "tourist industry."

The book is selective in its approach. The hard facts of publishing economics have made it essential to limit the size of the volume and it is no criticism of the publisher for the author to say that the limitations of space have proved to be among the most serious problems experienced in the writing of this book. Part One is devoted mainly to considerations of location theory, since this provides a basis upon which more detailed later studies can be erected, while an attempt is made to draw together the various points made in a study of the Industrial Region in Chapter VI. Considerations of space have made for sharp pruning even here. For example, the author expected initially to devote a full chapter to a consideration of quantitative location theories, but it became apparent as the writing progressed that this would make too great a demand upon the space available. On the other hand, it seems very desirable that even a beginner in this field of study should be aware of such contributions of workers such as Weber, and brief reference is therefore made to this writer in Chapter V. Other references are given to enable the reader to pursue the subject further if he wishes.

In Part Two studies are made of selected industries, an initial chapter on the small-scale industry forming a kind of antechamber to the main hall of study. The iron and steel industry almost chose itself as a subject of study since it is of fundamental importance in the modern world, while the cotton textile industry, one of almost world-wide importance, is offered next as a contrast. As opposed to these old-established interests, industries of more recent growth are represented by the manufacture of motor vehicles and by oil refining.

Part Three attempts a brief summing up of the present situation and makes mention of certain problems which accompany large-scale industrialisation and which cry out for solution. Thus an attempt is made to round off the work and give it some form of unity despite the gaps in it.

It was originally intended to include a chapter dealing with selected industrial regions. The growth and anatomy of these regions might then have been discussed in some detail. Shortage of space, however, forced the writer to abandon this idea, but the reader will find reference in some detail to certain industrial regions in the pages which follow; examples include New England (pp. 130-134), South-Eastern Australia (pp. 127-129 and 193-204) and South Wales (pp. 288-294).

In any work of this nature an author must rely extensively upon material already published, and the present work is no exception. Lists of material consulted by the author and which might with profit be studied by the reader where possible are given at the end of each chapter, but it might be fitting to record that the works cited by Boesch, Bengston and van Royen, Fryer, Estall and Buchanan, Jones and Darkenwald and E. W. Miller have been found particularly helpful. My thanks are due to these authors, and also to others who have helped me in a more personal way, especially to my wife, whose patience has made possible the labour necessary in the compilation of this work. Grateful mention should also be made of three of my former colleagues: Professor E. O. Hall, Professor of Metallurgy in the University of Newcastle, New South Wales, who offered very generous comments on an early draft of Chapter VIII; Professor David Miller from the University of Milwaukee-Wisconsin and Visiting Professor in Geography in the University of Newcastle, 1966, who gave very kind assistance with Chapter X; and Professor K. W. Robinson, Associate Professor of Geography in the University of Newcastle, who made very helpful suggestions with regard to Chapter XIII. Mr K. Bailey of the Department of Geography, University of Newcastle, gave much help with the drawing of the maps and diagrams, while Mrs D. Page of the same Department typed much of the manuscript.

I should also like to record grateful acknowledgement of the most helpful role played by the second-year classes in Geography at the University of Newcastle (the University College of Newcastle until 1964) between 1963 and 1966. These classes unwittingly acted as guinea-pigs since the present book is based on lectures given to them as part of their second-year course in Geography and upon tutorial discussions and conversations, many of which proved to be most stimulating and helpful.

H. R. J.

June 1969

Contents

Chapter		Page
	INTRODUCTION TO THE SERIES	v
	AUTHOR'S PREFACE	vii
	LIST OF ILLUSTRATIONS	xv

PART ONE: THE LOCATION OF INDUSTRY

I.	THE STAGE IS SET	3
	The Revolution in Power	4
	The Revolution in Transport	5
	The Increasing Use of Iron	8
	The Use of Machinery	9
	The New Technology	11
	Britain and the Industrial Revolution	12
	Some Effects of the Industrial Revolution in the U.S.A.	15
	A General Overview	19
II.	THE RAW MATERIALS	21
	Minerals and Industrial Location	21
	Types of Raw Materials	25
	The Mining Industry	26
	Conclusion	42
III.	FUEL AND POWER	44
	World Energy Consumption	47
	Some Characteristics of Energy	48
	Coal	51
	Electricity	57
	Mineral Oil	70
	Significance of the Newer Fuels	75
IV.	MAKING AND SELLING	77
	Labour	77
	Transport	85
	Markets	88

Chapter		Page
V.	OTHER LOCATIONAL FACTORS	92
	Water	92
	Climate	97
	Site	100
	Capital	101
	Government Activities	102
	The Personal Element	110
	Industrial Inertia	113
	Historical Accident	115
	Theories of Industrial Location	116
VI.	THE INDUSTRIAL REGION	121
	The Development of an Industrial Region	121
	The Growth of the Industrial Region (1)	127
	The Growth of the Industrial Region (2)	137
	Delimiting the Industrial Region	141

PART TWO: SELECTED INDUSTRIES

VII.	THE SMALL-SCALE INDUSTRY	151
	Domestic Industry	152
	Lock Manufacture at Aligarh	153
	Soya Bean Processing in the U.S.A.	155
	Fruit Canning in New South Wales	158
VIII.	THE IRON AND STEEL INDUSTRY	160
	Historical Development of the Industry	160
	The Modern Blast Furnace	166
	Raw Materials	168
	Modern Developments in Iron Production	171
	Developments in Steel Manufacture	173
	Modern Processes of Steel Manufacture	177
	The Integrated Iron and Steel Works	181
	Location of the Industry	182
	Iron and Steel in Communist China	189
	The Australian Iron and Steel Industry	193
IX.	THE COTTON TEXTILE INDUSTRY	206
	The Cotton Industry and Developing Territories	206
	Contrasts with the Iron and Steel Industry	207

	CONTENTS	xiii
Chapter		*Page*
	The Organisation of the Industry	209
	Structure and Location of the Industry	210
	World Production of Cotton Textiles	218
	The Cotton Textile Industry of the U.S.A.	223
X.	THE MANUFACTURE OF MOTOR VEHICLES	231
	Historical Development of the Industry	232
	Development of the Industry in Europe and the U.S.A.	235
	Functional Structure of the Industry	237
	Locational Factors	240
	Final Considerations	244
XI.	THE OIL-REFINING INDUSTRY	246
	Development of the Industry	246
	Production and Refining	253
	Further Considerations	261
	Siting of Oil Refineries	263
	Associated Industries	267
	Refining Techniques	270
	Present and Future Trends	271

PART THREE: THE PRESENT SCENE

XII.	SOME PROBLEMS OF INDUSTRIAL DEVELOPMENT	277
	Problems of a Technical Nature	277
	Living in an Industrial Society	286
	Problems of Development Related to Industrialised Regions	287
	Problems of Development Associated with Individual Industries	294
	Problems Associated with the Industrialisation of Developing Territories	302
	Mexico: A Case Study of a Developing Territory	314
XIII.	POSTSCRIPT	326
	Industry and Agriculture	326
	Problems of Capital Utilisation in Developing Territories	328
	Further Problems of Industrial Development in Developing Territories	332
	The Problem of Maturity	334
	INDEX	341

List of Illustrations

Fig.		Page
1.	Some aspects of the economy of the U.S.A. in the mid nineteenth century	16
2.	Areas of production of mineral oil in North-West Africa	28
3.	Iron ore ranges of Lake Superior	30
4.	Disposition of phosphate beds in Algeria and Tunisia	31
5.	A mineral vein	32
6.	The Frasch process of sulphur extraction	32
7.	Early methods of raising coal to the surface	34
8.	Iron ore in Sierra Leone	36
9.	Outlets from the Central African copper belts	39
10.	World energy output, 1850–1960	45
11.	World energy consumption, 1960	47
12.	World production of mineral oil, 1938 and 1966	50
13.	Composition and efficiency of selected varieties of coal	52
14.	Diagrammatic sections to illustrate varying coal-mining conditions	55
15.	The Kariba power project	58
16.	The Volta River project	62
17.	The Aswan High Dam scheme	63
18.	(a) The Kitimat project [A]	64
	(b) The Kitimat project [B]	65
19.	The Rheinau H.E.P. plant	66
20.	Thermal power stations in the Trent Valley	68
21.	Diagrammatic section showing a possible mode of occurrence of mineral oil	71
22.	Urban and industrial planned development	107
23.	A parkland city—Adelaide	108
24.	An example of isodapanes	116
25.	The locational triangle	118
26.	Examples of linked industries	124
27.	New South Wales: the central coastlands	128
28.	Manufacturing employment in the U.S.A.	135

xvi LIST OF ILLUSTRATIONS

Fig.	Page
29. Industrial regions of Western Europe	144
30. Belgium: industrial development	145
31. Belgium: agriculture	146
32. Belgium: population density	147
33. Lock manufacture at Aligarh	153
34. The processing of the soya bean in the U.S.A.	157
35. The blast furnace	167
36. Diagram of a battery of coke ovens	169
37. The Bessemer converter	175
38. The open-hearth furnace	175
39. The electric furnace	177
40. The basic oxygen converter and the Kaldo furnace	179
41. Flow diagram showing conversion of raw materials into iron and steel products	181
42. Diagram to show the raw materials needed to produce one short ton of ingot steel	183
43. The pattern of the iron and steel industry in Britain in 1966	184
44. The pattern of the iron and steel industry in Australia in 1966	194
45. Newcastle, N.S.W., and the iron and steel industry	195
46. Continuous casting machine	201
47. Uganda. Physical and economic features including cotton ginning	211
48. World production of raw cotton and cotton yarn	218
49. The cotton textile industry of the U.S.A.	226
50. The cotton textile industry of the southern states	228
51. World manufacture of motor vehicles, 1963	232
52. Ratio of people to motor vehicles in selected countries	233
53. The changing location of the oil-refining industry, 1938–64	253
54. World oil refining, 1938	255
55. World oil refining, 1964	255
56. (a) Total oil refinery output in the United Kingdom, 1938–66	257
(b) Sources of United Kingdom imports of petroleum, 1938 and 1966	257
57. Some features of the oil-refining and chemical industries of the United Kingdom, 1968	266
58. The Bristol region: industrial development	268
59. An oil refinery: the main processes and products	271
60. United Kingdom consumption of petroleum products, 1951, 1961 and 1970 (estimated)	272
61. Stream pollution in the Upper Silesian industrial district	285

LIST OF ILLUSTRATIONS

Fig. / *page*

62. Schematic plan of superblock housing layout — 287
63. The Abbey Margam steel works — 293
64. Changes in coal usage in the U.S.A., 1950 and 1960 — 297
65. (*a*) The vicious circle of poverty — 303
    (*b*) The spiral of developing production — 303
66. Mexico: agriculture and chief towns — 317
67. Mexico: minerals and industry — 320

PART ONE

THE LOCATION OF INDUSTRY

Chapter I

The Stage is Set

BECAUSE of the limitations of human nature it is very difficult for most of us to imagine conditions of life other than our own—at any rate in any vivid manner. We tend to assume that our particular conditions of life are "normal" and that any patterns of living other than our own are unusual. It requires for many of us a considerable effort of mind to comprehend that the present-day "Western" manner of life is a thoroughly unusual one, for nothing like it has ever before been known throughout the entire history of our planet.

Never before has there been a time when life has been so pleasant for so many people and when the various parts of the world have been in such close contact with each other as they are today. The horrors of wash-day, for example, are far behind for the modern housewife with her automatic washing-machine and spin-drier, while the husband can swiftly be transported by rail or can drive himself by car easily and comfortably to work. Evenings can be spent in the company of famous and distinguished people with the co-operation of the television network, while the death of a famous statesman in England can be known in the farthest parts of the British Commonwealth only a few minutes after it has actually happened —Australians can even see the funeral of the distinguished personage on their television sets just a few hours after it has taken place. Complex appliances such as Telstar and Syncom provide virtually instantaneous visual and aural links between one continent and another.

These are features of our modern life which we are already beginning to take for granted but which would have been far outside the wildest dreams of fancy in earlier times. And while we have so far mentioned only some of the more outstanding features, we should not forget that many more humdrum aspects of life with which we are so familiar are, in fact, comparatively modern innovations. Regular delivery of mail and newspapers, for example, and the certainty that if we have some reserves of money we can always purchase food, drink and shelter, are features no less peculiar to our modern age. Admittedly a minority of wealthy people have for long been assured of shelter and of supplies of food and drink, but for people at large to be similarly assured is new. In many territories even today there is no such assurance.

It is the purpose of this book to enquire briefly into some of the reasons why these modern amenities of life have become available. The sequence of events which has resulted in present-day patterns of life may be said to have begun in the latter part of the eighteenth century, though it is possible to distinguish contributory causes even before that time. It was, however, the series of developments which are collectively referred to as the "Industrial Revolution" which brought about the increasing mechanisation of our age, and in a sense this revolution is continuing even today; some observers, indeed, have claimed to recognise a second Industrial Revolution as they take note of the vastly accelerated rate of progress today in every field of invention and discovery.

In view of the importance of the Industrial Revolution as the nursery of our present industrial way of life, we ought to say a little more about the series of changes which were then set in motion, though we cannot institute anything like a systematic analytical study. For convenience our analysis will be arranged under five headings:

1. The revolution in power.
2. The revolution in transport.
3. The increasing use of iron.
4. The use of machinery.
5. The new technology.

THE REVOLUTION IN POWER

It is perhaps not unfair to argue that the basis of the Industrial Revolution was a revolution in power availability. One of the chief characteristics of life before modern times was the very limited amount of power available; indeed, almost the only power resources were those provided either by man himself or by animals such as the horse or donkey which could be tamed, though water power could be utilised in limited amounts at suitable points along river courses. Wind power was used at sea in quite early times, but it was infrequently used on land except in the windmill. If comparatively large amounts of power were required they could only be produced by using more men or more animals; for instance, rows of galley slaves could be compelled to provide the force needed to propel vessels across the oceans while four or six horses could be harnessed to a stage-coach.

The late seventeenth and early eighteenth centuries saw the first slow steps taken along the road which was destined to lead to the massive power plants of today, and the stimulus to development arose in the first place from the pressing need to drain water out of mines. Many mines in the latter part of the seventeenth century could only be drained at immense cost by horse-powered pumps; the first practicable steam pump, crude as it was, was devised by Savery in 1698 and was known as "The Miner's Friend," though it was cumbersome, inefficient and not without danger in operation. The next step forward came in 1710 when Newcomen invented his first steam engine (really an atmospheric engine since its operation depended partly upon atmospheric pressure), also a very inefficient piece of mechanism. It consumed about 20 lb of coal per horse-power hour and a serious drawback arose because of the enormous amounts of coal which were needed to keep it going. The important thing about it was that it provided a basis for future inventions though the transition to steam power was not easily made; much experiment was necessary. Wise, for example, mentions an interesting instance of a transitional stage in the progress from water power to steam. The Soho Manufactory established by Matthew Boulton in 1761* made use in its early stages of water power from the small Hockley Brook about two miles north of Birmingham, but this power proved insufficient to meet the demands of the growing factory. In 1767, therefore, a Savery engine was installed, followed in 1774 by a Watt engine; these steam engines did not directly provide power for the machinery, however, but were used to pump water from the stream below the factory back again into the mill pool, so increasing the available head of water.

In 1776 Watt patented several improvements to the old "fire engine" which greatly economised in the use of fuel; only about 5 or 6 lb of coal per horse-power hour were then needed. (By way of comparison we may note that in 1900 the corresponding figure was 5 lb and in 1925 1 lb.) A further development came in 1782 when Watt brought out his rotary (or planetary) movement which made it possible to translate the push-and-pull motion of the piston into a rotating motion; steam engines could then be used for driving machinery, and the future of the new form of power was assured.

THE REVOLUTION IN TRANSPORT

If one leading feature of the Industrial Revolution was the revolution in power resources, a second was the revolution in transport which was

* The Soho Manufactory attracted many engineers of skill, Watt among them. The Foundry for the manufacture of steam engine parts was opened in 1796.

made possible by the steam engine. We have seen that the application of steam to transport was not possible until after 1782 when Watt's rotary movement made it possible to turn shafts with the aid of steam power, and it was not until the early nineteenth century that George Stephenson began his series of inventions which made the locomotive a practical reality. In 1814 Stephenson successfully constructed a locomotive to draw loads of coal at Killingworth Colliery and it is interesting to observe that this locomotive had smooth wheels. This feature dealt the death-blow to the hitherto firmly held belief that a locomotive must have toothed wheels as smooth ones would slip on the rails! This point is mentioned here as illustrating a simple matter which we now take for granted but which was a hotly debated topic for some time.

Despite the success of the Killingworth experiment the new form of locomotion was slow in spreading, and the Stockton and Darlington Railway, famous as the first public railway in the world to use steam engines to carry passengers, was not opened until 1825. Between 1814 and 1825 steam traction was used for goods traffic but horses were used to draw passenger vehicles along the rails! The first railway seriously to challenge already existing forms of transport was the Liverpool and Manchester Railway, which was opened in 1830, and the £500 prize which was offered for the best locomotive was easily won by Stephenson's *Rocket*. The reason for the location of this railway between Liverpool and Manchester is not difficult to discover; it was to speed up delivery of raw cotton, for cotton often took longer on the journey between Liverpool and Manchester than it did between New York and Liverpool. This was because the canals had in those days a virtual monopoly of transport of such bulky merchandise as cotton, and the canal carriers therefore saw little reason to exert themselves to give good service.

The success of the Liverpool and Manchester Railway greatly encouraged the building of other railways, beginning with the Liverpool and Birmingham Railway in 1833. Acts of Parliament were thereafter regularly passed to authorise railway construction despite the strong opposition often encountered. Acts of Parliament were essential because it was necessary to authorise the purchase of land in long narrow strips along which the railways would be built. Without some compulsion landowners could have refused to sell the necessary land and in effect could thus have prohibited the building of the railways. The first attempt to pass through Parliament the bill authorising the construction of the Liverpool and Manchester Railway failed because a duke argued that such a railway would spoil his fox-hunting, while it became a popular idea that the coming of the railway would cause horses to die out; farmers would therefore be ruined as they would be no longer able to sell oats, hay or other fodder crops! Difficulty was often experienced by surveyors who were chased off by hostile landowners (surveys were necessary as details of the proposed route had to be included in the parliamentary bill concerned); on one occasion the implacable landowner was a clergyman and the necessary survey had to be carried out with expedition during the hour of worship which kept the clergyman busily occupied!

But the march of the railway could not be halted when once its efficiency and its superiority over other forms of transport had been demonstrated, and by 1848 no less than 5000 miles of railway had been constructed in Britain. From there it spread to almost every part of the world. Steam power also began to make its appearance on the world's ocean highways, though here again the change from sail was comparatively slow. Although the first vessel successfully powered by steam, the *Charlotte Dundas*, was working on the Forth and Clyde Canal as early as 1802, sail maintained its ascendancy on all major shipping routes until about 1860. It is true that an American ship, the *Savannah*, crossed the Atlantic with the help of steam in 1819 but the steam power was simply used as an auxiliary to sail and it was not until 1838 that the same ocean was crossed by vessels using steam power alone. By 1850 coaling stations were established at strategic points along the great trade routes of the world and the stage was set for the final change-over to the new form of power. An additional impetus came with the opening in 1869 of the Suez Canal on the tremendously important route linking North-West Europe and the Orient because large sailing ships could not successfully navigate the Canal.

As a result of all these developments in the sphere of land and ocean communications it became possible for the first time in the history of the world to organise economic affairs on a world canvas instead of on a merely local or national one. The new railways and steamships were able to transport mails and goods (even bulky and heavy goods) over long distances speedily, safely and regularly. This was something which had never before been possible. It became feasible to open up such areas as the pampa of South America, the interior grasslands of the U.S.A. and the prairies of Canada, and to link such regions with the economic and social life of North-West Europe. The wheat of Canada, the meat of the U.S.A. and Argentina, and the wool of Australia could all be sold in Europe and the manufactured goods of the new industrial nations could be sold almost anywhere in the world. The effect of this upon the newly developing industries was of the greatest importance.

THE INCREASING USE OF IRON

A third notable feature of the Industrial Revolution was the remarkable increase in the use of iron. In early centuries it took a very long time indeed before it was discovered that for the successful smelting of iron it is necessary to include the iron ore in the same container as the charcoal fire, and also to add some limestone which fuses with the slag so that this separates out from the iron (for further information, *see* Chapter VIII). Even when this was known, however, the way was not open to the widespread use of iron. For one thing, fuel shortages became desperate as the great forests of Western Europe began to disappear. Wood was used not only as a fuel but also in house construction, in the manufacture of such items as furniture, vehicles, various household vessels, waterwheels and windmills, and for the construction of the ships upon which the prosperity and the safety of countries like England depended. Even in the reign of Queen Elizabeth I laws were passed restricting the use for fuel of timber which was needed for the construction of naval and merchant vessels.

Another drawback lay in the fact that the iron produced from the smelting of iron ore is not pure iron. Pig iron, as it comes from the smelting process, normally contains between 2 and 5% of impurities, the main one being carbon, and it is the presence of these impurities which makes pig iron brittle and liable to snap under strain. This disability was itself a sharp limitation on the use of iron, for in order to guarantee sufficient strength, comparatively large amounts of the metal had to be used in manufacture, so that iron goods were usually cumbersome and very heavy.

Possible improvements were worked out in quite early times, and certain steels used for armour and swords in the Middle Ages (for example, Toledo and Damascus steels) became famous, but any secrets of manufacture were closely guarded (though *see* p. 162 below). It has, however, been general knowledge for a very long time that *wrought iron* can be produced from pig iron by stirring the molten metal for some time until the impurities are burned out. Unfortunately, while wrought iron is very malleable and does not snap under strain, it has not the strength of pig iron, and it tends to bend easily. Its use is therefore limited, though it is often used for "decorative" ironwork. The most useful form of iron is, however, the modified form known as *steel* which normally contains up to 1·4% carbon; further details are given in Chapter VIII. For the present we shall simply observe that it took many years of experiment before steel manufacture could be successfully undertaken following inventions such as those of Henry Bessemer (1855–56), Siemens-Martin (1862) and Gilchrist Thomas (1878). Thereafter, steel rapidly took the place of iron because of its more resistant and therefore better-wearing character.

Another important point was that as a result of the various improvements in technique, some of which we have touched upon, the costs of producing iron and steel were greatly reduced. This came about partly because of economies in fuel and in raw materials; for instance, the invention of Neilson's hot blast (*see* p. 165 below) in 1829 reduced the amount of coal required for smelting by more than one half so that, instead of about nine tons of coal being needed to produce one ton of pig iron, only about four tons were needed. We may note for comparison that by 1900 the corresponding figure was two tons, while in 1938 it was 1·8 tons in the United Kingdom and 1·3 tons in the United States (for later figures *see* p. 171 below). Similarly, the amount of coal required to produce one ton of steel fell from seven tons in 1850 to two tons in 1930.

Even this very broad outline may help to make clear why it was that the Industrial Revolution was largely a revolution in the use of iron and steel. Jones has remarked that the Victorians were lavish in their use of iron; structures such as the Crystal Palace and, in France, the Eiffel Tower demonstrate this, while the interiors of many Victorian public buildings were often embellished with wrought-iron decoration. More important, however, was the fact that steel could safely be used in the manufacture of machinery of all kinds.

THE USE OF MACHINERY

A fourth notable characteristic of the Industrial Revolution was the increasing use of machinery, a feature which was first prominent in the cotton textile industry—indeed, some accounts of the early developments of the revolution read like a simple account of the improvements brought about in this single industry. British exports of cotton goods rose dramatically during the eighteenth century, and, while the total value of British cotton exports was only £23,000 in 1701, it had risen to £5½ million by 1800.

It will be clear from these dates that the early inventions in the cotton textile industry came earlier than those in other industries. The first invention of the long series was that of the flying shuttle (1733), but the inventor, Kay, met with so much opposition that he had to take refuge in France. Thus began the pathetic history of militant opposition to the new inventions, an opposition born of fear of unemployment (and at times the fear was only too well justified) but which was doomed in the long run to failure. In 1767 came what is perhaps the best known of all the inventions, the spinning jenny of Hargreaves, named after the inventor's wife. The original jenny was a hand-machine and could be used in the workers' homes but it was quickly followed by the water-frame invented by Arkwright in 1768. The adoption of the water-frame which proceeded rapidly after 1785 was of particular importance as it inaugurated what was in fact a social revolution. This arose because the large-scale machine could not be used in the home; it had to be erected in a convenient place for water power and the workers had, as it were, to come to the machine. This pattern of working brought about the factory system on a scale previously unknown, and the hold of the system strengthened as it was realised that machines never tired and that they could go on working day and night. This naturally gave rise to the shift system of working and new urban growths began to develop around the cotton mills. Such a machine as the water-frame was, of course, out of reach of domestic workers, who could not afford the capital outlay required to set up one of the new factories.

By 1811 Britain had over 300,000 Arkwright spindles in operation but further improvements were already in being. In 1775 Crompton invented his mule, a combination of the jenny and the water-frame which could spin a yarn of a fineness hitherto impossible. The original mule was a hand-machine which could be worked in the home, but it was adapted for water power by 1792. Many ambitious men began operations with a single mule which they worked themselves, and as their businesses increased they added further machines and opened up factories so that they finally became wealthy. The mule soon superseded the Arkwright frame for the production of fine-quality cotton yarns.

It is perhaps easy to see why so much stress is sometimes laid on the textile industry, for it was in this industry that so much of the pattern to be followed by the new industrialism was worked out. The pattern included the strengthening of the capitalist as the means of production became more expensive and out of reach of the home worker; the concurrent development of the factory system as the new large-scale machines were operated in special buildings;* and the innovation of such working patterns as the shift system which in its turn greatly increased the numbers of people dependent for their livelihood on any particular mill. These developments encouraged the establishment of urban growths near the factories and helped to give birth to the industrial town.

* The use of the term "special buildings" is not meant to imply that in every case new buildings were specially built as factories. Many of the original factories were simply old barns and other buildings roughly adapted for the purpose.

It has for long been a matter for debate why there came this burst of activity in the cotton textile industry during the eighteenth century, and this is a question to which perhaps no final answer can be given. Certain possibilities may, however, be mentioned. In the first place, supplies of the only possible competitor, wool, were strictly limited. The extensive sheep farms of Australia were not opened up in the middle of the eighteenth century and supplies of imported wool were not great, while the home wool clip could hardly satisfy the increasing demands made upon it by a steadily increasing population (between 1750 and 1800 the population of England increased from about $6\frac{1}{2}$ million to 9 million inhabitants). Under these circumstances it was hardly practicable to envisage any great expansion in the woollen industry of the time, and some alternative form of textile was desperately needed.

Another factor which has been pointed out by Knowles is that the English people had during the eighteenth century become accustomed to cotton fabrics from India and the "taste" for this new cloth rapidly grew. In the middle of the century, however, the old-established Mogul Empire of India went to pieces and war was raging between Britain and France in the subcontinent. Under these circumstances merchants found ever-increasing difficulty in purchasing cotton fabrics from India, and this at a time when there was not only an expanding home market to satisfy but also a developing overseas one, particularly in West Africa. This is exactly the sort of situation which might be expected to lead to a vigorous search for other sources of supply, and it is surely no accident that during this period English inventors were busily occupied in the cotton textile field. It should be recognised, however, that this "textile revolution" came about in the face of great opposition from the woollen interests; wool, after all, had been the traditional cloth of England from very early times and the early trading strength of England had been founded upon it. Even today the Speaker of the House of Commons sits upon the "Woolsack."

THE NEW TECHNOLOGY

Despite many advantages which seem obvious enough to us today, the new power machinery was slow in working its way into the industrial world, and perhaps the main reason for this was the sheer difficulty of manufacturing moving parts with anything approaching precision. It is significant, for example, that after 1776, when Watt's engine superseded Newcomen's to pump water out of the Cornish tin mines, Watt was forced to live for a time in Cornwall, an area which he disliked intensely, simply because there was no one but he who could rectify the engine troubles which so frequently occurred. Even after Watt moved to Soho and was able to command the assistance of Boulton's best workmen, difficulties of manufacture were still legion. Sometimes cylinders were found after casting to be over one-eighth of an inch more in diameter at one end than at the other; yet this was after the finest workmanship then available! Expedients such as the use of chewed paper and greased hat packed around the piston were tried, but a great waste of efficiency was inevitable under these circumstances. It is perhaps not surprising that even in 1800 there were only 289 steam engines in the whole of England (84 of these were in cotton mills), while as late as 1835 there were just 2000 as compared with 1300 water-wheels. By the middle of the century, however, the "machine-tool" inventions of Maudsley, Bramah and Nasmyth had revolutionised the situation and thereafter steam power rapidly gained the ascendancy over water power.

BRITAIN AND THE INDUSTRIAL REVOLUTION

The question has often been debated why the Industrial Revolution first developed in Britain, an archipelago of only modest size and population; it has been estimated that the total population near the end of the eighteenth century was only about 9 million people, while that of France at the same time was 26 million. One suggestion made by Knowles is that the nation with the smaller population needed more machinery to produce goods for its export trade, but this line of thought does not explain why the export trade developed in the first place. It is more important to realise that France was showing considerable interest in the establishing of industry before 1789 but that the French Revolution was a tremendous economic disaster. Knowles talks of the "utter destruction of industrial and commercial life [in France] for ten years after the French Revolution" and argues that it took France until 1830 to get back again to the same pitch of commercial prosperity that she had enjoyed before 1789. The Revolution, then, probably removed what might have been one of Britain's chief competitors from the industrial and commercial race for many years.

The fundamental cause of the lead given by Britain to the world almost certainly lies in the robust and enterprising character of the people of the time. It was this character which had driven the British people to secure a greater degree of personal freedom by the eighteenth century than was the case in any other European country; it is difficult to imagine the French peasants, tied as so many of them were to the estates of often absentee landlords, migrating to find work in the factories as did their English, Welsh, Scottish and Irish counterparts. This same character had also sent British sailors and merchants overseas to fight and trade, and so the foundations were laid of family fortunes which were to help provide the necessary capital needed to nourish the newly developing industries. The case of William Miles shows us how a British merchant could prosper. William Miles walked into Bristol in the middle of the eighteenth century with $1\frac{1}{2}d$. in his pocket and worked as an unskilled labourer under conditions of considerable privation until he had saved £15. He then signed on as a ship's carpenter for a voyage to Jamaica, where he bought some casks of sugar which he took back with him to Bristol before selling them at a great profit. He then entered the "triangular trade" (see below), shipping English manufactured goods to West Africa and there purchasing slaves which in their turn were sold in Jamaica, and the money used to purchase sugar which was shipped home and sold at very considerable profit. Soon he was able to branch out with his own ships, and the affairs of the family so well prospered thereafter that when his son died in 1848 the Miles estate was valued at over half a million pounds sterling. And this example could be matched many times over.

No one can deny that such trading had its unsavoury aspects, but at the same time we must recognise that such determination as that envinced by Miles played a major part in making possible later developments in trade and industry. It is important to realise how much developments in one sphere of interest reacted upon associated fields. On the financial side, for example, profits reaped in trade were available for later investment in industry, while on the technological side developments in one industry stimulated developments in others. The invention of Watt's rotary movement, for instance, had an important effect on the engineering and iron-producing industries. Until Watt's time it was common for machinery to be made of wood, for the use of iron on a large scale was not possible in earlier years because of technical difficulties of manufacture. But Watt's invention made it imperative to discover ways of making iron available in larger quantities, and of improving its quality, because wooden machines simply were not strong enough to carry the strains and stresses imposed in a steam engine with rapidly moving parts. In a similar way the invention of textile machinery further encouraged the use of iron in the new machines, and all the time a search was carried on for new methods of production which would make possible the manufacture of precision parts (p. 12 above).

The whole Industrial Revolution was, in fact, cumulative, an advance in one branch of industry encouraging advances elsewhere. Basil Davidson had put forward the idea that it was the development of the slave trade which did more than anything else to stimulate expansion in industry in the eighteenth and early nineteenth centuries. This idea is not altogether new and we have already seen that the initial establishment of industry owed much to the availability of the capital accumulated by slave-traders, but Davidson also points out that the "triangular circuit" of trade created an unprecedented demand for cheap manufactured goods. This triangular circuit comprised a threefold trading system, the first stage of which consisted in the shipping to West Africa of goods that could be bartered with the coastal chiefs for slaves. The slaves were then transported along the second "leg" of the circuit, along the notorious "middle passage" across the Atlantic to be sold in the slave markets of the New World for work on the plantations.

The final stage in this unhappy trinity was the purchase of plantation products such as sugar, rum, molasses, tobacco, rice, indigo and cotton for transport back to Britain, so that as far as this country was concerned the overall result of the trade was the despatching of cheap manufactured goods, and the receiving in exchange of the plantation products of the New World which could be sold at great profit. It is clear that such a trade, with its insatiable appetite for such goods as cotton cloth, cheap jewellery, firearms, ammunition and spirits, must inevitably have stimulated industrial enterprise. Even if we do not agree entirely with Davidson, who seems to see in this trade the paramount stimulus to early industrial development, we can certainly agree that it must have exercised a powerful influence.

As this chapter is not meant to be a study in economic history but simply an introduction to the study of some aspects of the geography of manufacturing, it is not desirable here to go farther into the story of those early times. It is hoped, however, that the reader will now have some idea of how the slowly moving and predominantly rural world of the Middle Ages developed into the urbanised world of our own rapidly moving times. Even as late as 1815 40% of workers in Britain were engaged in agriculture; by 1900 the corresponding figure had dropped to 10%, while in 1960 it stood at 5%. In concluding this section we might note that progress in the industrial world has by no means halted; if anything it has been accelerated remarkably in recent years as is shown by the recent developments in the oil-refining and the associated petro-chemical industries, to take just two examples (*see* Chapter XI).

We might usefully bear in mind that such continuing progress is essential if an industrial society is to maintain its standards of living. There was a time, for example, when Britain could be assured of a comparatively affluent position in the world simply because she was the first in the industrial field, but today that state of affairs no longer exists. Other countries have followed her example of industrialisation and they have been able to benefit from her early experiments and mistakes. If a country like Britain is to maintain her standard of living and her world position she had to keep ahead of possible competitors in at least some fields of industry; it is not enough simply to maintain existing industries even at peak efficiency, since other countries are constantly catching up in productive capacity and imposing legal restrictions to protect their own growing industries. It is not surprising, therefore, that so much emphasis is today placed on research in industry, and the record shows that despite constant criticism British manufacturers have more than held their own in recent decades.* The position is broadly similar in New England, which has already lost her originally pre-eminent position in the cotton textile industry of the U.S.A. and which is developing instead such modern interests as the manufacture of electronic equipment. If New England did not do this she would fall rapidly behind as a major industrial region.

SOME EFFECTS OF THE INDUSTRIAL REVOLUTION IN THE U.S.A.

We have already emphasised the fact that the Industrial Revolution for the first time in history made it possible for economic affairs to be organised on a wider than local basis, and it might be appropriate at this stage to say a little more about this. The way in which new territories were opened up, for instance, can be illustrated from the U.S.A., where the interior regions were being settled during the nineteenth century. By about the middle of the century the humid forested eastern parts of the country had been settled and a domestic triangular trading system had developed (Fig. 1). This system included the despatch of raw cotton from the South to the mills of New England (the leading industrial region in the country at that time) while manufactured goods moved in the reverse direction. The main artery of movement was the sea. The two other legs of the triangle linked New England and the South respectively

* At the time of writing (1967) Britain has the largest electronic-computer industry outside the U.S.A.; she has pioneered the development of wireless, radar, hovercraft, the Mulberry harbour, the Bailey bridge and other devices; she is among the world's top producers of such items as ships, aircraft, motor vehicles (including tractors), bicycles, submarine cables, chemicals and machine tools. She produces considerably more nuclear power than the rest of the world put together, while her crop yields per acre are among the highest in the world (see also footnote, p. 334 below).

B

with the Middle West, a land of agricultural surplus, and part of this surplus found its way to New England and part to the South. In return, manufactured goods passed westwards from New England while imported goods were shipped northwards from New Orleans up the Mississippi.

The next phase in the opening up of the U.S.A. involved movement westwards from the humid and naturally forested east to the drier grassland areas farther west. Farmers began cultivation of the eastern fringes of the prairies in the 1840s though rapid development of these grasslands

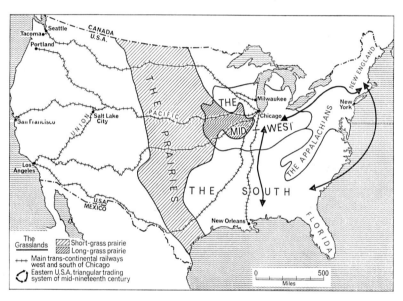

FIG. 1.—Some aspects of the economy of the U.S.A. in the mid nineteenth century.

did not take place until the late 1850s and subsequent years. This phase brought the settlers face to face with a radical change of environment. Pioneers of European origin were familiar with a forest environment; they knew that the forest yields building materials and provides a certain amount of shelter. They knew that when trees in the forest are cleared the underlying soils are comparatively easy to plough and are normally fertile and productive. But the open grasslands are quite different. There is little (if any) natural shelter and no wood for constructional purposes, while the thick sod and tall grasses of the "tall-grass (or long-grass) prairie" which lies immediately adjacent to the forest areas (Fig. 1) proved too difficult for the early farmers to cultivate; the wooden plough could not manipulate these soils and even the cast-iron plough found them very difficult to turn over. When we add to these difficulties the traditional belief that grassland soils are inherently less fertile than forest ones, we have little difficulty in understanding why it was that settlement for so long halted along the forest boundary. Philbrick well sums up the situation when he points out that the potentially rich grasslands of the interior of the U.S.A. were for long considered to constitute a great American desert. It is interesting to notice that settlement leapfrogged to some extent over the grasslands to the west (Pacific) coast, which was opened up before the prairies. Neither must we overlook what was in those days the oppressive isolation of the prairies, a verdant desolation given over to the bison and to Indian tribes.

The purpose of this analysis is to demonstrate how this unpromising outlook was altered, thanks to a changing situation brought about by industrial developments elsewhere. The two inventions which transformed the situation were the steel plough of John Deere and the railway. The steel plough was able to break up the matted prairie sod and therefore make cultivation possible, while railways, subsidised by extensive grants of free land, were built through the prairies during the middle part of last century; the first transcontinental railway, the Union Pacific, was actually completed in 1869 though it had reached the prairies considerably before this. After 1850 the isolation of the prairies was finally broken and settlement proceeded rapidly.

But this settlement was not something of local significance only. We have so far pointed to a fact which is fairly well recognised, that it was the products of the new industries (especially the plough and the railways) which enabled the prairies to develop; what is perhaps less well recognised is the extent to which the opening up of the prairies in its turn stimulated development and expansion of industry. Warren has pointed out in an analogous case the tremendous importance to the British iron industry of the railway boom in Britain in the 1840s, when a large proportion of the output of wrought-iron goods from British ironworks comprised iron rails for the new railways; similarly, a railway boom was under way in America by 1869. This boom stimulated the development of the iron industry in the U.S.A., and among the earliest products of the Chicago ironworks were iron rails. 28,732 miles of track were laid in the five years following 1869 and American ironworks were unable to meet the demand for rails; this demand therefore stimulated European industry as well, particularly the ironworks of Sheffield and Essen (Krupps). Sayers has pointed out that the tremendous burst of economic activity which took place in the U.S.A. after the Civil War "provided, by its appetite for British exports, a major thrust in the British boom." Railway construction was an important part of this burst of activity.

The first steel rails ever to be manufactured were rolled at Sheffield in 1861, and this was a most important event because the wrought-iron rails had shown themselves to be poor in quality; there were constant complaints of fracture and lamination and the rails had to be renewed frequently. Often they had to be replaced every four months though traffic was not at that time very heavy. The new steel rails proved far superior and this increased the transport potential of the railways since longer and heavier trains could be run over them.* This in turn stimulated production on the prairies because farm produce could be transported and sold more cheaply in the eastern markets, and this encouraged further settlement on the grasslands and brought about an increasing demand for industrial products such as ploughs and household goods. So industry went ahead in towns such as Chicago and Milwaukee which acted as outfitting centres for settlers moving into the prairies or for those already established there (note the particular importance of Chicago as a route centre from Fig. 1). It was undoubtedly the demands of the prairies for general household goods and for farming and transport equipment which proved to be the vital factor in the further development of industry, especially the iron and steel industry, in Chicago, Gary and Milwaukee.

This example of the development of the American prairies is a good one, for it serves vividly to remind us of the strong inter-relationships between the agricultural and the industrial revolutions. There is no doubt that the development of industry would have been very much slower than it was had concurrent developments in agriculture not been taking place. Today, of course, agriculture still needs the products of industry in the form of tractors, combine harvesters and other mechanical aids to cultivation which we are now apt to take for granted but which were hardly imagined a few decades ago; we should always bear in mind that, while for convenience it is necessary to carry on any study of geography in compartments, it is a mistake to suppose that the divisions between the various compartments (for instance, between agricultural and industrial geography) are clear-cut.

* Production of iron rails, still a flourishing industry in the U.S.A. in 1870, had virtually ceased by 1890. A comparable change in technology affected other industries, for example ship construction, for while the first steel ship was built in England in 1865, by 1888 more than 90% of all new ships being launched were made of steel.

A GENERAL OVERVIEW

The development of industry is apt to be a lengthy and complex affair; circumstances must be favourable, or, as the geographer might say, certain contributory factors must be present. Our study so far brings out the importance of such factors as the following:*

Raw materials. Nothing can be made unless the essential raw materials are available. Examples mentioned above include iron ore and raw cotton.

Fuel and power. The Industrial Revolution was very largely a revolution in power availability; this, in the early days, meant availability of coal supplies, but today electricity and atomic power are making substantial contributions.

The market. The sole purpose of industry is to make goods to sell, and if goods cannot be sold there is no point in manufacturing them. The early steam engines could be sold as they were needed to pump water out of mines, while the overseas market greatly helped the early development of the cotton textile industry.

Capital. Supplies of merchant capital helped to make possible early industrial enterprises, and capital is just as important today as it ever was in making possible industrial expansion. One of the great problems of the "developing" territories is that they lack domestic capital on the scale necessary to establish considerable industrial growth.

Labour. Supplies of unskilled labour are not in the modern world hard to come by except in areas with small populations, but modern industry needs *skilled* labour. Watt's early attempts at engine construction were constantly frustrated because he could not find workmen with sufficient skill to manufacture the precision parts he needed.

Location. It is very rare for a successful industrial undertaking to be located by chance; the choice of location is normally vital in any undertaking which needs to assemble its raw materials from widely spaced sources and to distribute its finished products equally widely. In many instances there are local advantages which have made all the difference in ensuring the success of an industry, while in other cases a poorly located industry is doomed to a low-level performance because of an absence of local advantages. Chicago was able to develop its iron and steel industries

* The reader will notice that the unsatisfactory term "geographical factor" is avoided. Simons ("What is a Geographical Factor?", *Geography*, July 1966) has drawn attention to the unfortunate nature of this term which is often narrowed down to include only factors which *in themselves* are of a geographical character—mainly relief, climate, soils and mineral deposits. We are here concerned with factors *which have geographical repercussions* and we must therefore cast our net of enquiry more widely than this.

largely because of its favourable location as a kind of gateway to the prairies.

The foregoing list of factors which affect the successful development of industry is not an exhaustive one, and in addition such features as the following might be mentioned:

Water supplies. Modern industry requires tremendous amounts of water, and access to large water supplies is not infrequently essential.

Climate. The importance of climate cannot entirely be overlooked.

Management. The success or failure of an enterprise may well be determined by the quality of its management.

Government activity. The role played by governments in encouraging and locating industrial undertakings is of increasing importance. This is perhaps inevitable as we move away from the lavish Victorian era, when there was not the need to conserve and allocate resources that there is today.

Miscellaneous. Some factors are difficult to classify as they may depend partly upon chance or upon the influence of one particular person.

The remainder of the first part of this book will be taken up by an examination of the foregoing factors and by a study of examples to show how they have influenced actual industrial developments in selected instances, while in the second part we shall examine selected industries as case studies. We shall conclude the survey with an examination of some of the problems which have come to the fore as a result of the industrial development of our time and we shall see that some of these problems are pressing indeed and are of much more than local significance.

SUGGESTIONS FOR FURTHER READING

Davidson, B., *Black Mother*, London, 1961.
Jones, E., *Human Geography*, London, 1964.
Knowles, L. C. A., *Industrial and Commercial Revolutions in Great Britain during the Nineteenth Century*, London, 1944.
Philbrick, A. K., *This Human World*, New York, 1963.
Sayers, R. S., *The Vicissitudes of an Export Economy: Britain since 1880*, Sydney, 1965.
Trevelyan, G. M., *Illustrated English Social History*, Pelican Books 1964, vols. 3 and 4.
Warren, K., "The Sheffield Rail Trade, 1861–1930," *Trans. Inst. Brit. Geog.*, June 1964.
Wise, M. J., "Some Factors Influencing the Growth of Birmingham," *Geography*, 1948.

Chapter II

The Raw Materials

THE importance of raw materials in manufacturing industry is so fundamental that it needs no emphasising. Indeed, the location of industrial enterprises is sometimes determined simply by the location of the raw materials concerned. The extractive group of industries based upon exploitation of mineral ores is an obvious case in point.

MINERALS AND INDUSTRIAL LOCATION

At the same time it must be obvious even to the most superficial observer that there is not always a clear-cut correlation between the location of raw materials and the location of industry—indeed, this is very commonly the case. In fact, such a correlation would often be impossible, even if it were desirable, because most industries use more than one raw material, and it is unlikely in the extreme that all the raw materials required by a particular industry could possibly be found in close proximity. Modern industry is so very complex that a wide selection of raw materials is often necessary for quite minor industrial processes, and we should bear in mind too that the raw materials of industry are not necessarily primary products; the raw material of one industry may well be the finished product of another. An obvious case in point is the pig iron produced by the smelting industry which is the raw material of the steel-making industry, while in a different field the cloth produced by spinners and weavers is the raw material of the finishing and clothing industries.

Apart from certain rather special cases, therefore, we cannot expect to find simple correlations between the location of raw materials and the location of industry. Many writers have in fact pointed out that manufacturing industry has in many cases not been based on local resources at all; examples of this include the engineering industries of towns such as Ipswich and Yeovil, the manufacture of jute goods at Dundee and the production of coal gas, which used to be found in almost every fair-sized town in Britain.*

To say that the situation is complex, however, is not the same as saying that no locational pattern at all can be discerned in the industrial field with reference to raw materials. Smith (1952), for instance, has pointed out that industries which owe their location primarily to the proximity of raw materials exhibit certain common characteristics. The weight of material per worker is usually high while the value both of the raw material and of the finished product per ton is normally low. Labour costs as a percentage of total costs are normally low in these industries. Examples which can be chosen to illustrate these points include the manufacture of cement and bricks, the concentration of low grade iron ore, and the extraction of foodstuffs from such agricultural products as sugar beet, cane sugar, oil seeds and corn (flour milling).

Here, then, we see the beginnings of a pattern of location, and we find that we are able to pursue the matter further if we take into account certain circumstances which are in some cases associated with the raw materials themselves. We might say, in fact, that the power of raw materials to attract industries varies in large measure according to the following factors.

Loss of Weight

The attractive power of a raw material is normally great if the material substantially loses in weight or bulk during the process of manufacture. This has been recognised since 1909, when Alfred Weber published his theory of the location of industry, a theory which has more recently been criticised as being incomplete (see, for instance, W. Smith, in his paper "The Location of Industry," *Trans. Inst. Brit. Geog.*, 1955, p. 2). Weber's weight-losing theory, however, has attracted much attention and is generally held to contain much truth, though Weber confused his case by including fuel as raw material.

Numerous examples can be given to show that the weight-losing theory possesses a considerable amount of validity. The crude sugar extracted

* Since the gas-producing industry was nationalised in 1949 great changes have taken place in the locational pattern of production. While 1050 gasworks were taken over in 1949 there were in 1963 only 307 works in operation, despite a 26% increase in productive capacity. Integration has been a deliberate policy of the Gas Council and it has been made possible by the installation of gas grids, by improved techniques such as the adoption of the German Lurgi process for the complete gasification of coal, by increasing use of "tail gases" which are produced in oil refining, and by the import of methane from the natural gas field of Hassi R'Mell in Algeria (Fig. 2) via Arzew and Canvey Island. For more details *see* Manners and Simpson.

from sugar beet, for example, weighs only about one-eighth as much as the raw materials involved (sugar beet, coal, lime), while the weight of butter, cheese, or other manufactured milk products is only about one-sixth of the weight of the materials which are used in their production in the milk factory. Pig iron produced in the blast furnace weighs about one-third of the total weight of the raw materials used in manufacture (iron ore, limestone, coal and scrap) if we exclude the enormous amounts of air and water which enter into production. These are examples of industries which tend to be attracted to their raw materials, though the tendency in the case of pig-iron production is not now as strong as it formerly was.

Perishability

The attractive power of a raw material is also normally great if the raw material concerned is perishable. We may note in this connection the way in which the canning of fruits and vegetables is normally carried on near to the areas of production (*see*, however, p. 158 below), while the processing of milk can also be included in this category as well as in loss of weight above. It is, of course, to be expected that many examples of industrial location can be classified under more than one heading.

Value per Weight Unit

The value per unit of weight of the raw material is of marked importance because a high-value material such as cotton or wool can bear heavier costs of transport than can a low-value one. If unit values of the raw material are high the transport costs involved* form a lower proportion of total costs and carry a lower proportional significance. Thus we find that industries using such materials as cotton or wool are not infrequently located at great distances from the sources of their raw materials; the cotton industries of Lancashire (and in earlier years of New England) and the woollen industries of Yorkshire and New England come readily to mind.

On the other hand, low-unit-value minerals such as copper ore and limonite are normally processed near the point of production. Much of the copper ore which is mined yields only between 1 and 3% by weight of pure copper, and it would not be practicable to transport an ore containing such a low payload over any appreciable distance. Limonite is a fairly low-grade form of iron ore which contains only about 30% of metallic iron, and it is noticeable that where limonite is mined (or quarried), as in the East Midlands of England or in Lorraine, the blast furnaces are located near to the source of supply.

* We should more properly consider the total transfer costs. *See* p. 85 below.

An interesting example which clearly shows the working of the principle under discussion is that connected with the processing of bauxite, the ore of aluminium, which occurs in nature as a clay-like mineral low in silica and rich in alumina. The bauxite contains the aluminium in the form of aluminium hydroxide, $Al_2O_3.2H_2O$, and it seems usually to have resulted from the weathering of igneous rock under tropical conditions. High-grade ores contain between 50 and 60% of aluminium oxide, while low-grade ores may show only 30%; there is clearly a large amount of waste involved and the unit value of the raw bauxite is low.

It is interesting to observe that under these circumstances it has become normal to process the bauxite in a preliminary way near the source of production. As much as possible of the valueless content of the ore is removed by washing and screening, and after this the residue is crushed, washed again and dried in rotary kilns. It is normally found that 100 tons of bauxite will yield 40 tons of the processed material which is known as alumina and this alumina is then transported to aluminium factories which refine the processed ore and extract pure aluminium. The alumina, of course, has a notably higher unit value than the bauxite. We may incidentally observe that it would be equally appropriate to use this example in the loss-of-weight context.

Possibility of Using Substitutes

The power of attraction of any raw material is greatly affected by the possibility or otherwise of using substitutes. An example which is often quoted in this context is the possibility of substituting scrap metal for pig iron in the open-hearth furnace (details regarding the manufacture of steel are given below in Chapter VIII), for enough scrap can today be used in the steel-making process to make the availability of scrap a very real consideration in the location of steelworks.

Numbers of Raw Materials Involved

Another point which is cognate to the previous one concerns the number of raw materials used in any given industrial process. It follows on purely mathematical grounds that, as the number increases, the influence of each separate one must diminish, though a material which is a big weight-loser may exercise a disproportionately large effect. An example of this general principle is seen in modern light industry, which includes the radio and electrical industries, for in this type of industry a very wide range of raw materials is needed. The result is that such industries are commonly situated with no reference to raw materials at all; these are the industries which are sometimes referred to as "footloose" because a very wide range of locations is possible within any fairly densely populated region.

Final Conclusions

The main point which seems to emerge from our study so far is that it is impossible to generalise about industrial location with reference to sources of raw materials. The matter is complex, as is shown by the fact that a single case may be held to illustrate more than a single theoretical point (*see* the cases of milk processing and bauxite refining referred to above). At the same time there are valid generalisations to be made, though it is important not to press these generalisations too far.

Smith (1955) had examined in some detail the whole question of the relationship between the location of raw materials and the location of industry and he comes to the conclusion that it is the weight of materials per operative that is the determining factor, though even here the relationship is not entirely clear-cut. But it may be laid down as a general principle that a large loss of weight linked with a large amount of material per operative ties an industry to its raw materials; conversely, a large gain in weight linked with a small amount of material per operative during processing frees an industry from them. Between these extremes lie various combinations of possibilities.

It is also true that changing circumstances may substantially modify an existing pattern. This may be illustrated by the case in which waste material produced during the process of manufacture acquires value in its own right when it may exert a pull sufficient to draw manufacture away from its raw material. When, for example, industrial coke was manufactured in the old beehive ovens the gas wastefully was allowed to escape, and under these weight-losing circumstances a location of the coke-manufacturing industry near the source of supply of the coal was the normal one. Today, however, the gas itself has considerable value and it is becoming much more usual for coking to take place at iron and steel plants where the gas can be used as a fuel. On the other hand, the value of the basic slag produced in the iron and steel industry is not sufficiently great to draw the industry near to agricultural areas where the slag can be used as a fertiliser.

TYPES OF RAW MATERIALS

We have so far assumed the availability of our raw materials and it might be useful at this stage if we turn our attention to some of the supply questions involved. In a broad sense it is possible to classify primary raw materials under two headings:

1. Those produced as a result of agricultural activities.
2. Minerals.

In order to complete the general picture we should add:

3. Raw materials which are themselves the product of industrial processes.

Of these types of raw materials we shall say little about numbers (1) and (3) above. A consideration of agricultural activities is outside the scope of this review and the reader is referred to other books for such a study, while an examination of the production of raw materials which are themselves products of industry is necessarily a study of industry itself, a study with which this whole book is concerned. We now turn, therefore, to a further consideration of the production of minerals, a class of primary raw materials upon which a very large segment of industry depends.

THE MINING INDUSTRY

Mining operations are usually expensive undertakings which demand large amounts of technical equipment, while complicated preliminary work such as the sinking of shafts may also be essential. It is, of course, true that under especially favourable conditions mining can be carried on comparatively cheaply, though the term "comparatively" should be stressed. When, for example, the mineral lies immediately beneath the surface *opencast mining* can be adopted and in this way the complicated business of sinking shafts can be avoided—but even when this is possible expensive equipment may still be required. One of the best-known examples of this type of mining is the quarrying for iron ore at the head of Lake Superior (Fig. 2), while low-grade limonite is obtained in a similar way in the East Midlands of England.

Because mining is normally an expensive operation* mineral exploitation is usually undertaken by large firms who go to great pains to assure themselves that the risks which are bound to be incurred are minimal before they commit themselves to the capital investment required. It is therefore appropriate at this point to turn to a consideration of the various factors which help to ensure that a mining undertaking will be

* Some mining undertakings such as surface digging for diamonds do not require expensive equipment and the small-scale worker is still to be found in such types of project.

successful, though it must be borne in mind that it will be rare for conditions to be uniformly favourable. Even in the most favourably located undertaking (and this is true of industrialisation in general as well as of mining in particular) some disadvantageous features will almost certainly have to be faced.

The Quantity of Mineral Deposits

The first point is quite simply that a mineral must be known to exist in economic quantities. This is an obvious statement but it may serve to remind us that even in these days of advanced technology prospecting still has to be carried out, sometimes under conditions of considerable difficulty. Nothing illustrates this point better than the dramatic way in which the continent of Africa, which for a long time was considered to possess only limited reserves of minerals (and almost no mineral oil), has now been shown to have very considerable primary resources.

Perhaps this point is demonstrated nowhere more spectacularly than in the north-western part of the Sahara Desert, an area which even informed opinion did not until very recently, credit with any great opportunities for economic development. It was in 1950, when prospecting for mineral oil was proceeding, that drills boring 4000 feet below the desert sands near Ghardaia struck unimagined amounts of water, artesian water which without warning suddenly leapt upwards into a fountain 600 feet high. This water is now being used for irrigation purposes. More recently artesian water has since been found at many other places in the north-western Sahara. At Hassi Messaoud it was found also at 4000 feet, and drilling at this location was continued to greater depths until oil was struck in 1956 at 11,000 feet; it has since been amply demonstrated that North-West Africa is an oil region of considerable magnitude (Fig. 2). Even more recently very large reserves of natural gas have been discovered beneath the North Sea and this discovery is likely to exercise a considerable effect on industrial patterns in Britain. Some observers are already forecasting a bright industrial future for parts of the east coast of England. Oil is today a raw material of very considerable importance as well as a source of fuel; we shall say more about this in Chapter XI.

In earlier years prospecting for minerals was an even more uncertain business than it is today, and perhaps one of the best-known discoveries in Africa was that of copper ore in the territory now known as Zambia, then Northern Rhodesia. Prospecting was undertaken for some years without success, but in 1926 a prospector relaxing from his duties on a hunting expedition shot a roan antelope. The animal fell partly on an outcrop of rock and as the hunter approached to retrieve his victim he saw that the rock bore a slight green stain—an indication of the presence of copper. It was this largely fortuitous discovery which led to the eventual establishment of the Roan Antelope Mine and to the development of the Copper Belt of Zambia. While this event had some unusual features about it, it does remind us that mineral prospecting, especially in the

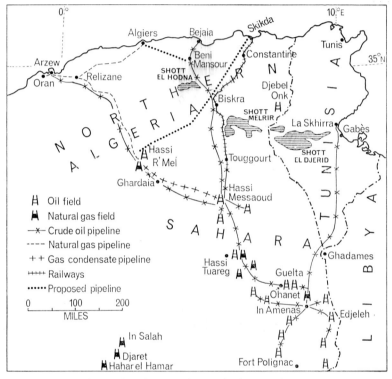

FIG. 2.—Areas of production of mineral oil in North-West Africa.

tropics where a lateritic crust of up to 20 feet in depth (in some instances even up to 100 feet) may blanket the surface, is by no means a simple matter.

The Quality of Mineral Deposits

The quality of the ore must be such as to justify mining. This does not of itself mean that the ore must be high grade though it is clearly of tremendous advantage if it is; the grade of ore must be taken into account along with other factors. For instance, low-grade limonite is mined extensively in the East Midlands of England because it is possible to establish ironworks near the centres of production and the ore does not therefore have to make a lengthy journey before smelting. It would be quite out of the question to mine ore of similar quality in a remote region unless the ore could be processed before despatch; without such processing the ore would need to be of high quality to justify mining. The iron ores of Schefferville near the Quebec–Labrador border and those of Kirunavaara in northern Sweden, both of which are mined, are of substantially higher grade than the East Midland ores (limonite about 30% iron content; Schefferville 50% or more, and Kirunavaara over 60%).

An example emphasising this point comes from Minnesota, U.S.A., where for many years very extensive reserves of taconite have been known to exist. Taconite is a low-grade ore containing between 20 and 30% of iron, and there is the further disability that it is a very hard cherty rock and not, therefore, easy to mine. Grains of magnetite and haematite are disseminated throughout the rock. Under these conditions it is not surprising that little thought was given to the possibilities of mining the taconite for many years; apart from the poor quality of the ore and its hardness there were two other reasons for this. The first was the presence of the greatly superior Mesabi ores near by (Fig. 3); it was only natural that while these ores were available they should attract the attention of mining interests. The second was the system of taxation adopted by the state of Minnesota. The difficulty about this system was the imposition of an *ad valorem* tax on exploitable minerals *in the ground*, and since the reserves of taconite were known to be very extensive it was realised that this poor-grade ore could not bear the heavy tax burden which would be imposed. As long as the taconite was not classed as an ore no tax could be levied.

This situation, however, was radically changed in the early 1940s for two reasons, the first one being the impending exhaustion of the rich Mesabi ore. The second developed as a result of a revision in the tax laws which enacted that actual production rather than proven reserves should constitute the basis of taxation. This revision, carried through in 1941, completely changed the situation, and exploitation of the taconite began in the late 1940s. There was a delay in commercial exploitation because even in the new circumstances it would not have been practicable to ship the crude taconite with its low payload of metallic iron to the smelters of Chicago or Pennsylvania; some method of treatment near the mine had to be worked out and considerable difficulty was experienced before a successful method was discovered. This is discussed below (p. 34).

There was one further circumstance which made exploitation of the taconite ore practicable, even desirable, and it was this: loading, shipping and handling facilities for iron ore had been set up many years previously for the handling of the Mesabi ores and others in the Lake Superior region (Fig. 3) and a very great capital loss would have accrued if it had not proved possible to continue using them. There was therefore every incentive to develop the taconite ore, which is at present providing more iron than the richer ores of the region (these ores are now almost worked out).

FIG. 3.—Iron ore ranges of Lake Superior.

This example is a good one as it shows us that it is not the quality of the mineral alone which counts—it is that quality in relation to other factors. Although we are here isolating for convenience the various factors which have relation to the production of minerals we should always bear in mind that possibilities of production are almost always related to more than one of these factors.

Mining Conditions

The general mining conditions at any given centre of actual or potential production must always be considered by mining interests. These conditions will depend on such factors as the depth of occurrence of the mineral; the thickness of seams or veins; the presence or absence of faulting; the amount of water encountered in the mine; and the prevailing temperature gradient.

Mineral deposits typically occur in seams or veins, according to the character of the rock concerned, seams being typical in sedimentary rocks and veins in crystalline rocks. When minerals occur in seams ideal conditions are approached when fairly thick deposits of ore lie almost or quite horizontally at no great depth; folding and faulting create difficulties when the ore moves rapidly from one level to another. This can be illustrated from the occurrence of the phosphate beds near the Tunisian-Algerian border (Fig. 4) which are partly horizontally bedded and partly folded. (The case of coal, which can sometimes be classed as a raw material, is examined below, pp. 51–56, and Fig. 14.)

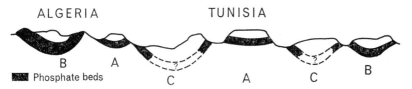

FIG. 4.—Disposition of phosphate beds in Algeria and Tunisia. Based partly on Thoman. (A) Almost horizontal strata—easy to mine; (B) steeply dipping strata—difficult to mine; (C) possible deep-seated strata—impossible to mine.

For an account of the formation of mineral veins the reader is referred to suitable textbooks on geology (see, for example, Lake and Rastall's *Textbook of Geology*, fifth edition, revised by E. H. Rastall, pp. 278 ff.). However, we might here remark that veins often develop when fissures in country rock are penetrated by hot solutions from below. If such solutions contain useful minerals these will separate out as the temperature falls to the appropriate level; commonly, the outside of the vein is taken up with *gangue* (useless rock) while metallic minerals occupy narrower central parts (Fig. 5). The degree of usefulness of such a vein is set largely by its thickness, by the proportion of useful contained mineral and by the angle of dip.

The leaking of underground water into mines is always troublesome and often dangerous and the presence of subterranean aquifers is therefore always a hazard. A further point is that in some parts of the earth's crust the temperature gradient is comparatively high and this can lead to uncomfortably high temperatures in deep mines.

The presence of beds of loose rock can hinder mining operations as shafts cannot easily be sunk through such rock without the provision of expensive linings. The mining of the sulphur deposits which exist near the Gulf coast of Louisiana, Texas and Mexico provides an example of this.

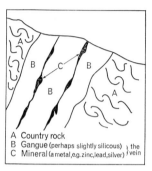

Fig. 5.—A mineral vein.

These deposits exist in connection with salt domes which have been thrust upwards towards the surface from very great depths, possibly as much as 30,000 feet; the sulphur is found just below the cap rock of some (not all) of these domes, but for many years it was not possible to extract it in any significant amount because of the unconsolidated nature of the overlying deposits. The situation was transformed, however, by the invention of the Frasch process of sulphur extraction (Fig. 6); this process was first used in

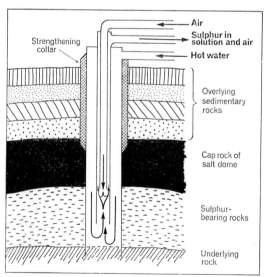

Fig. 6.—The Frasch process of sulphur extraction.

Texas in 1895 by Herman Frasch, and it involves the forcing by pumping of hot water and air into the sulphur deposits. The water melts the sulphur and the molten sulphur is forced to the surface by the pressure of the air, after which it solidifies. A great advantage is that sulphur produced in this way is virtually pure; further refining is not necessary.

Introduction of the Frasch process has enabled Mexico to become the second world supplier of sulphur since exploitation of the deposits near Minatitlan, on the northern slopes of the Isthmus of Tehuantepec, began in 1954. The fuel needed to superheat the water to a temperature of 320° F (160° C) is provided by an oil refinery at Minatitlan, and an interesting feature of one of the plants is that superheated water is led for a distance of one mile to the operating wells through insulated pipelines. Production in excess of domestic requirements is exported through the port of Coatzacoalcos.

The Level of Technology

Another factor which markedly affects the possibilities of mining is that of the level of technology available to mining engineers. It is easy enough to understand that a community with advanced techniques at its disposal is in a better position to make use of its mineral resources than a more primitive society, and we saw in Chapter I some simple examples showing how the invention of new machines affected mining operations. The most obvious case, perhaps, was the invention of the crude Newcomen engine which made it possible to pump water out of Cornish tin mines so that these mines could continue in production instead of closing down. Another early example which is not always given the recognition it deserves was the invention in 1839 by Andrew Smith of the iron-wire rope. Before this invention there were only two means of raising materials from the floor of a mine to the surface—by using an hempen rope or by using direct labour, usually female labour. Hempen ropes proved costly in practice as they quickly wore out under the strains involved and it was common practice, therefore, to employ women to trudge up inclined ladders with laden baskets on their backs (Fig. 7). The misery and drudgery involved in this system was quickly ended when the new iron rope became available, for it proved to be far more efficient and helped to increase mining output.

These are, however, examples from an age that is long past, though the underlying principle remains the same. As more recent examples we may note the success of the cyanide process for the separation of gold from its ore and of the flotation process which has lowered the grade of economically exploitable copper ore from that containing 2% of metal to that containing 0·75%. This has made possible the development of the low-grade copper ores in the mountain states of the U.S.A. and of the copper ores of the Canadian Shield.

The example of the mining of taconite in Minnesota to which we have previously made reference offers another example of this principle. In this very hard rock, grains of haematite and magnetite are rather sparsely disseminated, giving an ore with a metallic content of between 20 and

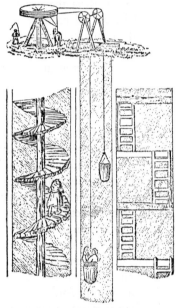

FIG. 7.—Early methods of raising coal to the surface.

30%, and some means of concentrating this ore was essential before it was practicable to commence mining. The difficulty of inventing such a process was increased by the fact that taconite is one of the toughest rocks known—it is said to be twice as hard as granite.* However, the concentrating can now be carried out. The ore is crushed to a powder, after which electric magnets draw out the iron-bearing magnetite from the gangue, though it is interesting to note that the toughest alloy steel linings

* The mining of this hard rock is difficult, for ordinary mining equipment barely scratches it. The technique now adopted is to sink blast holes by means of a 4300° F (2370° C) flame of oxygen and fuel oil, water sprays flushing out particles of burned rock. Blasting later breaks the rock into large blocks which can be removed for processing.

used in the ore crushers last for only one-sixth as long crushing taconite as they do when crushing other hard ores, while the teeth on the power shovels used to handle the ore, which are made of abrasion resistant steel, must be replaced at the end of every eight-hour shift. The separated magnetite is rolled into balls and baked into pellets about the size of a marble; these pellets are very hard and can withstand rough handling as well as bear the tremendous pressures met with in the blast furnace. About one ton of pellets containing an average of 63% metallic iron is processed from three tons of taconite. (*See* p. 171 below for further reference to pelletising.)

In the Lake Superior region the taconite ore is first crushed into smallish fragments (about three inches across) at Babbit (Fig. 3) before being railed to Silver Bay, 47 miles distant. There the ore is pulverised, concentrated and pelletised before shipment. Silver Bay itself has developed into a new town of between 4000 and 5000 inhabitants while Babbit has developed in a comparable manner. Taconite is playing an increasingly important role in the economy of the United States iron and steel industry; 20 million tons of pellets were shipped in 1959 but it is expected that by 1970 the figure will have risen to 40 million. It is anticipated that in less than twenty years' time American blast furnaces will be using more taconite than all other forms of domestic iron ore put together.

This example clearly illustrates the complexity of modern techniques; it is not the development of a single advanced technique which has made possible the mining of taconite, but the development of many. We need to bear in mind not only the invention of the crushing and digging machines but also the perfecting of the special steels which have been developed only after years of patient experiment.

Transport

It is rarely indeed that mineral ores can be exploited unless some form of transport is available to carry the ores or ore concentrates to the points where they are needed. And, as ores are usually both heavy and bulky, rail or water transport is normally needed. There are many cases where mining of rich ores has been declared impracticable because the necessary transport facilities do not exist.

A good example of this may be taken from Sierra Leone, a country which has proved to be very rich in mineral wealth. Possibilities of an iron-ore mining industry were first envisaged in 1926 when Dr Junner first discovered the rich Marampa ore near Lunsar (Fig. 8). The ore occurs in two hills which are virtually made of iron ore, but the ore was not discovered earlier because of the thick covering of laterite which mantles the whole area. The first practical step which had to be taken was to arrange for transport and shipment of the ore, about 50% of which goes to Britain while the U.S.A. and Western Germany each take about 25%. The necessary arrangements included not only the construction of a

FIG. 8.—Iron ore in Sierra Leone.

3 ft 6 in. gauge railway 52 miles in length to Pepel, but also the construction of loading equipment at the "port" (Pepel is perhaps better described as a loading point rather than a port; nothing is despatched from there except iron ore and nothing is imported). We might observe that more recently it has been found necessary to re-equip Pepel for the handling of larger quantities of ore and also to dredge the Sierra Leone River to provide access for large modern ore-carrying vessels; these facts remind us that the provision of adequate communications is not always a straightforward matter.

It was, then, the provision of transport facilities which helped to make possible the export of iron ore from Sierra Leone and exports of this ore have been running at more than 1 million tons annually for some time. The ore is exported in a concentrated form; as it is mined it carries about 50% of metallic iron but after concentration (which is carried out at the mining site at Lunsar) the proportion is increased to about 67% (95% iron oxide, Fe_2O_3).

But an interesting point now arises. Since the discovery of the original Marampa ore it has been shown quite definitely that even more extensive ore deposits exist in the Tonkolili District farther east. The village of Farangbaya, about 80 miles from Lunsar, lies in this ore belt, in which occurs laminated haematite together with some limonite extending over a distance of about 12 miles. Analyses have shown that the ore averages more than 60% metallic iron so that it is significantly richer than the Marampa ore.

For some years it was confidently expected that arrangements could be made to mine this ore and work even began clearing the proposed route along which the railway would run, but after prolonged negotiations with interested mining companies the whole scheme collapsed. The reason why it collapsed was largely because it was not considered economic to construct the necessary railway, which would have been in effect an extension of the existing Lunsar line. There were various reasons for this, one being the rugged nature of the terrain through which the railway must run; this mountainous region could have been pierced only with the help of expensive embanking, bridging and tunnelling. Another serious difficulty was that of stabilising the track. Ore trains are heavy (each train from Lunsar carries 1000 tons of ore concentrate in addition to its own weight) and they require solidly laid tracks. The problem (as in many tropical regions) is that an apparently firm rock surface may turn out on closer inspection to consist of a surface layer of lateritic crust underlain by a zone of amorphous clay, a clay which becomes very greasy and slippery when it is wet, as it normally is during the rains. Under these conditions the surface layer shows a tendency to slip under strain, and there have been cases in Sierra Leone of much lighter trains than the ore trains being wrecked because the lateritic crust over which they were passing suddenly gave way and slipped laterally. A further point is that severe storms commonly met with in tropical regions can cause very real difficulties for railway operation, and this fact could not be discounted. The railway between Katanga and Lobito, to take a somewhat parallel case (Fig. 9), is constantly subject to wash-outs and track movement during the storms of the rainy season, and these snags would certainly have been encountered on the proposed Tonkolili extension.*

Other examples of mining enterprises which are dependent upon transport are not hard to find. In another part of Africa, for example, the rich mineraliferous areas known as the "Copper Belts" of the Katanga and Zambia are tucked away near the centre of the continent many miles from the sea, and mining was only possible after the construction of railways linking the mining areas to tide-water ports. Figure 9 shows the general situation and the following notes may serve to emphasise the magnitude of the problem. The numbers in the following notes refer to the numbered routes shown in Fig. 9:

1. This route runs from Lubumbashi (formerly Elisabethville) southwards through Zambia, Rhodesia and South Africa to Cape Town. This is a lengthy route of 2305 miles and is rarely used.

2. From Lubumbashi via Zambia, Rhodesia and Portuguese East Africa to Beira: 1619 miles.

3. From Lubumbashi via Bukama northwards to Port Francqui, and thence by river steamer to Kinshasa (formerly Leopoldville); thence by rail to Matadi. Total distance: 1720 miles, but this is not directly comparable with an all-rail route owing to the breaks of bulk and the comparatively slow river passages.

4. From Lubumbashi westwards via Angola to Lobito: 1334 miles. This railway was completed in 1931 and provides the shortest and quickest route, the journey taking three days.

5. From Lubumbashi by a rather circuitous rail route to Albertville on the western shore of Lake Tanganyika. A service across the lake provides a link with Kigoma, which is in turn linked by rail with Dar es Salaam. Total distance: 1717 miles but the route suffers from the same disadvantages as route (3).

6. Since 1955 an additional rail outlet has been available via Zambia and Rhodesia to Lourenço Marques though so far this route has been little used.

7. The extension of the Mombasa–Uganda railway to Kasese has provided an additional outlet for the Congo Republic though it is rather too roundabout for Katanga minerals. It serves the gold-producing areas of the north-west of the territory.

* Another reason for the shelving of the Tonkolili scheme was the fact that overall iron-ore supplies were more than keeping pace with world demand. Mining activities in West Africa were being opened up and further developed in Mauritania and Liberia, so that additional Sierra Leone ore would have encountered severe competition in world markets.

8. Discussions and surveys are at present going ahead in connection with the proposed construction of a railway from Kapiri Mposhi (Fig. 9) to Tanzania in order to give Zambia an outlet for her copper independent of Rhodesia.

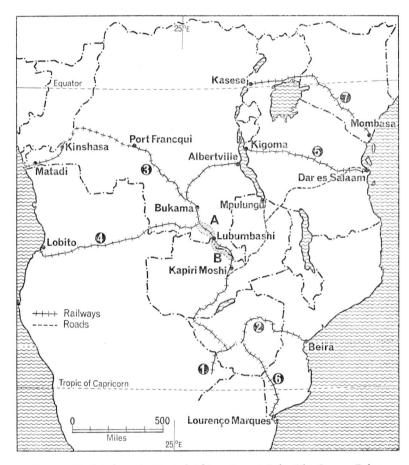

FIG. 9.—Outlets from the Central African Copper Belts. The Copper Belts: (A) Congo Republic; (B) Zambia. The numbers are referred to in the text.

Before 1965 most of the copper export from Zambia (about 60,000 tons a month) passed over Rhodesian railways prior to shipment, mainly from Beira. Following Rhodesia's Unilateral Declaration of Independence, however, other rail and road routes were urgently sought, and today barely 20,000 tons a month use the southern route. About 15,000 tons monthly is despatched via the Benguella Railway (4 above), while some goes by road to Malawi and thence by rail to Beira. But an astonishing total of almost 25,000 tons a month is sent by road to Dar es Salaam (Tanzania). Most of this travels all the 1100 miles by road, but some is shipped over Lake Tanganyika from the lake port of Mpulungu to Kigoma, whence it moves by rail to Dar es Salaam. The first steps are now being taken towards the construction of a railway from Zambia to Dar es Salaam. This example is instructive as it shows how very important is the provision of transport from mining areas.

It is not difficult to find examples from other parts of the world which demonstrate the importance of transport in the exploitation of mineral resources; the cases of the Kirunavaara iron ore in northern Sweden, the Schefferville iron ore of Canada, the silver–lead–zinc ores of Broken Hill (New South Wales), and the copper–zinc ores of Mount Isa (Queensland) come readily to mind. Lack of space forbids more than a passing mention of these.

Labour

Another factor which can influence mining developments is that of the available labour supply, and we should bear in mind that in this connection two distinct points arise—that of the quantity of labour available and that of the quality. The quantity of labour is in large measure governed by the population of a given area or of neighbouring areas, and a study of this would lead us into a consideration of population geography which is outside the scope of this book. We might properly observe, however, that it is not unusual for mining enterprises to encounter difficulties because sufficient numbers of workers are not available. In some cases, such as on the Rand of South Africa, special mining villages have to be established by the mining companies to attract workers, and, while the newest villages are often models to be emulated, the older "compounds" were frequently notorious for their poor living conditions. In some cases such as at Schefferville the problem arose of attracting an entire community as the barren region naturally supported almost no inhabitants before mining activities commenced.

The question of the quality of labour is in its way an equally important one, for mining operations involve the carrying out of tasks which can only be undertaken by skilled workers. The training of skilled labour can be a lengthy and expensive process and in territories where there is no tradition of technical skill the acquisition of the necessary skills is not an easy matter. Some "importing" of trained workers from more developed territories is normally essential.

Capital

We have stressed earlier in this chapter the fact that mining enterprises are normally conducted by large-scale firms. Only in some special cases such as panning for tin or gold, or digging for alluvial diamonds, can the small-scale operator carry on his activities. This means that supplies of capital are essential for mining successfully to be established, and where domestic supplies of capital are not forthcoming recourse must be had to foreign sources. This is a problem which has to be faced especially in the developing countries (*see* Chapter XII).

Markets

The existence of accessible markets is an essential prerequisite to any mining enterprise, for there is no point in producing a mineral which cannot be sold. In this respect the role of possible competition has to be taken into account (*see* footnote, p. 38 above). A striking example of this is offered by the case of Schefferville. Some of the most likely markets for the Schefferville ore lie in the great industrial regions of the U.S.A. which lie between the Great Lakes and the Atlantic seaboard, but until after the end of the Second World War these markets were amply supplied with iron ore from the Mesabi and near-by sources (Fig. 3). There was therefore no possibility of the Schefferville ores being developed until it was realised after the Second World War that the Mesabi ores were approaching exhaustion, and it therefore became imperative that an alternative source of supply should be opened up. It was at that stage that developments began at Schefferville and the railway, 357 miles long, was constructed to link Schefferville and the St Lawrence at the (hitherto) tiny village of Seven Isles. With regard to the questions of labour and capital it is intersting to note that a labour force of almost 700 men and financial outlay of 250,000 dollars were required before actual mining could begin at all. It was also during this same period that the St Lawrence Seaway was opened and the net result of all this activity meant that access to a large market was then available for Schefferville ores.

Political Factors

Political considerations can sometimes influence the opening up of ore deposits, an example of this being the development of the low-grade iron ores at Salzgitter, in Germany, during the Nazi régime. Not only did this provide iron for the Nazi rearmament drive but the location of the ore was such that in those days it was regarded as being outside the probable range of operations of an enemy force. We have previously mentioned the change in taxation policy on the part of the state government of Minnesota as being a potent factor encouraging the exploitation of the taconite ores near Mesabi.

Climate

There are occasions when the prevailing climatic conditions have to be taken into account by mining interests. In high latitudes, for example, the effect of prolonged and severe cold can be a sharp handicap. At Seven Isles, to take a case in point, the shipping season lasts for only about 240 days in the year, while in northern Sweden steam heating is necessary to thaw out ore dumps for loading and to keep rolling-stock moving. The Mesabi ores cannot be mined in winter as all navigation ceases on the Great Lakes and on connecting links such as the Soo Canals and the St Mary River. Enough ore must therefore be shipped during the rest of the year to feed the blast furnaces bordering Lake Michigan and Lake Erie as well as those in the Pittsburg and Youngstown areas throughout the twelve months. In low latitudes, especially in areas with a tropical climate, shortage of water during the dry season can cause severe difficulty. Even at Lunsar, where the average annual rainfall is over 100 inches, difficulty arises because of the long dry season; when the water shortage is most acute, in April and May, work can be carried on only because it has been found possible to recover 80% of the water used during the concentrating processes for re-use after treatment. The remaining 20% is made good by pumping 8000 gallons each day from the nearest river, an amount which the river is only just able to yield at that time of the year.

CONCLUSION

These in broad terms are the chief factors which affect the establishing and development of mining operations. In a highly competitive world such as this it is essential for a mining firm carefully to consider every possible factor which might affect the quantity and the quality of output, and also the possibilities of marketing the mineral concerned. The reader may well have noticed two significant points which develop from the part of this chapter which deals with mineral production:

1. It is in a sense a summary of the first part of this book. This is only to be expected because mining is an industry (though not a manufacturing industry) in its own right and the features which affect the establishing and

development of industry generally can well be expected to affect mining.
2. As mining is, in fact, an industry this chapter might be viewed as a study of a particular industry—the mining industry.

Considerations of space preclude any mention above of more than a very few examples and the serious reader is advised to follow up this chapter with studies of mining industries such as are to be found in the pages of such publications as *Economic Geography*, the *Geographical Review*, *Geography* and others.

SUGGESTIONS FOR FURTHER READING

Boesch, H., *A Geography of World Economy*, New York, 1964.
Highsmith, R. M., *Case Studies in World Economy*, Englewood Cliffs, New Jersey, 1961.
Jarrett, H. R., *Africa*, second edition, London, 1966.
Knowles, L. C. A., *Industrial and Commercial Revolutions in Great Britain during the Nineteenth Century*, London, 1944.
Lake, P., and Rastall, R. H., *Textbook of Geology*, fifth edition, revised by R. H. Rastall, London, 1956.
Manners, G., "Recent Changes in the British Gas Industry," *Trans. Inst. Brit. Geog.*, 1959.
Pounds, N. J. G., *The Geography of Iron and Steel*, London, 1959.
Simpson, E. S., *Coal and the Power Industries in Post-war Britain*, London, 1966.
Smith, W., *Geography and the Location of Industry*, Liverpool, 1952.
Smith, W., "The Location of Industry," *Trans. Inst. Brit. Geog.*, 1955.
Thoman, R. S., and Patton, D. J., *Focus on Geographic Activity*, New York, 1964.
Weber, A., *Uber den Standort der Industrien*, Tubingen, 1909. Translated by Friedricj, C. J., under the title *Alfred Weber's Theory of the Location of Industries*, Chicago, 1928.

Chapter III

Fuel and Power

THE point has already been made (p. 4 above) that it is very largely in the availability of inanimate forms of fuel and power that the world of the developed countries of today differs from that of years ago and from that of present-day developing territories.* Many of us today can hardly imagine the amount of time which used to be spent in producing simple forms of fuel (gathering firewood, for example, and cutting it up into forms suitable for domestic purposes), while great amounts of human energy had to be expended in such simple jobs as drawing water from wells and streams and in transporting it to the home. Today, the use of human and animal energy has largely been replaced by the use of inanimate forms of energy, and this revolution has helped to make life much easier for many thousands of people all over the world, though it is in the developed countries that the advantages of this change are felt most of all.

It was somewhere near the year 1800 that inanimate energy began to contribute noticeably to the world's output of energy, though estimates suggest that inanimate energy was the junior partner until about midway through the second half of the nineteenth century (Fig. 10). Since that time inanimate energy has passed animate energy in importance in first one country and then another, though it is interesting to note from Fig. 10 that the world's output of animate energy has steadily increased in actual amount right up to the present day; this feature is largely a reflection of the increasing population of territories such as China, India and Pakistan, which still rely largely upon muscular strength for the performance of many daily tasks which are undertaken by machines of various kinds in the more developed lands.

For a number of reasons Fig. 10 should be regarded as very tentative.

* When a source of energy is used to generate heat (to warm a room of a factory, for example) it is regarded as a "fuel"; when it is used to produce any form of motion (in driving any form of motor, for instance) it is regarded as "power."

In the first place, there is the impossibility of securing reliable statistics for all territories of the various forms of energy (especially of animate energy) recorded, and the amounts shown must be regarded as estimates only. Secondly, there is the difficulty of bringing the various forms of energy actually used to a common denominator for purposes of comparison, and it is doubtful whether a wholly satisfactory method of conversion can be devised. Comparison is sometimes effected in terms of the "coal equivalent" of each form of energy; it is estimated, for example, that mineral oil is 1·5 times as efficient as coal which is used under ideal conditions, while lignite is only 0·5 times as efficient. These are, however,

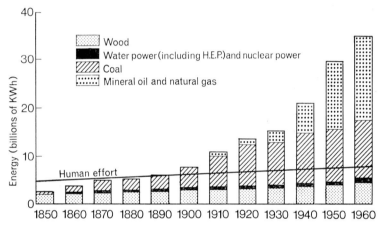

Fig. 10.—World energy output, 1850–1960. Based on information in "U.N. Statistical Papers," Series J.

average figures only and when used can give only a very general picture of the situation.

The equivalent used in the preparation of Fig. 10 is that of kilowatt hours,* and, although the warnings regarding the validity of the conversions used and the accuracy of statistics must be borne in mind, it is possible to argue that the diagram does give a good general representation of the actual pattern of fuel consumption over the past century and is therefore worth considering. The first obvious consideration is the remarkable way in which the use of inanimate energy has leaped ahead during the past one hundred years, and particularly during the present century. This is a point which we have already made but the diagram

* The usual equivalent used is that 100 kWh = 0·125 metric tons of coal burnt but other equivalents are sometimes used; the difficulty is that of selecting the most appropriate.

shows that more than this simple feature is involved, for a prominent characteristic of the changing pattern during this century is the way in which the earlier dominance of coal as an inanimate industrial fuel has been challenged, particularly by mineral oil. Oil, as a fuel, enjoys considerable advantages over coal, for it is easier to handle, it is cleaner, and unit for unit it is more efficient. It is not surprising that this changing pattern has coincided with a changing pattern of industry, the most obvious features of which is that industry is no longer tied to the coalfields as it once was. This topic will receive further consideration elsewhere (p. 56).

Another exceedingly efficient form of energy which today is being developed in increasing quantities is hydro-electricity, though from a global point of view the importance of this form of energy is still slight—perhaps only about 2%. This fact is apt to be obscured when we read of modern large-scale enterprises to harness water power. The importance of hydro-electricity is steadily increasing, however, and one great advantage of it is that supplies are renewable for an indefinite period; it does not destroy its raw material during production as does the production of energy from coal. It is not surprising that industrial location is being affected by the emergence of this new form of energy and we shall examine examples of this later.

In recent years the development of nuclear power has opened up possibilities undreamed of some years ago, and, while we are as yet in the very early stages of harnessing nuclear power, there is no doubt that it will bring about great changes in the pattern of energy consumption. In Britain electric energy has been produced by nuclear means for some years and the proportion of nuclear-produced electric energy will increase. Nuclear reactors have the tremendous advantage that the supply of raw materials presents almost no problem and the plant can therefore be situated with a close regard to areas of demand; it is likely that with increasing efficiency the consumption of one ton of uranium will produce an energy output equivalent to that derived from 10,000 tons of coal (Simpson, 1966), and even greater efficiency is theoretically possible.

Considerations such as these would seem to indicate that nuclear power has great possibilities in countries which at present suffer from a lack of energy resources, but unfortunately there are serious difficulties. Although running costs in nuclear stations are lower than in conventional power stations the capital costs involved are far higher, and shortage of capital is one of the chief problems of the developing territories. In order to make reasonable use of such a high degree of capital investment it would be necessary to use the installed power fully, but developing countries are not often in a position to consume such large amounts of power as are provided by a nuclear reactor. Reddaway argues that, since it is capital, rather than manpower, which is the scarce and expensive factor in developing territories, the emphasis in the drive for greater efficiency should be on raising the output from any given amount of capital rather than on increasing the output per man. Such a programme must weigh heavily against any installation of nuclear energy in developing areas for a long time to come, and it seems likely that it is in the more developed countries that this form of energy will be used in the foreseeable future.

WORLD ENERGY CONSUMPTION

These are some considerations which come to mind from a study of Fig. 10. We may now look at the question from a different angle, from

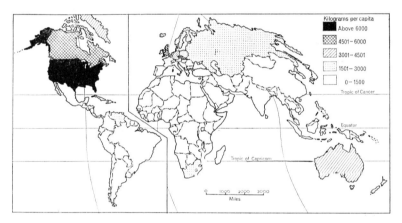

FIG. 11.—World energy consumption, 1960 (by territories).

the aspect of the geographical distribution of energy consumption in the the world generally. Figure 11 shows a classification of territories according to their *per capita* consumption of energy, and perhaps the most obvious features to emerge from a study of the map are the great disparities between the *per capita* rates of consumption in the low- and high-grade territories respectively, and the astonishingly large number of territories which even today record very modest rates of energy consumption. The map, indeed, could almost be taken for one showing a classification into "developed" and "developing" territories.

The highly industrialised parts of the world, especially North America

c

and North-West Europe, stand out fairly clearly on the map though we might note the low ranking of France, Italy and Poland, three countries which are very highly industrialised in certain areas. Africa and South America are both continents of low rating and territories like Argentina and Chile which have fair industrial development do not stand out, though the Republic of South Africa does to some extent. Industrialisation in India has so far made little impact on the life of the nation generally, judging from this map, while the big surprise may be the low ranking of Japan. This serves to remind us that industry in Japan is still strongly localised. Figures issued by the F.A.O. attempt to show the proportions of the inhabitants of different countries actively engaged in agriculture, and these figures show that in many developing countries such as India and most of the countries of tropical Africa, the proportion is as high as 70% or even 80%—sometimes higher. In countries such as as Japan and Italy which entered the industrial race late but which have made great efforts to overcome this disability, the figure lies between 40 and 50%, while in the U.S.A. it is 12% and in the United Kingdom it is as low as 5%. These figures give some indication of the tremendous differences in industrial development between the older industrialised nations and most others and throw further light on the situation portrayed in Fig. 11.

One point which the map does make clear is that most countries must have very considerable reserves of energy as yet unused, and this point is taken up below. Meanwhile it should be noted that the map can be misleading unless read with some care; for instance, Australia and Germany have the same ratings, though the former has very extensive underdeveloped areas while Germany is much more uniformly settled and developed. The map reflects the low overall population of Australia rather than extensive industrial development.

SOME CHARACTERISTICS OF ENERGY

Before turning to a consideration of the individual forms of energy we might briefly summarise one or two characteristics of energy which are of importance from our point of view. One most important feature, for instance, is that energy reserves can be classified from one aspect under two headings, "capital" and "income" reserves, and the point at issue here will become more and more important as time passes. Capital reserves include all reserves which are destroyed in the act of producing the energy, reserves such as coal, lignite, mineral oil and natural gas, as well as some less important forms such as peat and wood. These (except

FUEL AND POWER

wood) are sometimes termed "fossil fuels" as they owe their origin at least in part to fossil remains and at the present time they account for by far the greater share of energy produced in the world (Fig. 10). Uranium also falls into this group.

Income fuel reserves, on the other hand, are derived primarily from the power of the sun and the force of gravity, as these act upon the atmosphere and the hydrosphere, and they are renewable for an indefinite period. Easily the most important up to the present is the force of running water in streams and rivers. While the overall importance of this form of energy is not yet very great it is increasing notably and must continue to do so. Of small but increasing importance is the energy derived from the tides, the wind, and from direct solar energy.

Another point to bear in mind is that while in theory one form of energy may be substituted for another, such substitution is not always possible in practice; sometimes such a change would involve heavy capital expenditure as would be the case if a coal-fired energy system were changed to an oil-fired one. Sometimes the substitution is not possible at all; for instance, there are now available in India small stoves which work by focusing the direct rays of the sun on to the cooking utensil, and no other form of energy can be used in such appliances. As a further example we may note that no practical alternative to coke has yet been found as a fuel for the blast furnace, despite careful search and experiment.

The available supplies of energy are very unevenly distributed over the surface of the earth. For example, it is estimated that North America has about 38% of the world's coal reserves and that Australasia has about 1%, while it has been suggested that two-thirds of the estimated world potential of water power may be located in Africa. This may overstate the case, however, and it is possible that a more realistic estimate is the 41% cited by Kimble. Jones and Young (in *Science*, 5 March 1952) have given the following estimates of the world's potential and developed water power (w.p.):

	Capacity of installed w.p. plants*	Potential w.p. based on normal minimal flow*	Developed power as %age of potential power
Africa	715	250,000 (39%)	0·28
Asia	14,296	156,000 (24%)	9·22
Europe	48,516	64,000 (10%)	75·80
N. America	46,430	90,000 (13%)	51·59
Oceania	1,778	23,000 (4%)	7·73
S. America	3,962	62,000 (10%)	6·40
WORLD	115,000	645,000 (100%)	17·95

* Thousands of h.p.

In the case of mineral oil reserves the situation is even more uncertain, for new discoveries, some of them in areas traditionally considered to be barren of oil supplies, have greatly changed the world picture. "Proven" reserves of mineral oil in 1951 amounted to 14,450 million metric tons but by 1961 the figure had risen staggeringly to 40,000 million tons and by 1965 to 48,000 million. Actual production has risen from 280·5 million metric tons in 1938 to 538·33 million tons in 1950 and to an estimated 1,697,200,000 metric tons in 1966, while, as Fig. 12 shows, the general world pattern of production has changed greatly over the same general period.

We might at this point turn our attention more specifically to the main sources of energy and examine them separately, and in this connection

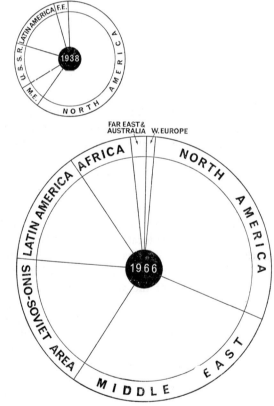

Fig. 12.—World production of mineral oil, 1938 and 1966. *From information supplied by the Petroleum Information Bureau.*
World total production: 1938, 280,500,000 metric tons; 1966, 1,697,200,000 metric tons.

we shall commence our study with coal, the use of which in the first place enabled the Industrial Revolution to get under way.

COAL

It is well known that the energy value of coal varies in accordance with its constitution, higher proportions of carbon and hydrocarbons producing coals of superior quality. Most coal contains ash which is not combustible and is therefore waste matter,* while all coal contains some moisture. Not only is the moisture of no calorific value but it also absorbs heat during combustion as it is evaporated away, thereby further reducing the efficiency of the coal; furthermore, evaporation from a comparatively moist coal during storage or transport may lead to crumbling and a consequent lowering in the quality of the fuel.

Coal may be classified under several headings as follows:

Lignite, or brown coal, has a high moisture content (sometimes as high as 35%) and a low content of fixed carbon (between 30 and 50% though values near the lower end of this scale are more usual). In some areas, however, large amounts of lignite occur and much use is made of them despite the comparatively poor quality of the fuel; this is the case in Germany, for example, where to ignore the extensive lignite deposits available would be to ignore a substantial national resource. Lignite is often used as a fuel for the generation of electricity and for processing into synthetic petroleum and petroleum products, while experiments designed to produce coke from lignite have met with some success.

Sub-bituminous coal is also of comparatively poor quality, partly because of a high moisture content and partly because the fixed carbon content is fairly low (between 35 and 60%). During combustion it tends to throw off showers of sparks, a feature which can under some circumstances be dangerous. Use has been made of these coals in Germany by burning them underground for the production of gas.

Bituminous coal is probably the most widespread form of coal and the form which is most generally useful. The fixed carbon content ranges between 60 and 80%, and the quality therefore varies considerably, different types of bituminous coal being suitable for different purposes. One important group comprises the coking coals, coals which form a hard coke suitable for use in the blast furnace, while steam coals are relatively ash-free and smokeless and are efficient producers of heat.

* If the ash produces clinker during combustion the value of the coal is considerably impaired for most industrial purposes.

Anthracite coals burn with great heat; they throw off little smoke and leave little ash. The fixed carbon content may lie between 80 and 95% or even higher. Although productive of great heat such coals do not burn as easily as the other forms of coal and they require specially constructed furnaces or stoves which produce a strong draught. They are very heavy, bulk for bulk, and expensive, and are mainly used for domestic purposes.

It will probably be clear from the figures given above showing the average fixed carbon content of the various types of coal that considerable subdivision is possible. The classification given above is simply a skeletal one and can be greatly extended; steam coals, for example, are

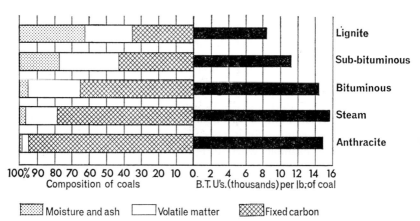

FIG. 13.—Composition and efficiency of selected varieties of coal.

often recognised as constituting a separate category. For further information on this and for details regarding the origins of coal the reader is referred to books such as Lake and Rastall, Holmes, or Fearnsides and Bulman which deal with physical geology.

Figure 13 attempts to give some indication of the varying composition of typical varieties of coal and also of their relative efficiencies in terms of British Thermal Units. An interesting point is that the efficiency of the grade with the highest fixed carbon content (anthracite) is slightly less than that immediately below (steam coal). The inferior quality of lignite is clearly brought out.

The distribution of world reserves of coal is very uneven, a point which is brought out by the following table (the figures are in millions of metric tons):

	Anthracite, bituminous and sub-bituminous	Lignite	Total	%age of world total
N. America	1,390,617	519,857	1,910,474	38
U.S.A.	1,325,564	495,265	1,820,829	36
Canada	65,053	24,592	89,645	2
U.S.S.R.	998,000	202,000	1,200,000	24
Asia	1,096,575	4,255	1,100,830	22
Communist China	1,011,000	600	1,011,600	20
India	62,143	2,833	64,976	1
Europe	572,045	87,890	659,935	13
Germany	279,516*	56,758†	336,274	7
United Kingdom	172,200	‡	172,200	3
Africa	69,734	210	69,944	1
Republic of S.A.	68,014	—	68,014	1
Australasia	13,957	39,689	53,646	1
Australia	13,900	39,200	53,100	1
South America	13,733	4	13,737	
Colombia	10,000	—	10,000	
World	4,154,661	853,905	5,008,566	100

Source: U.S. Geological Survey Circular, no. 293.
* Mostly in West Germany.
† Mostly in East Germany.
‡ Negligible.

Note that owing to a rounding off of the percentages in the final column and to the omission of the comparatively small total in South America the continental totals add up to 99% instead of 100%.

This uneven distribution is bound to favour industrialisation in the more favoured countries and to place less favoured ones under a considerable handicap, and it helps to explain the dominance of Europe and North America in industry in past years. It also shows that countries such as the U.S.S.R. and India which started late in their industrial development are well favoured as regards their overall coal supplies, while other territories such as those of Africa and South America must in the main base any industrialisation on sources of energy other than coal.

On the other hand, we must not try to read into the table a significance which is not there. As we have noticed earlier, all coals are not equally suitable for industrial (especially for coking) purposes, while the disposition of the coal within a territory may present sharp difficulties if it is difficult of access. Furthermore, its disposition within the earth's crust may not be entirely favourable for large-scale exploitation, and this is a question to which we might devote a little more space.

Coal normally occurs in seams comparable to the strata of the sedimentary rocks with which it is often associated, but the seams can vary widely in thickness, in horizontal extent, in depth and in continuity. As regards thickness, a seam may vary from a fraction of an inch to over 100 feet but usual measurements lie between 1 and 12 feet. Seams between 6 and 8 feet in depth are perhaps the least wasteful to work, for underground galleries cannot exceed 6 feet in height by very much, and if the seam is much thicker coal will simply have to be left above and below the galleries as it will not be safe to work it. On the other hand, if seams are less than 6 feet in thickness there must be much wasted effort in its recovery, because coal can be won from only part of the working-face and much waste material will have to be moved as the underground workings are extended. It is not always realised what a comparatively low proportion of coal can in some cases safely be worked from underground deposits; in some areas it is laid down that as little as 25% of the existing underground reserves can be removed in order to ensure as far as is possible that roof falls and collapsing of workings will not occur. This is, however, a low figure, and under more favourable conditions the proportion may rise to 75%.

Seams vary in horizontal extent very widely, for some deposits extend over only a few square feet while others may extend over many square miles; one of the best-known seams in this respect is the Pittsburg Seam of West Pennsylvania, which underlies a surface area of over 14,000 square miles. The depth of a seam below the surface is important, for while great economies of working can be secured with the help of opencast or drift working, deep seams can be reached only with the help of expensive shafting and pit-head equipment.

Seams may be continuous or they may be fractured and displaced vertically (and sometimes laterally) by severe faulting, and this is always a troublesome feature. Even sharp folding, such as that which occurs in the anthracite field of eastern Pennsylvania, causes inconvenience as the level of underground working is constantly changing under these conditions.

Two examples have been selected and are illustrated in Fig. 14 to show (*a*) a case in which the disposition of the coal seams is such as to make possible comparatively easy working, and (*b*) a case in which the arrangement of the seams is such as to make the working more difficult. Figure 14(*a*) illustrates the disposition of the coal seams near Pittsburg, Pennsylvania, and for convenience we might summarise the features which have helped to facilitate coal mining in this region:

1. Individual coal seams are remarkably undisturbed over a very wide area which extends from Kentucky and West Virginia northwards into

FUEL AND POWER

Pennsylvania (they are faulted and folded farther south). This makes for ease of operation and has also facilitated the use of coal-cutting machinery at the coal-face.

2. The coal seams are productive, varied and of high-quality coal. Coking coals, steam coals and coals suitable for domestic use are all well represented.

3. The seams are of convenient thickness for mining. The Pittsburg Seam, itself, for example, is about 6 feet throughout.

4. The seams are within convenient reach of the surface. The Pittsburg Seam, the most important coal horizon, is rarely more than 100 yards below ground level and, as Fig. 14(a) shows, it is not infrequently exposed at the surface. Where the seam is below ground level, shafts can be sunk with comparative ease and where it is exposed opencast mining or adit mining is comparatively simple.

5. Where rivers such as the Monongahela have incised themselves

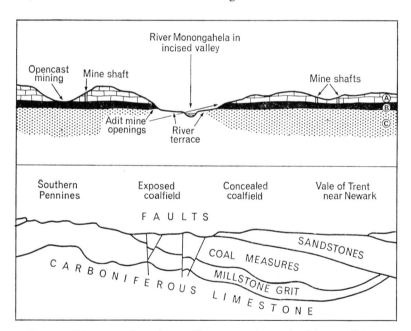

FIG. 14.—Diagrammatic sections to illustrate varying coal-mining conditions:
   (a) Diagrammatic section to illustrate mining conditions in West Pennsylvania.
      (A) Limestone; (B) Pittsburg Seam; (C) Sandstones and shale.
   (b) Diagrammatic section to illustrate mining conditions in the South Derby and Notts Coalfields.
Note in (b) that only parts of the coal measures are productive of coal. The coal seams occur at varying depths "sandwiched" between non-productive strata.

down into the Allegheny Plateau, the Pittsburg Seam commonly is exposed along the higher edge of a river terrace. This terrace is well situated to take buildings (including industrial establishments), roads and railways, while the river itself acts as a major line of transport. In West Virginia crustal uplift has been greater and the degree of river incision is therefore correspondingly greater, and the seams outcrop higher up the valley sides.

It is not surprising, under these favourable conditions, that coal mining rapidly developed into a major industry in West Pennsylvania and West Virginia, especially in view of the overall strategic location of the coalfields. The second example shown in Fig. 14 illustrates conditions in an important coal-mining region in England, but a region in which the physical conditions are by no means as favourable for mining as they are in West Pennsylvania.

Figure 14(b) shows that the coal measures outcrop at the surface immediately to the east of the Pennines, but farther east again they dip below the surface under sandstones of Bunter and Keuper ages. Furthermore, the coal seams are both folded and faulted so that seams are not infrequently displaced by several feet. Although coal outcrops in the exposed coalfield, the seams fairly rapidly dip beneath the ground so that opencast mining is possible in only a very limited area. Shafts are generally necessary even in the exposed field, while deep shafts are needed in the concealed coalfield.

The industrial strength of the developed territories of the world has been built up very largely on their coal resources and Fig. 10 shows that larger amounts of coal are being produced today than ever before. Despite this, the coalfields no longer exercise the power over industrial location which they once did, and there are two chief reasons for this. The first is that coal can today be used with greater efficiency than in years past so that the *effective* value of a ton of coal is greater than it ever was; this makes it more possible to take coal to industry and it is no longer essential for industry to gravitate to coal.★

The second reason for the loss of attraction of the coalfields is to be found in the increasing use of more easily handled and more efficient forms of fuel, and we now turn to an examination of one of these.

★ The complex situation facing this country with regard to future exploitation of coal reserves is well set out in Simpson, E. S., *Coal and the Power Industries in Postwar Britain*, London, 1966.

ELECTRICITY

The significant use of electricity dates only from the 1890s, though it has been only during the last four decades that really substantial use has been made of this relatively new form of energy. It was in 1879 that Edison completed his first incandescent electric light and in the early 1880s the same inventor's first electricity station in New York was set up and was able to transmit low-voltage direct current for a distance of rather less than one mile. In 1891 electric current was transmitted over a distance of 108 miles in Germany with a loss of about 25% of the original energy and this may be taken as the beginning of the possible use of electricity for general purposes. Transmission distances did not normally exceed about 30 miles even in the 1890s, but short-distance transmission was quite practicable. For instance, in 1894 the new power station at Niagara Falls was in operation and was transmitting current to Buffalo, 20 miles away, and in 1895 a new electro-process industry which included the reduction of aluminium was attracted to the Niagara site.

By 1910 the use of electricity was fairly widespread though current could be transmitted economically only to a compact and fairly densely populated area, but soon after 1910 long-distance transmission became practicable and a new stage of industrial progress was possible. Today, to take just one example, the generators at Harspranget, in northern Sweden, produce one-tenth of the total power output of the whole country and they transmit this power from the location on the Lula River, which lies north of the Arctic Circle, to southern Sweden over a distance of 600 miles with only a 7% loss of power in transmission. It is said that this probably represents a peak level of efficiency for the transmission of alternate current but that direct current can be sent over distances up to 1000 miles. (*See* p. 300 below for a further observation on a modern transmission technique (EHV).)

We might at this point begin to take note of the various factors which make possible the production of electrical energy on a large scale, and it is probably fair to say that the fundamental factor is the existence of a powerful enough demand. The level of technological achievement in these days is such that, given this demand, means can usually be found to provide the electricity. In some cases, indeed, particularly in the Owen Falls Scheme of Uganda and the Volta River Project of Ghana, steps have been taken to provide the energy *in anticipation of* the demand; the idea is that if the power is there industrial development must surely follow.

A good example which shows the power of market influence is that of the famous Kariba Scheme in Central Africa (Fig. 15). The demand for power in Central Africa came from many sources but the strongest source was the Copper Belt of Northern Rhodesia (as it then was; the territory is now the independent state of Zambia). We have previously noticed (p. 23 above) that copper ore as it is mined is normally of very low grade and that some form of processing at the mine is essential

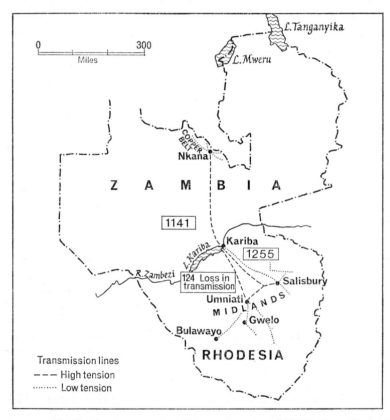

FIG. 15.—The Kariba power project. The total output of H.E.P. from Kariba is 2520 million kWh. Of this, 1141 kWh are transmitted to Zambia and 1255 to Rhodesia, while 124 kWh are lost in transmission.

because the crude ore cannot bear the costs of transport generally involved. Now considerable amounts of energy are needed for the concentrating process and for a very long time the Copper Belt had the greatest difficulty in securing sufficient for its needs. It relied mainly on coal imported from the Wankie mine of Southern Rhodesia (now Rhodesia) but the single railway line which even now forms the only connecting link proved incapable of meeting all the demands which were made upon it. When coal supplies therefore failed from time to time the mines had to fall back upon wood as a fuel! An agreement was signed in 1956 permitting the Copper Belt to import electricity from the (then) Belgian Congo but unfortunately a serious drought in the years following made it difficult for the Congo to produce enough electricity even to supply its own needs, while the troubles following the granting of independence to that territory in 1960 showed clearly enough the dangers which can arise when one country depends upon another for essential supplies. It was obviously too dangerous for the copper interests to rely upon their northern neighbour for any vital necessity.

It was amid these general circumstances that the Kariba Scheme was born, and it is not difficult to see why the Copper Belt was one of the strongest interests behind the whole project. Work actually began at Kariba in 1955 and the first electricity was produced in 1960. At present there is a single underground power plant producing 600 megawatts but a second plant generating 900 megawatts is to be constructed within the north bank. Of the power now produced the Copper Belt takes most of the 48% which is consumed within Zambia while the 52% which goes to Rhodesia is utilised mainly in the "Midlands" industrial belt which extends along the railway between Bulawayo and Salisbury through such towns as Gwelo, Que Que and Gatooma.

Unfortunately, the unilateral declaration of independence (U.D.I.) by Rhodesia in 1965 could adversely affect subsequent developments, and plans for the future are very uncertain. It is understandable that Zambia is not anxious to remain dependent upon Rhodesia for vital resources and she is endeavouring to press on with the north bank scheme, originally planned to become operational after 1970. There is also the possibility that she may go ahead with another H.E.P. scheme centred upon the lower Kafue Valley, which lies entirely within Zambia. This valley is gorge-like in its lower reaches above the confluence with the Zambezi and a scheme comparable to that at Kariba could be developed there. Politics and economics are very closely interwoven in this region, as, indeed, in most parts of Africa.

The importance of the market in helping to promote the provision of electrical energy is thus underlined, and the second factor of which we should take note is capital. Modern generating plants and installations are expensive; the Kariba Scheme, for example, cost about £80 million in its first stage and is expected to require £113 million in all. These are very large sums and capital on this scale can generally be supplied only by large-scale agencies such as governments or the International Bank for

Reconstruction and Development. In the case of Kariba this bank supplied almost 36% of the capital costs involved in the first stage, the copper companies supplied 25%, and the Colonial Development Corporation almost 19%. Other sums were contributed by the federal government of the now moribund Central African Federation, the British South Africa Company, the Commonwealth Development Finance Company and various banks interested in the project.

Hydro-electricity

We need at this point to make a distinction between electricity which is produced from water power (hydro-electricity) and that produced from coal (thermal electricity), because the factors involved are different. We shall consider first the production of hydro-electricity, and the most important determining factor for the generation of this is the presence of a moving mass of water such as is provided by a river. The only essential requirement is that there shall be a sufficient flow of water, preferably throughout the year, to provide the force necessary to drive the generators. Such features as a break in slope or a sharp gradient which result in a rapid flow and a strong head of water are not essential though they may well be helpful in individual cases. The important thing is that enough water shall flow to provide the necessary power, and breaks of slope are not essential for that. It is, of course, true that a greater *vertical* distance between the upper surface of the water reservoir above the dam and the point at which the water enters the turbine produces greater power than a smaller fall of the same amount of water would do (the pressure per square inch on the turbine blades is about 0·433 lb per foot of vertical distance through which the water falls), but in lowland areas the handicap of a small vertical fall is not infrequently offset by the greater volume of water available; this makes it possible to install several penstocks, each one of which leads water to a turbine, and the total amount of energy so produced may be very great.

A reliable flow of water is maintained as a result of numerous factors, a well-distributed rainfall being one of the most important. A large catchment area is helpful as local variations in rainfall are of less general account when the overall catchment area is extensive. Other contributory factors include temperature (high temperatures normally mean increased evaporation which must be made good by increased precipitation, while low temperatures may result in a freezing of the river and a consequent cessation of flow) and the nature of the underlying rock (water moving over porous rock is likely to disappear underground and be lost).

A further feature of great importance is that a suitable site for the necessary dam must be found. Helpful physical features include a narrowing of the valley and stream course; strong bedrock to withstand the tremendous weights imposed by the dam itself and the impounded water; and impermeable rock both at the dam site and beneath the lake which will form (it is possible to seal the floor of a reservoir so that water will not be lost through underlying permeable rock, but such sealing imposes heavy additional costs). It is also desirable that the area concerned shall not be subject to earthquakes and shall be reasonably accessible.

The initiation of a large-scale hydro-electric project may have important repercussions upon the human geography of a region. It was realised, for example, in the Kariba Scheme that the interests of riverside dwellers upstream from the dam had seriously to be taken into account, and in some cases this could entail a lengthy and costly removing of such dwellers to other favourable locations. Two other classic instances of this are provided by the Volta River Project and by the construction of the High Dam at Aswan (Figs. 16 and 17).

The history of the Volta Project in Ghana goes back a long time, but the essential features behind its implementation are modern enough: an increasing need for power to permit industrial development and the need felt by a developing country for a large-scale project for reasons of prestige. The immediate urge behind the scheme was the plan to build an aluminium smelter, originally planned for Kpong but finally located at Tema, which is expected to produce perhaps 135,000 tons of aluminium ingots a year. Clearly, very large amounts of electricity were essential if such a plan was to come to fruition.

But a scheme of this magnitude is bound to have far-reaching repercussions. The building of the Akosombo Dam, 370 feet high, and the consequent formation of Lake Volta, 200 miles long and at 3275 square miles in area covering more than 3% of the total area of Ghana, means that people originally settled along the lower Volta and its tributaries have had to be moved. The town of Keta Krachi has been submerged and the 5000 inhabitants have been settled in a new town; altogether fifty-two new villages have been established to replace those which have vanished beneath the waters of Lake Volta. A new port has been constructed at Tema to handle the increased traffic, while the new smelter, upon which the whole scheme initially depends, has been constructed in the same town. It is confidently expected that other industries will take root now that power from Akosombo is available.* Another point is that

---

* Transmission lines run from Akosombo to Tema, and thence to Accra and along the coast to Sekondi. A large circuit then runs from Sekondi roughly along the line of the railway northwards to Kumasi and then south-eastwards to Koforidua which is

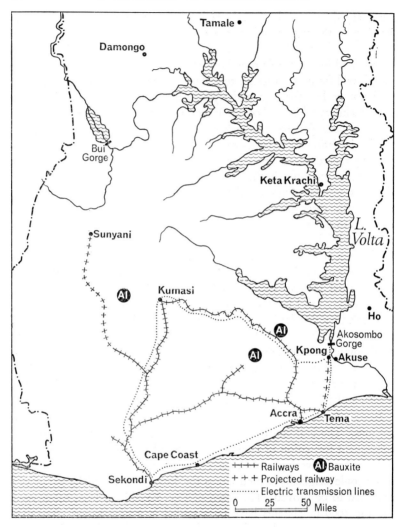

Fig. 16.—The Volta River project.

although Lake Volta will constitute a formidable barrier to east–west communications it will be navigable and it will supply water for irrigation purposes, while it is also anticipated that it will supply much-needed protein in the form of fish to Ghana's daily diet.

---

also linked directly to Akosombo (Fig. 16). It is proposed to construct a further transmission line from Akosombo to Cotonou (Dahomey).

The Aswan High Dam in Egypt is being built at a point four miles south of the original Aswan Dam (Fig. 17) and it is expected that the dam will be completed by 1974. A large storage reservoir, however, has already built up behind the coffer dam and when this is full it will have a high-water mark 200 feet above that of the old reservoir. It will be 250 miles in length (180 miles within Egypt and 70 miles in the Sudan Republic), while it will vary in width between 5 and 10 miles.

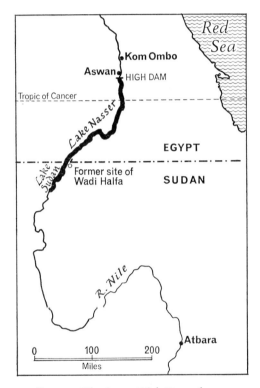

FIG. 17.—The Aswan High Dam scheme.

An inevitable result of the flooding of so much riparian land was that the former inhabitants, including the townsfolk of Wadi Halfa (which is now under water), had to be moved elsewhere. The Egyptian people were moved northwards to a 50-mile stretch of the river near Kom Ombo, a sugar-refining town where a scheme is in hand to increase the amount of cultivated land (sugar cane is the chief commercial crop) by 250,000 feddans (1 feddan = 1·038 acres). The Sudanese people concerned had to be moved very much farther to Khasm el Ghirba, a town on the Atbara River about 600 miles to the south-east (further details of the scheme are given in Jarrett (1966)).

This project has been mentioned because it brings out another point at issue. In these days it is becoming increasingly usual for hydro-electric schemes to dovetail in with other projects, notably for the provision of irrigation, so that the *multi-purpose project* is becoming a typical feature in developing territories. In the case of Aswan an important part of the scheme is the provision of electricity, and it is anticipated that the amount of energy which will be generated will be sufficient to supply all industrial and domestic needs in Upper Egypt. But this is not all; equally important is to be the provision of water for irrigation, with a consequent increase in the available cultivable acreage and the amount of crops grown. Another multi-purpose scheme is under construction in Nigeria at Kainji, upriver from Jebba, on the River Niger, while in Asia the Damodar Scheme in India is well advanced. Work has also commenced on the Hwang-ho Project in northern China. It is likely that we shall hear more about schemes of this nature.

Although we have given above a summary of factors which encourage the establishment of hydro-electricity plants, it is surprising under what apparently unfavourable conditions such plants can be set up if the incentive is strong enough. A good example of this is to be seen in the famous Kitimat Scheme for the production of aluminium from imported alumina in the coastal mountain region of British Columbia (Fig. 18(*a*)).

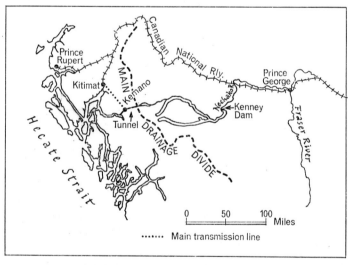

FIG. 18 (*a*).—The Kitimat project (A).

FUEL AND POWER

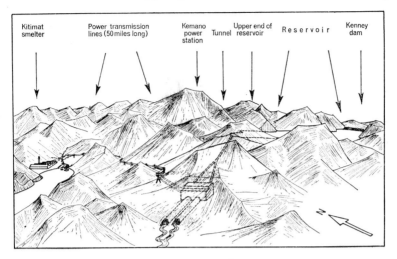

FIG. 18 (b).—The Kitimat project (B).

The original settlement of Kitimat was an Indian village situated on a flat alluvial area at the head of Kitimat Arm, a typical fjord; it was remote and without any regular contact with the outside world.

To the south-east of Kitimat the River Nechako flows *eastwards* to join the Fraser River at Prince George, but today the Nechako has been dammed by the Kenney Dam and the overflow from the resulting lake is led *westwards* to the Kemano power station (Fig. 18(b)). An unusual feature of the scheme is not only that the outflow of water is led underground to Kemano but that the power station itself is also underground, 10 miles to the west of the lake and 2600 feet below it; there is therefore a considerable head of water. One great advantage of this arrangement is that the power-house is protected during the winter and spring from snow and ice, as well as from avalanches.

The Kemano power is led 50 miles to the aluminium smelter at Kitimat. The power line ascends from sea-level to an altitude of almost *one mile* before it descends again to sea-level; great care has been taken to safeguard as far as possible against damage from avalanches as a continuous supply of power is essential to the smooth working of the Kitimat smelter.

This example is remarkable in many ways. It shows how a region which has a very limited selection of natural resources to offer and one which suffers from many handicaps, can support industrial growth if the incentive is strong. The Kitimat area is remarkable for its wild terrain, its harsh climate and its remoteness, while the region contains no native ore and no market. What then, was the secret of the success of this enterprise? Basically, it lay in the astonishing measure of control established over the unpromising environment. A totally inadequate catchment area was greatly extended by the reversal of flow of the Nechako, while the underground working offers protection against climatic excesses. The

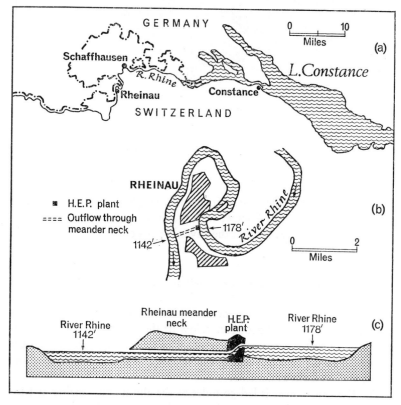

FIG. 19.—The Rheinau H.E.P. plant. *Based partly on Boesch.*

navigable Kitimat Arm provided an undoubted advantage as did also the flat delta upon which the smelter and the modern town of Kitimat have been built. This tide-water setting makes it possible to import alumina for smelting from Jamaica and to export the refined aluminium, mainly to the U.S.A. and to Europe (*see* p. 24 above). The ingots which are produced at Kitimat are 99·65 pure aluminium.

In order to correct the widely held idea that a considerable fall of water such as that found at Kitimat is necessary for the production of hydro-electricity. Fig. 19 illustrates the siting of the Rheinau plant situated on the River Rhine near the German–Swiss frontier. The diagram shows that in this case the head of water is slight and has been secured by cutting across a meander neck.

Thermal Electricity

The various factors which govern the production of thermal electricity are quite different in character from those we have just been considering, and for convenience we will tabulate the main points:

1. The location of the thermal electric station should be such that large amounts of coal are available comparatively cheaply. Sixty per cent of the works cost of generating electrical energy comprises fuel costs, so the importance of this factor will be clearly understood. Poor quality coals are not infrequently used in power stations and to secure the maximum economic benefit the coal should not have to travel long distances.

2. The location of the power station should be fairly near to the consuming area in order to minimise costs of distribution and losses due to transmission. It is, however, true that this factor is becoming of less importance in countries where a grid system of transmission has been adopted.

3. The site* must have easy access to very large supplies of cooling water. This is a most important point. When the cooling is direct each 60 mW unit of operating capacity needs about 2 million gallons of water *hourly*, while a 1000 mW station needs 39 million gallons an hour. When this water is returned directly to the source of supply it is between 10° and 20°F warmer than it was originally, and this fact can itself exercise severe limiting effects. For instance, the balance of aquatic life can be greatly disturbed by the influx of such large amounts of heated water, while it is not normally possible to site power stations very closely to one another along any single river.

When large amounts of water on the scale mentioned are not available recourse must be had to supplementary cooling techniques, and cooling towers then become necessary. Even so, the amounts required are considerable; the daily requirements of the Hams Hall "B" station near Birmingham of 321 mW capacity are 10 million gallons and this amount is supplied by the modestly sized River Tame. Cooling with the help of cooling towers is more expensive than by the direct method but it must be used if very large amounts of water are not available.

* Notice that the term "location" refers to the general situation of a town or factory and the term "site" to the specific situation at a certain point.

4. The site must be fairly extensive to allow for the generating plant and cooling towers (if there are any), for coal-handling plant and coal dumps, for railway sidings and for ash dumps (even if these are temporary). When we realise that a single large power station may excrete up to one and one-third million cubic yards of ash a year we shall realise that the disposal of waste upon this scale constitutes a most pressing problem, even if ultimately the disposal is more widespread than on the actual site. The Castle Donnington station (Fig. 20) covers 190 acres of ground and larger stations may need up to 1000 acres.

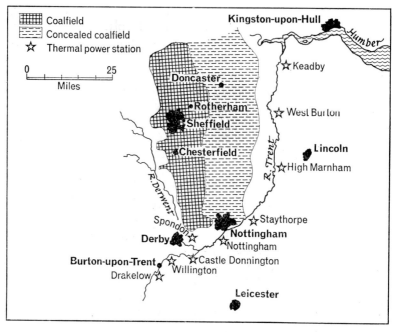

FIG. 20.—Thermal power stations in the Trent Valley.

5. The mention of the disposal of waste, especially ash, reminds us that such disposal is an important consideration in its own right. An interesting example comes from the Trent valley, where fuel ash is finally dumped in gravel pits so that the general level of the river terrace from which gravel is excavated is maintained. Waste gases emitted by the plants are obnoxious and care must be taken to ensure that these gases are removed as far as possible from settlement; one Trent station, for example, emits 160 tons of sulphur dioxide daily, and to disperse this

chimneys up to 600 feet high are needed. Unfortunately, chimneys of this size can constitute a very real hazard to aircraft and it is not therefore desirable to construct them in busy flying areas. There is also the criticism that high chimneys may simply throw the waste over adjacent areas on to more distant locations.

6. In view of what we have said about the necessity for an extensive site, we might add that land which is comparatively cheap and of a low rateable value is desirable. This need is of more importance than easy access to a large labour supply as labour requirements are modest in relation to the other factors concerned (*see* note on fuel cost, item 1 above). This is one reason why power stations are typically located outside town and city limits (another reason for this is the question of waste disposal previously referred to).

It will be clear from the foregoing that suitable locations for large-scale electric power stations will not be numerous. It is no accident that in England the River Trent region (Fig. 20) between Burton and the Humber has developed into one of the leading suppliers of thermal electricity,* and Rawstron (1951) has examined the reasons for this. The following are the main points:

1. The area is well placed to supply power to the industrial districts of the Midlands and to those of the North, especially to the very important region which lies near the southern and eastern flanks of the Pennines.
2. The necessary supplies of coal lie within economic distance. The chief supplier is the extensive Yorkshire, Derby and Notts Coalfield, the largest and most productive in England, while the Midlands coalfields are also within reasonable reach.
3. The River Trent is a large river by British standards and for the greater part of the year it carries a sufficient flow of water to supply the water needs of one 400 mW station every 10 miles along its course below Burton.
4. Railway communications are good. The valley is crossed by main lines which can cope with the necessary assembling of fuel.
5. For the most part the river flows through non-urban areas so that the problems connected with cheap land and waste disposal are not acute. We have previously mentioned the unusually favourable circumstances for the disposal of ash.
6. This part of the course of the River Trent is the only substantial length of river in England which satisfies all these conditions.

* See map in Simpson (1966), p. 148. This region is sometimes known as "Killowatt Valley."

MINERAL OIL

The third of the sources of energy with which we are concerned in this chapter is mineral oil. Like electricity, mineral oil is a comparatively new energy source and its effects have become notable within the past few decades (Figs. 10 and 12). In the United Kingdom, for example, the consumption of petroleum products increased fivefold between the years 1938 and 1960, the most significant increase being in the use of fuel oils.

It is not easy in these days, when the use of mineral oil products is so widespread and so much a part of the lives of so many people, for us to realise that it was only in 1859, little more than a century ago, that the first oil well was drilled at Titusville, Pennsylvania, amid much scoffing (how could it be possible to secure liquid oil out of hard rock?). It is true that hand-dug wells had been producing even before that in Burma and in Romania but their output was extremely limited. Opposition to the use of the new fuel was not unknown; in 1864, for example, a petition was presented to Congress in which the petitioners prayed that:

> "A stop may be put to the irreverent and irreligious proceedings of various citizens in drawing petroleum from the bowels of the earth—thus checking the designs of the Almighty, who has undoubtedly stored it there with a view to the Last Day, when all things shall be destroyed."

We have come a long way since 1864!

Origin of Mineral Oil

There is no certainty as to the origin of mineral oil. At one time it was believed that oil was inorganic in origin and that crude oil was formed by purely chemical processes. Laboratory experiments certainly show that oil can be made in this way, but opinion slowly swung round to the belief that such processes were unlikely to have occurred in nature on a scale sufficiently great to produce the very large amounts of mineral oil which have already been discovered. For some time, therefore, opinion generally was inclined to accept an organic theory of origin, according to which an extremely slow accumulation of fragments of past plant and animal life was supposed to have collected in the mud flooring of ancient seas and lagoons, and to have provided the "raw material" for mineral oil. Slow chemical change taking place as a result of the action of anaerobic bacteria on this raw material (possibly with some chemical interaction with salts in the overlying sea-water) converted the organic remains to oil, and the oil entrapped within the shale changed its location as earth movements deformed the strata (the original mud was normally changed to shale by pressure exerted by overlying rock laid down later, by pressure of overlying water and by earth movements and compression). The oil is thought then to have migrated into more porous rocks and to have become trapped in structures such as anticlines (Fig. 21). Further details can be found by the interested reader in textbooks on physical geography and geology, and also in Bengtson and Van Royen, but it might be useful to point out that when petro-chemical material migrates underground the constituent parts to some extent may separate out according to their relative densities. For example, natural gas, the lightest constituent, commonly occupies the upper horizons and it typically occurs in the roofs

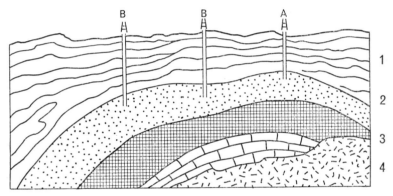

FIG. 21.—Diagrammatic section showing a possible mode of occurrence of mineral oil. (A) Well which might produce natural gas; (B) Oil wells. (1) Surface rocks; (2) oil-bearing stratum arched into the anticline which forms the oil trap; (3) unconformity; (4) underlying crystalline rock.

of anticlines (Fig. 21), while the mineral oil itself occupies a lower horizon.

In recent years a duplex theory of the origin of mineral oils has been advanced by Sir Robert Robinson, a director of the Shell Chemical Company and a former Nobel prize-winner. He considers that both organic and inorganic processes may have played a part in the formation of crude oil, and for the inorganic basis he would go back to a very early stage in the history of the globe when the atmosphere may well have consisted mainly of methane, carbon monoxide and ammonia. The action of sunlight on such an atmosphere may well have produced hydrocarbons, a large proportion of which may have floated for long periods on the surface of early seas. Such hydrocarbons could have been washed inshore gradually by ocean currents and tides and there mixed with organic matter formed in a manner comparable to that already described. Such an origin would require very lengthy periods for completion, but, as available evidence seems to show that mineral oil deposits date back for as long as between a million and 400 million years, this is not a weighty objection.

The game of guessing the origin of mineral oil has a serious side, for the more we know about the origin of this valuable source of energy, the more we shall know about the most likely places to search for it, and the more we can deduce about the extent of possible reserves. Prospecting based on the organic theory, for example, has been most successful in leading to the very productive exploration of recent years; this theory also provides a reason why oils from different regions vary widely in character from the very light crude oils of North Africa to the unusually heavy ones of Venezuela and the Netherlands. It seems reasonable to suppose that different source materials could explain such variations. The duplex theory would explain the variations on similar grounds. The argument continues, however, and Russian scientists now appear to support an inorganic theory of origin.

Early Developments

It was in 1859, then, that Colonel Drake (who, ironically enough, died in poverty) sank the first oil well with the help of a new technique—the cable-drilling method. The maximum depth reached with the help of this technique was 7350 feet but the rotary-drilling method (which was in use by 1900) permits drilling to depths of 25,000 feet and even more. Transport facilities have also been enormously developed and it is a far cry from the *Elizabeth Watts*, a 224-ton brig which in 1861 sailed the first cargo of kerosene from the Delaware to the Thames, to the *Gluckauf*, the first tanker, constructed in 1886 and of 2307 gross tons, and to the giant supertankers of today. A tanker of 12,000 d.w.t.* was considered a reasonable size just after the Second World War but tonnages have risen sharply since that time and a tanker of 131,000 d.w.t. was built in Japan in 1961. In 1966 the 150,000 d.w.t. *Tokyo Maru* was the largest

* Ship tonnages are designated in various ways. *Gross tonnage* refers to the capacity of a ship measured in ton units of 100 cubic feet each with certain authorised deductions; *net tonnage* is similar but allowance is made for space occupied by the crew and any passengers; *measurement tonnage* is similarly arrived at but the unit capacity is 40 cubic feet; *displacement tonnage* refers to the weight of water (in long or metric tons) displaced by the loaded or unloaded vessel; while *dead-weight tonnage* (d.w.t.) refers to the cargo-carrying capacity of a vessel expressed in tons (long or metric). The latter form of designation is useful when the capacities of ships carrying similar cargoes are to be compared, but in other circumstances it is more usual to use the net or displacement tonnages.

vessel of any kind afloat, but by the end of that year a 205,000-tonner was launched. Tankers of 300,000 d.w.t. will shortly be in use, while it is likely that even larger vessels will follow (p. 248 below). More details regarding the distribution of mineral oil will be found in Chapter XI.

Advantages of Mineral Oil

Mineral oil has very considerable advantages as a source of energy, and these have been touched upon earlier in this chapter (p. 46 above). We might now examine this point a little further, noting first of all that unit for unit fuel oil is notably more efficient than coal; the calorific value of one ton of mineral oil averages about that of one and a half tons of bituminous coal: substantial fuel economies can therefore be secured by using the newer form of fuel. It has been pointed out that in the heavy clay industry productivity has been increased by as much as 40% where kilns are fired with oil rather than coal.

Another great advantage of oil is its mobility. As it is a liquid it can be comparatively easily handled in bulk without the dirt and dust which characterise bulk handling of coal. Neither is there any risk of damage in transit. There is an economy of storage space, too, for a ton of oil occupies less space than a ton of average-quality coal, while in terms of calorific content the saving is even greater. A further saving occurs as oil contains no ash, while the quality can be standardised as the quality of coal cannot be. These economies of space operate with regard to transport as well, for the costs of moving any given number of British Thermal Units will be less if the units are locked in oil rather than in coal.

Consumption of Mineral Oil

It is not surprising under these circumstances, that consumption of mineral oil has greatly increased in recent years. This is borne out by Figs. 10 and 12 and also by the following table:

World Energy Consumption

	Coal	Oil	Natural gas	Nuclear power and H.E.P.
1938	73%	20%	6%	1%
1948	63	27	8	2
1962	47	36	15	2
1963	44	39	15	2
1965	43	39	16	2

Source: Statistics supplied by the Petroleum Information Bureau.

We should bear in mind that the increases in the proportion of fuel oil and natural gas shown in the above table do not tell the whole story since the amounts of all fuels (including coal) actually used have increased very considerably during the present century. This is shown in Figs. 10 and 12, and in the case of mineral oil in the following table:

Production of Mineral Oil

	1938 Thousands of metric tons	%age	1966 Thousands of metric tons	%age
N. America	171,600	61	510,010	31
Middle East	16,200	6	471,240	23
Sino-Soviet Area	37,780	13	291,570	18
Latin America	44,230	16	239,600	15
Africa	—	—	128,740	8
Far East and Australasia	10,000	4	33,320	3
Western Europe	690	—	21,140	2
WORLD TOTAL	282,500	100	1,697,200	100

Other Considerations

Effective production of mineral oil is dependent upon factors other than the mere occurrence of oil reserves, essential though these are. The quality of the crude oil, for instance, can be important and may become more so. Venezuela has for many years held a leading world position in the supply of mineral oil, most of her exports going to the U.S.A., but Venezuelan oil suffers from a high sulphur content. This will become a very important point with the passing of anti-pollution laws in the U.S.A. (p. 278 below), for sulphurous gases emitted into the atmosphere are not only unpleasant but also toxic. The cost of sulphur extraction from the crude oil is comparatively high—about 80 cents a barrel on oil costing roughly $1.60—and this must be a handicap on the Venezuelan product (which is, however, at present the cheapest oil available along the American east coast).

Political considerations are bound to be important when so much of the world's crude oil comes from developing territories. There is a natural tendency for host governments to try to squeeze the last ounce of revenue out of the oil companies, and this can lead to periodic uncertainties surrounding production in individual countries. This may be particularly the case in such territories such as Venezuela where general elections periodically fall due, for no political party which aspires to power can afford to relax terms offered to oil companies; political kudos will rather accrue to the party which promises greater returns from these companies. In an age of emerging nationalism it is also true that political pressures are constantly urging host governments to attempt to secure a greater measure of direct control over the exploitation of what is perhaps their only major natural resource, and such governments are becoming less willing to leave the taking of executive decisions to foreign companies based in overseas centres such as London or New York. The days when companies were awarded concessions which gave them the right to explore, exploit and export mineral oil for an indefinite period according to their own judgements are on the way out. We shall almost certainly see more moves like that of Aramco, which has moved its headquarters from New York to Dahran in Saudi Arabia, and more governmental participation in the production of mineral oil is certainly to be expected.

SIGNIFICANCE OF THE NEWER FUELS

The increasing use of the newer forms of fuel mentioned in this chapter has had a profound effect on the location and development of industry. Without the additional fuel supplies which they afford the development of industry on the scale which has actually taken place could never have been achieved; coal supplies could never have been sufficiently expanded to meet such a vastly increased demand and this would have been a severely limiting factor upon industrial growth in general. Furthermore, the flexibility and mobility of the newer fuels (especially electricity and mineral oil) have helped to liberate industry from the coalfields in a way which could not have been foreseen at the beginning of the century. Electricity and fuel oil can be distributed with far greater ease than coal, and increasing use of them has permitted industry to become far more orientated towards markets. This has been a powerful factor behind the movement of industry to the south of England and to the South in the U.S.A.

We might finally notice that the development of new forms of energy during the present century has led to the parallel development of a wide range of completely new but associated industries. It is obvious, for example, that the growth of industries related to electricity has been phenomenal and includes such various activities as the manufacture of sparking plugs for cars and of diesel-powered engines, as well as the extensive wireless and television interests. The electro-plating industry is dependent upon electrical energy as is the smelting of aluminium from alumina. Industries based upon mineral oil and natural gas are today bewildering in number and complexity, and a selection of some of the main products derived from crude oil after refining will demonstrate this.

76 A GEOGRAPHY OF MANUFACTURING

Raw material	Manufactured products
Gases	Ether, solvents, lacquer, explosives, saccharin, dyes, varnishes
Light distillates	Petrol, dyes, varnishes, turps, substitutes, kerosene
Intermediate distillates	Diesel fuel oil
Heavy distillates	Heavy oils, waxes (used among other things as a base for detergents), lubricating oil
Residues	Heavy lubricating oils, petrolatum, asphalts and bitumen, roofing material. From petrolatum are manufactured gear and axle grease, metal-coating compounds, cable-coating compounds, petroleum jelly, cosmetic creams, rouge, lipsticks

Further reference to this point will be made in Chapter XI.

SUGGESTIONS FOR FURTHER READING

Bengtson, N. A., and Van Royen, W., *Fundamentals of Economic Geography*, London, 1957.
Boesch, H., *A Geography of World Economy*, New York, 1964.
Dury, G. H., *The East Midlands and the Peak*, London, 1963.
Estall, R. C., and Buchanan, R. O., *Industrial Activity and Economic Geography*, London, 1961.
Fearnsides, W. G., and Bulman, O. M., *Geology in the Service of Man*, Pelican Books, 1945.
Holmes, A., *Principles of Physical Geology*, London, 1945.
James, J. R., Scott, S. F., and Willats, E. C., "Land Use and the Changing Power Industry in England and Wales," *Geographical Journal*, 1961.
Jarrett, H. R., *Africa*, second edition, London, 1966.
Kimble, G. H. T., *Tropical Africa*, New York, 1960.
Lake, P., and Rastall, R. H., *Textbook of Geology*, fifth edition, revised by R. H. Rastall, London, 1956.
Mountjoy, A. B., *Industrialization and Under-developed Countries*, London, 1963.
Rawstron, E. M., "The Distribution and Location of Steam-driven Power Stations in Great Britain," *Geography*, 1951.
Rawstron, E. M., "Changes in the Geography of Electricity Production in Great Britain," *Geography*, 1955.
Reddaway, W. B., "External Capital and Self-Help in Developing Countries," *Progress*, 286, 1965–66.
Simpson, E. S., *Coal and the Power Industries in Post-war Britain*, London, 1966.

Chapter IV

Making and Selling

WE have now considered the more important natural bases of industry, the raw materials from which manufactured articles can be made and the sources of energy which make the manufacture possible. But the road which must be travelled by the various materials between the time they enter the factory and the time they reach the consumer as manufactured goods may be a very lengthy one, and we shall in this chapter examine three vitally important factors which make the journey possible; these factors are labour, transport and markets. The importance of the labour element is obvious since nothing can be made without it, but we might also bear in mind that the organisation of labour (management) can be just as important in its way as the contribution made by the man on the factory floor. Transport is an essential element because it is rare indeed that manufactured goods are sold at the factory site, and arrangements must normally be made to transport the goods to potential consumers. And, finally, we must realise that the attractive power of the market can itself be an important locational factor.

LABOUR

One of the most important features regarding labour is that it must be paid for, but the comparative importance of labour costs varies greatly from one industry to another. Figures in the *Annual Survey of Manufactures*, *U.S.A.*, 1955, suggested that in that year the wages of production workers accounted for about 37% of the total value added by manufacturing industry[*] over the whole range of industry in the country, but this

[*] It is not easy to compute with any exactness the total contribution made to any economy by a manufacturing establishment or industry. Ideally, if a figure could be calculated it would exclude all value created elsewhere and include only the wealth produced by the application of land, labour and capital within the manufacturing establishment or industry. The closest approximation to such a figure which is normally used is the *value added by manufacture*, and this is calculated by subtracting expenses incurred in the purchase of raw materials, component parts, containers, fuel

average figure masks wide variations. Estall and Buchanan have shown that labour costs can vary between 62%* in the clothing and related industries and 29% in the chemical industry; in the fabricated metal products industries they work out at 43%. The reasons for these variations are not hard to find. The clothing industry, for example, includes the high-cost fashion interests and these interests are tied geographically to high-wage fashion centres such as London, Paris and New York; location at these centres confers much of the glamour and prestige which constitute an essential feature of these industries. Since the labour costs involved are necessarily unusually high, because of the insistence upon top-grade workers at every stage from designing to modelling, it is easy to see why labour costs in the world of *haute couture* are well above the 62% average for the clothing industry as a whole.

In this branch of the clothing industry location is established despite high labour costs and the same is to some extent true of the bespoke interests which grow up wherever there is a market of such a size to warrant it. The important feature is the close personal relation which exists between the customer and the producer, a relationship which the customer values so highly that he is willing to pay extra for it. On the other hand, ready-made clothing manufacturers do not have to consider such relationships and they are free to locate their factories with more regard to other factors such as local skills and wage differentials.

In the case of the chemical industry mentioned above, the situation is completely different because the total labour force normally employed is comparatively small. We have noticed a comparable case in the production of thermal electricity (p. 69 above), while the oil-refining industry is another case in point; the large Milford Haven refinery employs only 330 men. Under these circumstances the industry concerned can be located with a considerable disregard for problems of labour supply—a marked contrast to the case of the textile industry (*see* item (3), p. 225 below).

Mobility of Labour

Having established the point that relations between labour and industry vary very considerably from one group of industries to another, we might proceed to examine the relationship from the point of view of

---

and power purchased externally, and other related items from the value of the finished product. Certain expenses such as taxes, costs of insurance and transport costs are normally not regarded because of the difficulties encountered in trying to allow for them.

* Of the total value added by manufacture.

the labour itself, and the first question we will raise is that of labour mobility. This question has at least two distinct aspects, that of mobility in a geographical (spatial) sense and that of mobility between different industries, and the present state of opinion regarding labour mobility in the spatial sense seems to be very divided. Estall and Buchanan talk about the "relative immobility of labour" though they admit that the "impediments to geographical mobility of labour are much less serious in advanced than in under-developed areas." Miller, however, argues that "in the modern period labour is exceptionally mobile" and that the unskilled worker is rarely tied to a particular locality by ownership of property or by strong social ties, a statement that appears to be at variance with observations made elsewhere which suggest that unskilled workers and their families often display a very strong attachment to their local community. On the other hand, Miller seems to assume that skilled and specialised labour is often concentrated in particular localities and draws industry to itself, and he cites the examples of Naugatuck Valley, Connecticut, which manufactures about half of the output of brass of the whole U.S.A., and of New Britain, which is noted for its hardware. Other examples could include such cases as the pottery industry of North Staffordshire, the small metal trades of Sheffield and the craftsmen gunsmiths of Birmingham. While these examples are real enough it seems dangerous to generalise, though experience generally would seem to indicate that skilled labour is, in fact, relatively mobile, while unskilled labour is noticeably less so. It may be that the real answer to the problem rests upon the distinction between the short run and the long run of the economist. In cases where an industry is strictly localised, perhaps because of its historical setting, the appropriate skilled labour will be correspondingly localised in the short run and the necessary skills will only with difficulty (if at all) be found elsewhere, but in the long run such a monopoly or quasi-monopoly is bound to be broken and labour skills will become more widespread. This has happened in the case of the textile industry, to the great discomfiture of such old-established areas as Lancashire and New England.

We should note that labour migration in these days is not infrequently international in character, as has been shown by the considerable post-war migration of industrial workers from Mediterranean Europe to such countries as France, Western Germany and the United Kingdom. A case remarkable for the distances involved is that of the migration of large numbers of workers from Britain, Italy and Malta to Australia; much of the post-war industrial development of the latter country has been made possible by immigrant labour. In the mid 1960s, for example, almost 50% of the workers at the Port Kembla iron and steel plant in

D

New South Wales were migrants from southern Europe, while of the 4000 employees of the B.H.P. iron and steel plant at Whyalla, South Australia (*see* p. 203 below), more than 1000 came from continental Europe. Over 25% of the total population of the town of Whyalla were of British origin.

The case of mobility between industries is not a simple one. While unskilled workers can often move without much difficulty from one industry to another, the case may be quite different for the skilled worker. It is true that certain skills are needed throughout a wide range of industries but it is equally true that other skills are of value in one type of industry only. No general point, therefore, can probably be made regarding the mobility of skilled workers, for so much depends upon the type of skill required.

Labour Availability

No one can deny that the prior existence of a labour force is attractive to industry unless there are strong reasons to the contrary (some are mentioned later in this section). Notice that there may well be a double point here, since the existence of a considerable labour force often implies the concurrent existence of a market, though the existence of a market in general does not necessarily imply a market for a specific product. Dury has shown that the two main factors involved in the locating of an aluminium-milling industry at Banbury were (*a*) the general location of Banbury with respect to markets (London, Birmingham and Coventry) and (*b*) the existence of a local labour supply in an agriculturally depressed area where there were few competitors for labour. The Birmingham Chamber of Commerce (*The Times*, 28 May 1968) argues that availability of labour is a prime consideration for industrialists, and that in some cases shortage of labour has led to costly withdrawals from industrial enterprises, and in some cases even to complete closures of companies. The first man-made fibre factory in Australia, at Newcastle, New South Wales, was attracted to its site partly by the availability of female labour in an area notable for its mining and its heavy industry. This point reminds us that male and female labour are generally found in joint supply, and, as in Newcastle, there can often be a surplus of female labour in an area in which the main industries are those which need male labour.

An industrialist will normally have serious doubts about the wisdom of establishing an industry in an area devoid of labour, though industry has been established under such apparently unpromising conditions as is shown by the examples of Kitimat and, to a lesser degree, Milford Haven. The case of Milford Haven, with its oil refineries, is in this respect somewhat comparable to that of Kitimat, for the location was "far from any great concentration of population" (James, Scott and Willats) and the refinery was attracted there by advantages which had nothing to do with labour (*see* Chapter XI). It is only fair to point out, however, that these industries are examples of those in which labour costs are comparatively small and the numbers of workers comparatively few.

The Wage Factor

The question of differential wage rates is one which must be referred to, for it is well known that wage rates can vary considerably, even from place to place within the same country; notably they are higher in large towns and cities and lower in small towns in the countryside. In 1960, for instance, the average hourly earnings for production workers in the U.S.A. varied between $2.62 in California and $1.54 in North Carolina, while the figure in Tennessee was $1.84. Perhaps the main reason for the variations is the sharp competition for labour in California, where industry and agriculture are booming, as opposed to the lack of competition in North Carolina and Tennessee. A similar situation is illustrated by the fact that a survey conducted in 1959–60 showed that the wages earned by Class A welders in the U.S.A. varied between $1.83 in Dallas, $2.26 in Pittsburg and $2.43 in Newark and Jersey City. It is quite possible that within the same country, wage differentials of this kind occur in the short run only and are evened out in due course as labour redeploys itself and as industry locates, or relocates itself, to take advantage of the differentials.

We should not make the common mistake of supposing, however, that cheap labour is necessarily low-cost labour. The reverse can well be true, for labour of poor quality is usually high cost because it is inefficient, however low crude wages may be. What is significant is the wage rate: labour-output ratio, and, where this ratio is high, labour costs are in a sense always low, however high crude wage rates may be.

It is sometimes assumed that low wages constitute an attraction to employers *per se*, but this is not always true. Labour rates in India, for example, are nominally low yet we find that in many cases employers are not anxious to take advantage of this fact but install advanced automatic machinery which actually economises on labour. Part of the reason lies in the poor quality of the very limited supply of skilled and semi-skilled labour available, but there is the additional reason that firms may suffer what is in effect a heavy tax on labour, as they may be required by law to provide fringe benefits in the form of housing and other services which do not normally fall on employers in more highly developed countries. There is under these circumstances a strong incentive for a firm to keep its labour force as low as possible, despite the unfortunate social repercussions which such a policy may have. On the other hand, there seems to be little doubt that the amazing rate of industrialisation in Japan can in part be accounted for by the low overall labour costs which generally obtained during the earlier years of industrial growth (*see* also the case of cotton spinning and the American Piedmont, p. 228 below). It is interesting to note that wage rates have now risen in Japan so that she in turn is feeling the pinch of competition from low wage countries such as China and India (p. 222 below).

Labour Relations

Perhaps just as important as wage differentials are labour relations, and it is often true that newly developing industrial areas have an advantage in this respect. The reason is that with development of industry over a fairly lengthy period of time workers often come to adopt what may perhaps not unfairly be described as "fossilised" attitudes towards their work, and such attitudes can have unfortunate results. What begins as a reasonable working practice can fossilise into a custom which must not be broken, even when changing circumstances make change desirable, while constant attempts are generally made to increase wage rates and fringe benefits and to decrease hours of working. It is significant that a nominal reduction in hours of work agreed upon frequently means that just as many hours are worked after the agreement as before, but that more of them are classed as "overtime."

"New" industrial areas, therefore, in which workers' attitudes have not had time to harden (or fossilise) frequently have an advantage over older ones, and this is held to be one of the reasons why industry is moving south both in England and in the U.S.A. McLaughlin and Robock have examined the reasons why eighty-eight firms made location decisions which led to the siting of new factories in the U.S.A., and they discovered that in almost every case the firm concerned made a careful study of labour–management relations in areas considered suitable for development on other grounds, and that in some cases a history of poor relations led to the rejection of locations which were considered suitable in other respects. It has been argued that one reason for the startling development of the Illawarra District of New South Wales (*see* below) has been the fact that labour relations in the earlier-developed Newcastle and the adjoining coalfields districts have been very poor, and that an influential firm such as the Broken Hill Proprietary Ltd. (B.H.P.) preferred, therefore, to expand many of its activities in the newer district. It has many times been argued that, while the decline in the Lancashire cotton textile industry was probably inevitable, the decline was hastened by restrictive practices and other forms of fossilised attitudes displayed by the workers.

The question of the relationship between labour and management as evidenced in wage rates and in general attitudes one to the other is clearly a complicated one and simple generalisations are not easy to draw. Even the apparently straightforward matter of wage differentials is not simple on closer inspection, for in addition to the vital question of relative labour efficiency there is the question of relative costs of living in different areas to be considered. While it may be true that wages in one area may be lower than in another part of the same country it may also be true that the cost of living is lower in the low-wage area and that the worker therefore, does not suffer in real terms (*see* p. 97 below). It is even possible that it is only the fact of lower wages in such an area which enables a firm to continue operations there, because the lower wages offset higher operating costs in other directions.

Food Supplies

It might be useful to introduce at this stage a point which will receive further consideration later, that industrial populations must be fed from non-industrial sources; indeed, the question of maintaining an adequate food supply may be critical in particular instances (note what was said about North America above, p. 18, and what is said about Spain and Mexico below). Professor Arthur Lewis pointed out some years ago that in developing countries it may be more important to develop food supplies than to encourage industry. If food supplies are not adequate and industrial growth takes place it can mean that labour is lost to the land, that less food is therefore produced, and that the money wages earned in the factories simply cause food prices to rise in an inflationary trend.

Managerial Skill

We have already made mention of the importance of managerial skill, and it seems especially important to emphasise this in this egalitarian age. The blunt fact is that without far-sighted direction any industrial economy will sooner or later (probably sooner) run into serious trouble, and it is another blunt fact that a community which is unwilling to accord to managers and entrepreneurs a favourable position economically is likely to find the supply of such people slowly diminishing. Perhaps one may pause to marvel at the suicidal tendencies evinced by certain societies today, which begrudge economic advantages and benefits to their economic leaders, but which are happy to lavish them upon winners of football pools and pop singers.

It is sometimes said that managerial skill is usually mobile, but the mobility is evident only within fairly sharply defined limits. It can, of course, move only within or between industrial areas though potential industrial areas may attract members of the managerial class if prospects are sufficiently attractive. Questions of climate, of social prospects, of educational opportunities for children, and of general living conditions are all relevant, and it might be argued that one reason for the emergence of the south of England as a major industrial region is that it can offer conditions of living which are said by many to be preferable to those typical of the North. Garwood has shown that the post-war migration of industry in the U.S.A. to Utah and Colorado, is to be explained in part by the attraction of the climate to owners of industrial enterprises and to the workers.

Managerial skill can only blossom in the right environment; to take an extreme case, it is clearly hopeless to look for this type of skill among communities which have no industrial development. While there may be potential Henry Fords and Lord Beaverbrooks among the Amazon Indians, they will remain strictly potential, since there will be no opportunities for them to put their inherent abilities to good use for a very long time to come. This is one reason why industrial development feeds on itself, for, when the opportunities are presented by co-operation between inventors, capitalists and workers, the entrepreneurs to spearhead further expansion have an opportunity to emerge. Developing territories are naturally in an unfavourable position in this respect as their potential managers have had little or no opportunity to develop their skills, and organisers must therefore be sought elsewhere. This may give rise to difficulties since resentment can easily be bred from this enforced dependence upon the human resources of other nations and other races. It is also true that managers are generally not easily persuaded to take up employment in developing territories where prospects are often very uncertain, though a potent persuasive force can be exercised by unfavourable economic and social conditions in their homelands. A combination of attractive forces operating from overseas and repellent forces at home can be very powerful and is largely responsible for the "brain drain" from Britain today.

The importance of the quality of management is clearly shown when an area has to compete in a broad market, possibly an international one, to sell its products. Outstanding examples of this are to be found in England and in New England, long-established industrial regions which in earlier years enjoyed a near monopoly of manufactured products, the one on an international scale and the other in the developing markets of the U.S.A. As other countries and other regions entered the industrial race, however, these older developed areas discovered that they were no longer able to sell at will, and they had to give considerable thought (a) to bringing established industries up to date, and (b) to the necessity of scrapping some industries altogether and developing new ones. Now such policies involve far-sighted appraisals of future trends, and the taking of decisions which may in the short run be painful, since old-established industries cannot be run down and new ones established, without at least temporary dislocation in the labour force (which means crudely that many workers will almost certainly suffer periods of unemployment). Constant development of new technical skills is also vital. Questions of management are clearly heavily involved here, though imagination and skill on the part of the workers are also quite essential. Professor Sir John Baker has lamented that "it is staggering to find among otherwise perceptive people a lack of appreciation of the imagination, originality, scientific knowledge and courage which is called for in designing and making an engineering product," and it seems an essential point that a community which does not evince the qualities enumerated by Baker cannot hope to develop and maintain its industrial strength to any marked degree.

TRANSPORT

We might at this stage turn to a consideration of situations which typically present themselves as goods are moved from the factories to the markets. In this case, too, the situation is not as simple as it may at first seem. For instance, it is generally realised that transport costs can exercise a considerable effect on the location of industry, but it is perhaps less often realised that the essential feature is not that of crude transport costs, but that of a less tangible element to which the term "transfer costs" is normally applied. It is true that transport costs enter heavily into transfer costs but it is also true that other elements are involved. Such elements include clerical costs which may be heavy and which may, in the case of a large scale enterprise, involve the maintenance of a complete department to supervise the receiving and despatch of goods; insurance costs; costs involved because of damage to or deterioration of goods in transit; and the indirect capital costs which arise during transit when goods are locked up, as it were, and effecting no return to the manufacturer. The reality of these capital costs is clearly demonstrated by the working of bills of exchange and by the discount at which these bills can be purchased before maturity.

The Determinants of Transport Costs

Transport costs do not vary in any direct manner, certainly not according to mere distance. In many instances, indeed, short hauls can prove comparatively expensive as costs of loading and unloading have to be met whether the intermediate haul is one of five miles or five hundred, and the total cost of the longer haul is unlikely therefore, to rise in proportion to the distance involved. Martin has remarked that in Lorraine the high charges imposed by the French railways for short hauls may raise ore costs for works "at a moderate distance" from iron-ore mines by as much as 30%, and this places works in the Northern and Nancy districts at a comparative disadvantage.

The efficiency of the means of transport available is a further important factor in determining costs of transport. In pre-railway days in Kenya, for example, the cost of head-portering one ton of produce from Uganda to Mombasa worked out at about £300, but the rail freight today on one ton of cotton lint moving over the same distance is £5. Russell Smith points out that when the Erie Canal was opened in 1825, the freight rate between Buffalo and New York dropped from $100 to $5 a ton, while the time taken over the journey dropped from twenty days to eight; it has been this of course, which has so greatly assisted the continuing development of New York and the Erie Canal zone as areas of manufacturing industry.

Another point to bear in mind is that the well-known principle of charging "what the traffic can bear" can completely upset any possible correlation between distance and total costs of transport, for the possibility of competition has constantly to be borne in mind by the transport operators. It is not surprising, for example, that the opening of the St Lawrence Seaway in 1959 brought about a rapid reduction in railway freight rates between Lake and Atlantic ports, but perhaps it is rather less to be expected that freight rates between Chicago and New York are lower than those between Chicago and Philadelphia, although the goods bound for New York from Chicago actually *pass through* Philadelphia! The point is that there is an alternative (water) route via Buffalo and the Erie Canal available to goods passing between Chicago and New York, and rail costs have to be adjusted to allow for this, but there is no alternative route between Chicago and Philadelphia.

Similar instances to the above are not difficult to find. Some Atlantic coast industries of the U.S.A. are not able successfully to compete with rivals in the Mid-West because of high transfer costs but they can compete with these same rivals along the Pacific coast as their goods can be sent by sea, an advantage not available to the Mid-West competitors. In Canada freight rates between Montreal and Calgary (2240 miles) are higher by one-third than those between Montreal and Vancouver (2880 miles) because the latter rates have to be competitive with water transport via Panama. It was for a time a bone of contention in England that cases of Danish butter despatched by rail from Hull to London, were charged less than butter sent by English farmers from the Midlands to London, but it has been pointed out that the two situations are not strictly comparable. The Danes despatched their butter, conveniently packed in boxes, regularly and in such quantities that full trainloads could be made up, while the shippers concerned made themselves responsible for the unloading at the London railway terminals. The English farmers, on the other hand, despatched smaller amounts less conveniently packed at less regular intervals, and the railways handled the butter at the terminals. These less efficient transport arrangements therefore placed the English butter-making industry at a decided disadvantage.

Breaks of bulk and artificial restrictions invariably act to the disadvantage of some areas and may produce a stimulus to industry elsewhere. For example, the standardisation of through freight rates within the territories of the European Coal and Steel Community has had important effects in aiding the freer flow of fuel, iron ore, scrap iron and steel across the international boundaries concerned and will certainly affect the future localisation of industry in Western Europe, while the now abandoned Pittsburg Plus system in the U.S.A. has had very important repercussions on the location of the American iron and steel industry.

Briefly, the system worked like this. All steel made in the U.S.A., wherever it was made and whatever were the costs of production, had to be sold in any U.S.A. market at the Pittsburg price *plus* the freight costs to the market concerned from Pittsburg. Any comparative advantages which might have been enjoyed by a location such as that of Birmingham, Alabama (*see* p. 182 below), were therefore wiped out and the development of new production areas was greatly handicapped. The system was in operation between 1900 and 1924, and between 1924 and 1948 in a modified form. It was finally abandoned in 1948 and it is significant that since that time, the importance of other areas of iron and steel production such as the north-east coast and Birmingham, has greatly increased. Admittedly there was more to this system than simply a juggling with transport costs, but such manipulation formed an essential part of the scheme.

The Present Significance of Transport Costs

While it must be conceded that in some cases transport costs can make the difference between the success or failure of an enterprise, it is also true that such costs today form in general a small proportion of the selling price of most manufactured goods. This situation stems from the greatly increased efficiency in transport today as compared with that obtaining before the Second World War. There are several causes of this increased efficiency, some of them technological (for example, more efficient and more powerful haulage units which can pull greater loads over long distances more speedily than ever before), but one of the most important undoubtedly, is that it is a reflection of the increasingly efficient use made today of raw materials generally. Manufactured goods are now made with notably less raw material than formerly, and this means in effect a greater unit value for each ton-mile per hour of raw material transported. This situation applies to transport equipment as well as to other manufactured commodities. Lorries, railway trucks and ships all today weigh notably less in relation to their ton-mile per hour transport capacity than in past years, when transport systems were characterised by comparatively low efficiency and high costs.

In many industries it is true to say that a smooth and uninterrupted flow of raw materials or component parts is a more vital factor than low freight cost differentials. Motor-car assembly plants, for example, frequently operate with only a few hours' supplies of component parts on hand, and any interruption to the inflow of further parts (for instance, such as may be caused by a workers' strike) can quickly halt production. This requirement makes it essential that assembly firms are located in such positions that a free flow of incoming parts is assured, and not in positions where traffic complications or other factors are likely to hold up supplies.

One other feature connected with transport is worthy of note. When industrial establishments are located along major routeways they are able immediately to take advantage of the excellent advertising facilities which such a location makes possible. Advertisements on hoardings or factory buildings reach a wide range of people when they face a motorway or a main railway line, and most firms are not slow to take advantage of this.

MARKETS

We finally turn to a consideration of the market, noticing first that it is becoming more and more true that industries are seeking locations as near as possible to their markets; indeed, it has been remarked that market attractions are now so great that a market location is being increasingly regarded as the normal one, and that a location elsewhere needs very strong justification.

In order to examine in a fairly simple manner the relations between industry and market we shall proceed by way of a tabulation of the main points, some of which have received attention earlier:

1. Sometimes there is a considerable material increase in weight, bulk or fragility during the process of manufacture, and in such cases industry tends to be market orientated. The baking and brewing industries are cases in point as the products are large in bulk and fragile (in the case of bread a considerable increase in bulk takes place during the baking process, while in the case of beer and soft drinks the fragility is a result of the bottle containers).

2. When the product exhibits a high degree of perishability, industry is usually market orientated. Baking falls into this category of industry, while in a sense printing does, for the value, of, for example, local newspapers depends to some extent upon the freshness of the news in them.

3. Where personal contact is necessary the industry concerned must be located near its market. The bespoke-garment industry is often quoted as an example of this form of control, while manufacturers of containers of different kinds often position themselves within easy reach of their best customers. The production of local newspapers also falls within this category as well as within (2) above.

4. If the manufactured product is comparatively cheap and bulky, transfer costs incurred in moving it will be proportionately heavy, and this provides an incentive to locate the industry concerned near the consumer. The manufacture of bricks and tiles may be offered as an example in this connection.

5. The feature normally known as "linkage" can be a very important one, though it is of wider application than in the sphere of market relations with which we are at present concerned. It will be dealt with further at a later stage in this study (pp. 123-125 below). It is relevant to notice, however, that in cases where the market for a manufactured product is another industry, a close juxtaposition of the two sets of industries commonly occurs. The production of iron and steel, steel plate, and of steel-plate products (such as cans for foodstuffs) not infrequently is carried on within the same general area, while textile machinery is commonly produced in textile-manufacturing regions such as New England and Lancashire. The production of sparking plugs at Rugby and of tyres at Birmingham has

a clear connection with the manufacture of cars in the Midlands and the south of England.

6. It is worthy of note that large-scale market regions often are suppliers of raw materials in substantial amounts, an outstanding example in this respect being the ability of manufacturing regions of certain kinds to supply large quantities of scrap iron. Areas like north-eastern England, New England and Le Nord produce large amounts of scrap from industries such as ship-building and the manufacture of motor-cars. One of the reasons for the establishment of new iron and steel industries along the north-eastern seaboard of the U.S.A. was the accessibility of markets in the megalopolis region of that seaboard and also of those along the Gulf Coast, but it has also been remarked that the megalopolis region supplies large amounts of scrap to feed the steel furnaces.

7. Another reason why market regions often attract industry is that such regions by their very nature are well populated, which is another way of saying that they can offer considerable labour resources. It is a trite observation that earning power is spending power but it is nevertheless true, and it is important to bear in mind that there is no necessary distinction between the workers in any industry and the consumers of its product. The simple fact is that large centres of population are potent attractive forces both from the labour and from the market aspects and it is not perhaps surprising that market locations, therefore, are becoming normal locations for industry. Where other locations are chosen there is usually a clear advantage in some other respect which involves a substantial saving in costs.

8. The tendency to establish industry near its market is greatly accentuated today because of the practice of levying higher freight charges on finished manufactured products than on the raw materials concerned. In many cases there is a perfectly good reason for this, for raw materials such as iron ore and raw cotton can be handled expeditiously in large amounts and without any expensive measures being necessary to protect the material. Manufactured products, on the other hand, generally need more careful handling and packing, while they are more wasteful of transport space. A truck can be filled completely with ore so that no space is wasted but a cargo of motor-cars must be loaded with care, space being left to minimise the risk of damage (*see* Chapter X, p. 239 below). In a sense, therefore, the available freight space has to be used to less advantage.

In another sense, however, the practice is simply an extension of the principle already mentioned of charging what the freight will bear. Manufactured products have greater value unit for unit than raw ma-

MAKING AND SELLING 91

terials, and transport costs are therefore likely to form a lower proportion of total costs; it therefore becomes possible to charge more highly for transporting them. (*See* pp. 130 and 239 below for further examples of this principle.)

SUGGESTIONS FOR FURTHER READING

Baker, Professor Sir J., "Education and Recruitment," *Progress*, 1964.
Dury, G. H., *The East Midlands and the Peak*, London, 1963.
Estall, R. C., and Buchanan, R. O., *Industrial Activity and Economic Geography*, London, 1961.
Hoover, E. M., *The Location of Economic Activity*, London, 1948.
Jackson, Commander Sir R., "The Volta River Project," *Progress*, 1964.
Lewis, W. A., *Report on Industrialization and the Gold Coast*, Accra, 1953.
McLaughlin, G. E., and Robock, S., *Why Industry Moves South*, New York, 1949.
Martin, J. E., "Location Factors in the Lorraine Iron and Steel Industry," *Trans. Inst. Brit. Geog.*, 23, 1957.
Miller, E. W., *A Geography of Manufacturing*, Englewood Cliffs, New Jersey, 1962.
Smith, J. Russell, and Phillips, M. O., *North America*, New York, 1942.

Chapter V

Other Locational Factors

W E have now made a general survey of some of the more important bases upon which the development of industry depends, but still we cannot pretend to have covered the full field of enquiry. In this chapter we shall examine some further factors which notably influence the location and development of industry. These factors are water availability, climate, site, capital, government activities and certain others which can conveniently be gathered together under the vague and umbrella-like category of "miscellaneous."

WATER

Members of communities who live in comparatively well-watered parts of the world have for so long simply accepted without question the water which is available to them that they do not easily reverse this traditional pattern of thinking. We are, however, slowly beginning to realise that water is a very precious commodity indeed, and that large areas of the earth's surface do not have enough of this natural resource to satisfy the many and growing demands which are today being made upon it. Even areas which have until recently assumed that their water reserves are sufficient to supply all their needs are having to readjust their thinking in this regard. Miller has reminded us, for instance, that parts of the English Midlands would produce better pastures than they do now if the fields were irrigated—a salutory corrective to that line of thought which conceives of Britain as a rain-swept archipelago with virtually limitless supplies of water.

The importance of this question is further brought into relief when we consider the very large amounts of water required by industry today. As an overall estimate it has been suggested that in the U.S.A. in 1900 an average of 40,000 million gallons of water were used each day, while in 1955 the corresponding figure was 240,000 million; it is estimated that the 1975 figure will be 450,000 million. While by no means all of this water is used for industrial purposes a large proportion is (an estimated 55% by 1975), and a few examples will underline the modern need of industry for water. The processes of manufacture leading to the production of a single ton of steel, for example, may require 65,000 gallons of water, while the production of one ton of sulphite paper demands 64,000 gallons, and it is equally important to realise that in this case the water is fouled during the manufacturing process and is therefore not available for any other purpose. Seven hundred gallons of water are consumed in the refining of one barrel of oil and 100 gallons in the production of one pound of rayon.

Other examples of the great demands of modern industry for water are not hard to find. Average figures tell us that in the U.S.A. 38,000 gallons of water are needed to manufacture one ton of paper from wood, 220,000 gallons to produce one ton of rayon fibre and 8000 gallons per kWh of thermal electricity (*see* also p. 67 above). In England, five chemical plants between Immingham and Grimsby on the south bank of the Humber estuary consume on an average 10 million gallons a day; this enormous amount is derived from the underlying chalk. One synthetic-fibre plant in Liverpool uses as much water as a town with a population of between 50,000 and 100,000 people, while a single electric-power station in south Lancashire has a total consumption of water (most of which is used for cooling purposes and is derived directly from the Mersey estuary) greater in amount than that supplied to the Liverpool County Borough! The steelworks of Shotton use no less than 66,600,000 gallons a day, of which more than 60 million gallons come from the River Dee. Miller reminds us that as long ago as the latter half of the nineteenth century the industries of Lancashire and Yorkshire used an amount of water equivalent to almost 10 inches of rainfall over the extensive gathering grounds of the Pennine watershed. And it is important to notice that as usage for industrial purposes increases, and as industrial populations increase, the amounts of water concurrently needed for domestic and general purposes also sharply mount. It is not perhaps surprising that it has been suggested that one criterion which may be used to differentiate between developing and developed countries is that displayed by the *per capita* consumption of water. In developing countries the figure is commonly of the order of 5 gallons a day while in the U.S.A. it is 1300. It is a significant commentary upon the American scene that in the north-east of the U.S.A. 81% of all water consumed is for industrial purposes, while in the south-east it is 67%. In the west, on the other hand, almost 94% of all water used is for irrigation, and only 2% for industrial purposes.

Quality of Water

Modern industry, therefore, demands very large supplies of water, but it is equally important to notice that the quality of the water is also a factor to be taken into account. It is sometimes forgotten that the substance lightly called "water" may vary greatly from place to place in composition and in its properties. Important considerations to an industrialist may well include the corrosive power and the scale-forming propensity of any given water, while the amounts of inorganic matter carried in suspension and of organic matter present can markedly affect the value of water for the manufacturer. It is generally true that water derived from underground sources is of better quality than surface water, and because this is so it is sometimes worth while to draw on subterranean sources though higher costs (capital and operating) are involved in securing it. It is, of course, possible to improve the quality of water by processing it, but when very large amounts are required the consequent costs may prove too heavy a burden for industry reasonably to carry.

In cases where water is used for cooling purposes the quality is often not of prime importance though some prior treatment may be necessary if the water is corrosive. Water used for steam raising, on the other hand, must be very pure or problems of scaling and corrosion will soon upset the working of a modern high-pressure boiler. This means not only additional expense involved in periodic cleaning and servicing but also a heavy capital loss when such an expensive item of equipment as a modern boiler plant is out of use.

The importance of the quality of process water will vary considerably according to the processes concerned. Process water used for washing the product in a steel works need not be of high quality, but the manufacture of photographic equipment, paper and many textiles requires very pure water. It is also true that in cases where the water is used as a raw material and enters into the manufacture of foods, drinks, drugs and medicines generally the highest standards of purity are essential.

The following table illustrates some of the points made above:

Water Use in Industry

Industry	%age treated before use	Cooling	Steam	Process
Primary aluminium	4	93	7	—
Radio manufacture	12	44	45	11
Manufacture of motor vehicles and accessories	16	22	37	41
Pulp milling	54	18	12	69
Textile (exc. wool) finishing	56	11	12	77

Source: Information in the *U.S. Census of Manufacturers*, 1954, vol. 1.

The first industry in the table shows how little water is treated before use when the main use is for cooling purposes. The second and third industries are representative of those which use proportionately more water for steam raising and more of this water is treated before use. The fourth and fifth industries show that much more water is treated in the case of industries which use large amounts of process water.

Water Supplies and Industrial Location

It is probably increasingly true that modern industry is being located with careful regard to available water supplies. Although water is generally viewed as a cheap commodity this is the case only because supplies are fairly widespread and industrial location has been able to take advantage of this fact. Water has to be cheap for industry since such large quantities are required, and it would not generally be practicable to transport massive supplies over considerable distances. Langdon White considers that water plays a dominant role, for example, in the iron and steel industry, partly because water transport can be invaluable in transporting both raw materials and finished products, partly because of the large consumption during the various processes of manufacture (for every 3·2 tons of solid material fed into the blast furnace 57 tons of water are used), and partly because proximity to large water bodies facilitates the disposal of waste. It is therefore not surprising that iron and steel plants are typically located alongside large rivers and lakes or on the sea-coast. Areas which are not well blessed with water are finding themselves in ever-increasing difficulties in the industrial world. In Australia, for instance, it is no accident that most industrial development has taken place along the relatively well-watered eastern and south-eastern seaboard, while Cole points out that the rapid rate of industrial development on the Rand of South Africa has placed a great strain upon the available water supplies. It has, in fact, been recommended that "no industries requiring large quantities of water and which do not for geographical reasons require to be placed there should be permitted to establish themselves in the Vaal River basin" (*The Vaal River. Report on the Water Supplies of the Vaal River in Relation to its Further Development.* Union of South Africa, Natural Resources Development Council, 1953).

Water Economy

One result of the increasing importance of water is the development of techniques designed to economise in its use, generally by some form of treatment after use so that it can be used again. While considerable costs are involved in such treatment they are well worth while if they make it practicable for industry to continue in operation, or even to expand, instead of limiting production or closing down. The following table shows the degree of success which has been achieved in the U.S.A. in selected cases:

Range of Water Use in the U.S.A.

Industry	Unit of product	Gallons of water used per unit	
		Max.	Min.
Primary metals	Ton of finished steel	65,000	1,400
Chemicals	Ton of sulphuric acid	7,270	780
Pulp and paper	Ton of dry ground wood pulp	60,820	4,860
Petroleum	Gallon of crude oil	44·5	0·8

The possibilities of water economy are heavily underlined by this table and further advances along these lines can be expected. It is, for example, notable that the fairly new American iron and steel works located at Sparrows Point on the brackish Patapsco River originally used water from wells for most purposes, but these wells became contaminated as a result of over-use. Since 1955, therefore, the plant has processed and made use of almost the entire flow of sewage of the near-by city of Baltimore—about 150 million gallons each day.

A problem which is often associated with water supply is that of the disposal of waste, particularly of liquid waste. The problem is particularly acute in built-up industrial areas where some streams and small rivers have been reduced to the status of sewers. Gibson gives an example of a straw-pulping mill on the Irwell ("the hardest-worked river in the world") which until 1957 discharged 600,000 gallons of kier liquor daily into the river. This liquor is the waste from the pulping process and is a dark-brown, strongly alkaline liquid containing a large amount of organic matter. It produces a brown scum and froth which persists some miles downstream from the point of discharge, though various measures have since 1957 lessened the discharge and the consequent pollution. The discharge from artificial-fibre industries is not noxious in itself but has the property of absorbing oxygen. When fed into rivers, therefore, it robs them of their oxygen and therefore kills all aquatic life; since the natural purifying processes are therefore inhibited, the water becomes offensive in appearance and odour. (The question of stream pollution is examined in Chapter XII, pp. 278–281 below.)

The need of industry for large water supplies for various purposes is increasing, and it is not perhaps surprising that a Presidential Commission in the U.S.A., reporting in 1952, gave as its opinion that by 1975 access to good water may have become the most important single factor to be considered in the location of new industries. Estall and Buchanan make the cautious comment on this claim that it could well contain more than a grain of truth. It could indeed!

CLIMATE

The influence of climate on manufacturing industry is often indirect though very real. Certain avowed correlations of the past, e.g. that which was for long supposed to have existed between the climate of Lancashire and the establishment of the cotton textile industry in that county, have for some time been shown to be excessively generalised, but at the same time it must be realised that correlations undoubtedly do exist in many cases. Climate may, for instance, affect labour costs; it has been stated as a case in point that a working man in the South of the U.S.A. can receive a wage of 10 to 20% less than that of his New England counterpart and still enjoy a comparable standard of living. This is because of heavy savings in the South on such items as winter clothes and heating costs, while the variety of locally grown foods in the South can be bought at appreciably less cost than is possible farther north. These points have proved to be important contributory factors to the southward migration of industry in the U.S.A.

An obvious link exists where the industry concerned is locally based upon agricultural production, since the production of crops is closely linked with climate. It is natural to find sugar-processing industries located in areas of sugar-cane or sugar-beet production (*see* p. 23 above), while fruit-canning industries are typically situated in areas suitable for the large-scale production of fruit (but *see* p. 158 below). The manufacture of silk and silk products may to some extent be determined in this way as the rearing of the silkworm is largely determined by the suitability of the area for the growing of the mulberry tree.

Climatic Extremes

Climatic extremes can be as little favourable to industry as they are to agriculture. From what we have already said about the need for adequate water supplies it will be realised that excessive dryness must militate against industrial development; this is a problem which South Africa is already having to face (p. 95 above) and other territories such as Australia will increasingly face in the future. Extreme heat can be troublesome, partly because human efficiency falls off when temperatures are very high; Stamp has remarked that he has found a definite lowering of mental ability when the dry-air temperature has reached 100° F (37·8° C) and at a much lower temperature when relative humidity is high, and has suggested that under normal conditions temperatures ranging between 70° F and 75° F (21° C and 24° C) are most favourable to mental exertion and slightly lower temperatures for physical effort. Mountjoy refers to the poorer performance of dockers at the Queensland port of Townsville as compared with those at Brisbane and to the evidence of fatigue and inertia experienced among professional people in Queensland. The experience of the present writer would confirm that a definite lowering of physical and mental ability does take place in hot conditions, and to the extent that this is true industrial prospects for hot climatic areas must be the poorer.

It has been observed that periods of extreme weather have proved detrimental to industrial effort in the U.S.A. For instance, work on shipbuilding in the north-eastern U.S.A. is normally interrupted during the bitterly cold spells to which that region is subject during the winter; men cannot work out of doors under such conditions, especially when they are dealing with metal objects. On the other hand, work may be held up during the summer under very hot conditions when the men are not able to work with metal under the hot sun. We may notice in this respect that in the more temperate north-west of the U.S.A., where such climatic extremes are very rare indeed, shipbuilding is equally rarely interrupted either by extreme winter cold or by undue summer heat, and it is not, perhaps, surprising that a marked increase in shipbuilding activity has taken place in the north-west during recent years.

Sometimes climatic extremes may be directly unfavourable to industrial enterprise. It has been asked, for example, why Ghana (a leading producer and exporter of cacao) does not establish a chocolate-manufacturing industry. While many factors are involved it is reasonable to point out that chocolate rapidly melts and deteriorates under hot conditions and that cool conditions are therefore essential for the successful manufacture of this commodity. It is conceivable that extensive air conditioning might provide the means for getting round this difficulty but the costs involved would be a heavy additional burden for the industry to bear.

Extremes of temperature can at times seriously interfere with transport, and when such extremes are of normal occurrence due allowance must be made for them. Snow and frost can cause considerable inconvenience, while excessively high temperatures can cause railway tracks to expand and buckle with a consequent danger of derailments. Flooding can constitute another major hazard to roads and railways.

Industry and Climate

The aircraft industry of the U.S.A. provides an interesting example of the adjustment of an industry to climatic factors. The manufacture of aircraft engines and propellors is concentrated in the north-east of the country, and this may be an off-shoot from the early industrial development of this region, which has long been a producer of petrol engines and associated products (*see* Chapter X). Airframe production and the final assembly of aircraft, however, take place in the south-west of the U.S.A., a feature of the industry which well shows the effect of climatic control. Aeroplanes can be very large final products and very extensive workshops are necessary to assemble them; it follows that very substantial heating costs are saved in an area like the south-west U.S.A. where winters are mild and sunny. And, since there is little rain to be expected and the effects of frost and snow are little known on the lowlands, equipment can be stored in the open air without danger of deterioration from climatic circumstances. This, again, represents a very real cost saving. There is even a further point; the clear skies which are normal in the region permit flying and testing of aircraft all the year round, and this in itself is a great advantage and leads to further substantial savings in costs.*

In connection with the aircraft industry Garwood has noted that the migration of the manufacture of aluminium pistons from the Great Lakes region to Utah effected an economy of production which was entirely unforeseen. In the humid Great Lakes area a fairly high proportion of completed pistons (70% on days when humidity was high) was rejected because of the occurrence of gas pockets in the finished product, but under the low humidity conditions of Utah the proportion of rejections from this cause fell very sharply.

There is often a connection between climate and type of industry. For example, it is no accident that Sweden has come to the fore as a manufacturer of high-quality heating equipment, while the manufacture of agricultural implements at Chicago is in some measure a reflection of the agricultural productivity of the Mid-West, a productivity that is linked with climate.

A final point which we might make in connection with climate is the simple fact that people often prefer to live in congenial climates, and this

---

* Other advantages of the south-west for the aircraft industry include a comparatively high degree of strategic security, cheap land and plenty of labour. The *Grapes of Wrath* drought of the 1930s induced substantial migration of labour into California from the interior plains so that abundant labour was available for industrial expansion.

applies to factory owners and entrepreneurs as well as to other folk. Garwood states that in his survey of the reasons which took 116 firms to Colorado between 1946 and 1951, he discovered that 25% of these firms stated that climate was for them the primary reason why they chose to set up in this state. Another American firm which advertised exactly similar posts both in the north-east of the U.S.A. and in Florida received only a few routine enquiries about the posts in the north-east but numerous applicants enquired about the posts in Florida.

SITE

Site requirements for industrial development can in themselves be of considerable importance, and we have earlier noted (pp. 68–9) the fact that thermal electric power stations need extensive, cheap, sites. Sites commonly need to be reasonably flat and to be served by adequate transport facilities (we have mentioned the important part played by water transport in the iron and steel industry). The Abbey Margam steel-rolling mills established near Port Talbot cover four square miles and the availability of unused flat marshland along the coast was a prime reason for the location of the works (Fig. 63). It was a similar set of features (flat areas of marsh and dunes fronting an extensive water mass) which encouraged the location of the enormous Chicago-Gary iron and steel industry near the southern tip of Lake Michigan, while the extensive Ford plant at Dagenham, Essex, stands on 500 acres of formerly little-used land well located to take advantage of facilities for water transport provided by the Thames estuary.

In some cases industry may be attracted to particular sites because of the prior existence of useful facilities. Two of the largest firms examined by Garwood, for instance, were located in Colorado because facilities were made available to them in former government installations. The availability of general services can be a most important factor with regard to the suitability of particular sites, for a site which is located within a developed area is likely to be within reach of public services such as gas, electricity, sewerage, roads and railways, while houses, schools, churches and other buildings may be available within reasonable distances for the workers concerned and their families. The situation is very different if these features are not available (p. 84 above). So-called "greenfield" sites are much more expensive to develop than are established ones; an American estimate suggests that extensions to an existing works designed to secure a given additional output of steel would involve a cost per ingot ton for the extra output only one quarter of that which would be necessary if the additional capacity were secured as a result of the erection of a new works on a greenfield site.

It is appropriate to point out that in some cases sites which might have been suitable for industrial development on general grounds have had to be discarded because of government requirements. It is usual in developed territories to find that competition for land is strong, and in order to ensure that the land available is used to the maximum advantage some degree of government (or local government) planning is essential. If it is decided that land in a particular locality is best suited for crop production or for purposes of recreation (or even for strictly residential purposes), then industry may be barred from it, however suitable particular sites in it may be for industry in other respects (*see* pp. 102–110 below for further discussion of this question).

CAPITAL

The capital requirements of modern industry are great and in a study of this nature they need little elaborating. Figures illustrating the large amounts of capital required by industrial enterprises are given elsewhere in this book, but we might observe that this is properly a study for the geographer as well as for the economist, because industrial locations may well be affected by the availability or the lack of capital reserves. There is, for example, the obvious contrast between developed countries and developing ones; the first group has, in general, considerable supplies of capital at its disposal but the second has not. Industry is therefore more likely to advance in the first group than in the second, while there may well be selective development in the second group as a result of assistance granted by members of the first group. Actual selection of developing countries to be granted capital aid may be influenced by a number of factors, including political as well as economic considerations.

The influence of capital equipment (as distinct from capital goods) may be strong since most types of such equipment are relatively (some even completely) immobile. Heavy plant cannot easily be moved elsewhere, and this is not infrequently a reason for the immobility of industry. If the moving of an industry means the abandonment of capital it will not be undertaken lightly and this may well lead to industrial inertia (*see* p. 113 below).

While capital in the form of money is theoretically extremely mobile, in practice the mobility may be sharply lessened by various limiting factors. Any form of uncertainty with regard to the possible safety or productivity of capital invested is likely to restrict sharply the supply of available money capital, and this is particularly true where a possible movement of capital across international boundaries is concerned. While it may be natural for inhabitants in developing territories to resent their enforced dependence upon outside help, and tempting for them to abuse the overseas investor who may appear to wax fat at the expense of the territory concerned, such feelings if overtly expressed may do much to discourage the entry of further capital which may be badly needed; indeed, it may induce an outflow of existing capital. Lewis put the matter clearly in 1953 when he wrote that "the Gold Coast needs the foreigner more than he needs the Gold Coast. . . . The Gold Coast cannot gain by creating an atmosphere towards foreign capital which makes foreigners reluctant to invest there." This is a principle of wide application.

The availability of capital is markedly affected by the presence or absence of financial institutions which specialise in assembling capital and making it available to industrialists and others who are in need of it. Developing countries suffer badly from the lack of such finance houses, which means that it is not always easy for the prospective capitalist to find avenues of investment. This may be one factor which encourages wasteful spending. The attitude of banks can vary widely between one part of a country and another, and ultra-cautiousness to lend money may discourage industry from taking root in a region. Garwood observes that Colorado banks were alleged to be ultra-conservative, and in this case industrial location appears to have been established despite this unfavourable factor. It is less easy to discover whether firms were actually prevented from establishing themselves in the area because of this.

GOVERNMENT ACTIVITIES

The importance of government activities in affecting both the general development and the location of industry is steadily increasing. This is inevitable in a world in which population increase is proceeding at such a staggering rate as is the case at the present time, for as population increases there is more competition for the available land, and a point is reached when some degree of planning is essential if the best use is to be made of the land which is available. We can do no more in this short section than touch upon some of the main points involved, and the reader who wishes to learn more about this subject is advised to consult other works which examine the topic in more detail.

Direct Government Assistance

Government action may be either positive or negative in its approach to industry. It is positive when there is direct encouragement to industrial development, and examples of this were given in Chapter III in connection with the Volta River Scheme and the High Dam at Aswan. These are schemes which could never have come into being without the direct interest of the governments of Ghana and Egypt respectively, and the object of these schemes is in part the production of electricity, much of which is to be used in industrial projects. Such schemes may give direct encouragement to industry as was the case in the Owen Falls Scheme in Uganda. This project is similar in many ways to those already mentioned and the power available has stimulated industries such as textile manufacture and food processing at Jinja and other centres, while it is also interesting to note that the cement industry at Jinja, which was originally established to produce cement for the construction of the dam, has continued in being since the dam was completed. It should be remarked, however, that the development of industry in Uganda as a result of the project has not fulfilled expectations.

These are examples of large-scale government encouragement, but we might well observe that direct action can be on a much more modest scale and still be effective. We mentioned earlier in this chapter, for instance, Garwood's example of two firms who established themselves in Colorado because former government installations were made available to them.

Effects of Taxation and Control

In other cases encouragement is offered by governments, but is of a more direct kind. Taxes, for example, may be remitted for a period of years to industries newly establishing themselves, as has been the case in some West African territories. This kind of policy may also affect the localisation of industry within a particular territory under a federal form of government; for instance, especially attractive financial prospects are offered in Australia to firms establishing themselves in the Northern Territory. The success of this kind of policy depends, however, not so much upon the action taken by the government concerned as upon the prospects of success in any undertaking as seen by possible participating firms. In the case of West Africa these policies have met with some degree of success, though political factors such as fears of appropriation of assets and limitations on the export of profits have in some cases worked in the opposite direction and have discouraged the entry of new firms, but in the case of the Northern Territory the chances of industrial success are at present so slender that the action of the government has met with no response from industrialists.

Government action may be negative, and this occurs when any form of restrictive control is imposed upon industrial undertakings. We mentioned in the last paragraph financial measures taken by some West African governments which have encouraged firms from overseas to establish themselves in the territories concerned. Such measures do not normally affect the location of industry within a territory very much except under a federal government, when different states may enact different laws so that conditions may be more favourable to industry in one state than in another. A well-known example comes from the U.S.A., where it was illegal until fairly recently to employ women after 6 p.m. in New England, while in the southern states women could be employed on night work. It seems fairly clear that these circumstances directly encouraged some industries, especially the cotton textile industry, to move southwards from New England.

The imposition or the manipulation of taxation can exercise powerful effects upon the development and the location of industry, and we might observe in this regard that the expectation of taxation can sometimes be just as powerful a force as the actual imposition of it. This was shown by the establishment of an iron and steel industry at Duluth, near the western tip of Lake Superior. This industry was set up in order to avoid a tax on ore exports which was threatened by the state government of Minnesota, a threat which never materialised. It is significant that this iron and steel industry has never flourished (for further details, *see* p. 186 below). It has many times been observed that taxes are unusually heavy in the state of Massachusetts, and this has gained the state a poor reputation among U.S. industrialists though, as Estall and Buchanan point out the tax structure does much to mitigate the full force of the taxation and renders Massachusetts in some ways more favourable to industry than competing areas. The bad reputation of the state, however, remains and is said actively to discourage many prospective industrialists.

Selective taxation can operate adversely upon particular industries and slow down their development. An example of this kind of effect may be seen in a purchase tax which artificially increases the price to be paid for selected commodities and therefore may decrease demand for them; the production of substitutes for these commodities may thus be stimulated, and so the industrial structure of a country can be affected. Manufacturers often argue that such a discriminatory tax, with consequent fears of restricting domestic purchases of their product, will lessen their ability to make use of maximum economies of production. So, they argue, the domestic purchaser has to pay even more for the product while the manufacturer's ability to compete in overseas markets is reduced. To the extent that this is true even a domestic tax can have wide international repercussions.

On the other hand, governments may, as we have seen, offer tax concessions which may have the effect of stimulating industry in selected areas. A good example of this is that of the establishing of the iron and steel industry at Hamilton, Ontario, the town which is now the largest producer of iron and steel in Canada. It is a generally accepted fact that the Lake Peninsula offers several suitable locations for the establishment of an iron and steel industry, but the industry was attracted to Hamilton because of tax concessions initially offered to the iron and steel company by the city council.

A tax concession which came about with a change in tax law in 1941 has led to large-scale exploitation of the extensive taconite deposits of Minnesota. Before 1941 tax was chargeable on all exploitable mineral reserves in the ground and this effectively inhibited the development of the taconite. As long as the mineral was not considered to be commercially exploitable it was safe from taxation but if it were classified as an ore it would have been taxed, and the incidence of the tax would have been extremely heavy because the reserves are so enormous though the ore is low grade. In 1941, however, the state legislature enacted that taxation should thenceforth be assessed upon actual production and not upon reserves, and it was not long before the taconite was being mined and processed in Minnesota (*see* p. 29 above).

In many instances countries attempt to stimulate industrial production within their borders by the imposition of taxes on competitive imports. The argument is that such tariff barriers offer protection to a young developing industry and the barriers can be removed when the industry is sufficiently developed to stand on its own feet. Unfortunately, the time seems uncommonly laggard in most instances where the barriers can be removed; the reason is probably twofold. Firstly, the removal of the barrier would result in a sharp drop in revenue for the government concerned, and secondly, industries become accustomed to operating within a given framework of costs and prices, and a sudden change can present very awkward problems for them. A removal of tariffs could mean increased competition and a need to reduce prices, a policy not often popular with manufacturers nor, incidentally, with workers, who may fear cuts in wages or in fringe benefits as a result.

A good example of an industry developing very largely as a result of protection is to be seen in the rise of the automobile-manufacturing industry in Australia. Extensive tariff protection has made possible the development of the very large General Motor Holden Company which now manufactures most of the cars today used by Australians. This company originated in Adelaide in 1948 though today it has very extensive works at Melbourne, while in more recent years other firms, notably Fords, Mercedes, Renault and Volkswagen, have all set up assembly and manufacturing centres. Not every part of the car concerned is necessarily made in Australia but the tendency is to extend the range of Australian-made parts until the car is 100% locally made. The Holden car was wholly Australian from the start. The significant point is that despite the protection offered by tariffs since the inception of the industry, and despite the natural tariff protection conferred by the situation of Australia with respect to the other main areas of automobile manufacture, there is still strong insistence from the industry that tariffs must be strongly maintained; despite the financial protection afforded, Australian manufacturers are increasingly feeling the strength of competition from as far away as Japan!*

Planning

It is becoming increasingly common for planning authorities (including governments) to attempt directly to control industrial location and development, and this takes place locally and on a national scale. Local examples arise when local authorities limit the development of industry to specific areas and forbid it elsewhere. Closely allied to this practice is the idea of the "Green Belt" or the "Green Wedge," zones which are looked upon as areas within which urban, including industrial, developments are generally forbidden (Fig. 22). The Green Belt comprises a continuous area around a town or city, and the idea is that such a belt can offer the facilities of the countryside to the townsfolk. Unfortunately, such a belt can severely restrict building which may be desperately needed to accommodate an increasing population, and the result will almost certainly be that building will leap over the belt and commence on the other side of it. Outstanding examples of cases where this has happened are offered by the Parkland Towns of South Australia, Adelaide being a case in point (Fig. 23). The journey to work for those who live beyond the park is greatly lengthened in these cases.

* There is a good economic reason for this. The size of the Australian market is very restricted judged by world standards and by the standards of large-scale modern industry, and Australian firms find that they cannot avail themselves of large-scale economies of production to the same extent that firms elsewhere can. This puts up unit costs and places Australian car manufacturers in a weak position competitively.

The Green Wedge attempts to overcome this difficulty by permitting natural urban development along radiating roads but forbidding it in the wedge-shaped areas in between them. Obviously, the localisation of industry will be affected by such a policy. We might observe that the Green Wedge, like the Green Belt, suffers from certain disadvantages. For example, the wedge-shaped rural areas can become very awkward to handle from the farmer's point of view and the near-by urban populace can become a very real additional hazard for the stock rearer and cultivator.

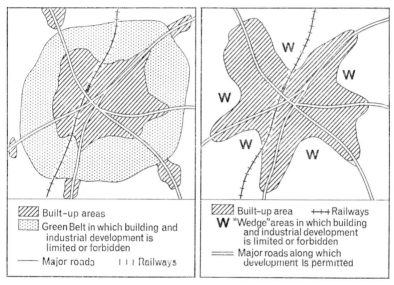

Fig. 22.—Urban and industrial planned development: (*left*) Green Belt; (*right*) Green Wedge.

On a national scale governments at times enter into industrial planning in a fairly direct manner. In the United Kingdom the location of the iron and steel plant at Ebbw Vale was decided at government level in order that work should be brought into a "special" area, *i.e.* an area suffering from a high rate of unemployment, while similar considerations prompted the choice of location for the Nowa Huta plant in Poland. The iron and steel industry, developed near Hannover in Germany, which is based on the Salzgitter-Wattenstedt iron ore, was established because of the intervention of the Nazi government which wished to extend the manufacture of iron and steel as part of the armaments drive, while one marked result of the wartime invasion of Soviet Russia by the

Nazi armies in the Second World War has been the rapid development of industrial potential in areas east of the Urals.

Sometimes a government may encourage industrial development mainly for reasons of prestige, and a marked case of this mentioned by

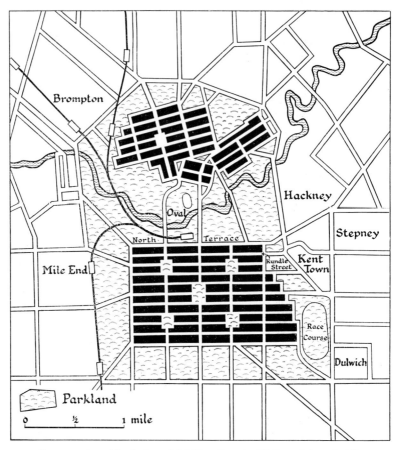

FIG. 23.—A parkland city—Adelaide. *Courtesy H. Rees: "Australasia".*

Mountjoy is the establishment of an iron and steel industry at Helwan, near Cairo, Egypt. Such a plant in Egypt is bound to incur heavy transport costs on raw materials, for there is no coking coal at all in the country though iron ore of fairly good grade (50% iron) exists in the Eastern Desert near Aswan. The actual location chosen for the industry lies more than 100 miles from the ports which import the necessary coke and about 500 miles from the ore deposits, so that assembly costs are high.

The plant is a small one, which means that the iron and steel produced carries a high unit cost of production; in a freely competitive world the Helwan works would rapidly be put out of production. Despite the small size, however, the initial cost involved was £E28 million. It seems reasonable to say that there is little justification for this enterprise; it is simply a costly prestige symbol which owes its inception to government backing.

It should not be imagined, however, that governments always act in such a way, and to close this part of the chapter we will examine a case study from southern Italy. This is done, not only because of the intrinsic interest of the example, but because the study shows something of the many different ways in which a government can help to create a favourable "climate" for industrial development.

A Case Study

For reasons which cannot detain us here, southern Italy has traditionally been a backward and undeveloped area. The region known as the Mezzogiorno (southern Italy, Sicily, Sardinia and other islands) is inhabited by about 38% of the 50 million people of Italy but it produces only about 25% of the gross national product. It is not difficult to imagine that in such a region, with its dense population, its impoverished soils, and with small hope for the future, trouble was always likely, and in 1950, following disturbances which threatened to become very serious, the Italian government took responsibility for a ten-year plan aimed at encouraging general economic development in the Mezzogiorno. Schemes involving land reclamation, water supply and general agricultural development were set in motion, while encouragement was given to industrialists to interest themselves in the region. The following are the main inducements offered to industry developing in southern Italy:

1. Duty-free importation of machinery and building materials necessary for establishing new factories.
2. Exemption of industrial firms from income tax for ten years after initial establishment or after major expansion.
3. A considerable reduction of turnover tax on new machinery and equipment.
4. Appreciable reduction of local taxation or exemption altogether as local authorities may decide.
5. When an existing Italian firm, say from northern Italy, invests part of its profits in an enterprise in the Mezzogiorno, these profits may enjoy a 50% reduction in taxation.

6. Stamp duties and registration fees are greatly reduced.

7. Investment in the Mezzogiorno qualifies for bank loans at reduced rates of interest.

8. Grants may be made towards "social costs" involved (the provision of roads, sewerage, electricity, etc.).

9. Financial assistance may be available towards the cost of machinery and other essential equipment.

10. Transport rates incurred by firms in the south may be reduced and land and existing installations may be purchased from local authorities at reduced rates.

It is only fair to say that industrial development in the Mezzogiorno has not been greatly stimulated by these measures, one reason for this being the high costs involved in importing energy supplies, but this situation may change following recent discoveries of methane deposits. So far, however, the case has been somewhat comparable to that of the Northern Territory of Australia; unless, in a free society, conditions for industrialisation are reasonably favourable on economic grounds, a government can temptingly display the water trough of economic persuasion under the nose of the industrial horse, but it cannot make the animal drink. It has, unfortunately, to be recorded that some industrial horses have drunk at the trough of economic subsidy only to find the water rudely snatched away after a short period, and such examples do not encourage faith in governmental promises. A case comes to mind of a manufacturer of motor vehicles who was persuaded to establish his factory in the north of England, though such a location far removed him from the main suppliers of his raw materials in the Midlands and south. (*The Economist* 1955, p. 936). The deciding factor was a heavily subsidised rent, but when this rent was suddenly almost trebled "without discussion or redress" the manufacturer found himself in a highly disadvantageous situation. Delivery charges on his raw materials, for example, came to about £20 per single chassis and he had to absorb these costs himself as he could not pass them on to customers in a highly competitive market. It is not surprising that the manufacturer stated that if he had to face the problem of location again he would choose a situation in the Midlands or south and not in the north of England.

THE PERSONAL ELEMENT

The personal element in industrial development arises from the fact that industries are established by people, and exceptional people may leave their mark indelibly stamped on the industrial landscape. Estall and Buchanan, for example, remind us that the fact that people like Ford, Olds and Haynes lived in Michigan has undoubtedly had much to do with the rise of that state into a leading area for the manufacture of automobiles, while the fact that Oxford has also become a centre of automobile manufacture is in part due to the circumstance that at a critical period the school which Lord Nuffield's father attended as a boy came up for sale, and it was purchased for use as a factory by Lord Nuffield largely on sentimental grounds. The development of Essen in particular and the Ruhr in general owes a very great deal indeed to the Krupp family, who in the early part of last century established the famous iron works which bears the family name. It is true that circumstances greatly helped the development of what has virtually become an economic empire; for instance, we may recall that in the early years of the enterprise British competition was virtually eliminated by Napoleon's blockade; that later in the century the war with France provided a market for arms; that developments overseas were helpful (Krupp made rails for the new railways which were being laid not only in Germany but also in the U.S.A. and even in Britain); and that during the present century the demands for war goods have resulted in a market of unprecedented scale in addition to the demand for normal peacetime iron and steel goods. Yet when we have made all these allowances we have still to admit that a great deal of the credit of developing the Krupp interest inevitably must go to the Krupp family itself.

Other examples of the importance of the personal element are not hard to find. A simple case is that of the British Paints factory which is located at Bankstown, in Sydney, New South Wales, because the site ultimately selected reminded the Englishman who was sent out by the parent factory to explore the various possibilities of the site of the parent factory in Britain! Smith and Phillips record a very interesting case from New England. A manufacturer set up a small mill to grind lenses and to make other optical equipment at Southbridge, Massachusetts, using the water power produced by a small stream. In fact, the stream was so small that it often dried up altogether in the summer and the operator then used a horse for power; on one occasion he even paid a Negro ten cents an hour to take the place of the horse! From these very modest beginnings the great American Optical Company has grown and today the extensive plant of this company is situated on the same site as the original tiny mill. The city of Troy, in New York State, is the leading centre in the U.S.A. for the manufacture of men's detachable collars, to the extent that two-fifths of its workers are employed in making shirts and collars. This interest goes back to 1829, when a Troy housewife realised that detachable collars would greatly reduce the labour of washing and ironing and made such a collar for her husband. Neighbours copied the idea, a retired Methodist minister began to make and sell detachable collars, and the husband himself later set up a full-scale collar factory. In this way the collar and shirt industry of Troy began, an industry which was greatly stimulated by the use of the sewing machine in 1852.

More recently, Scargill has shown that "individual initiative became the chief factor affecting industrial growth in the town," the town in this case being Halifax and the individual initiative deriving its force from two families, the Akroyds and the Crossleys. There was a period in the first quarter of the nineteenth century when the fortunes of Halifax showed a marked decline relative to those of Bradford, and Scargill shows that this situation was changed by the two families mentioned. The Akroyd family was responsible for bringing the manufacture of new kinds of worsted cloth to the town, while the Crossleys were responsible for the introduction, about 1850, of a new carpet loom, which led to a great expansion in the carpet-producing industry.

It is perhaps worth mentioning again that individual enterprise of the kind which we have been considering can work only within certain limits. We have noted earlier (p. 84) that managerial skill can operate only where other conditions are favourable and there is a point beyond which even the most skilful and determined individualist cannot go. It is straining credulity to argue that, if Ford had been living in Dakota, that state would have become to the automobile industry what Michigan is today, or that, if Morris had had connections with Dolgelly, that admirable beauty spot would have developed as an industrial centre as Oxford has done. The Huntley and Palmers biscuit factory at Reading is not only a monument to the capabilities of the two families concerned but also a testimony to the value of locating a factory within easy reach of raw materials (wheat) and markets. For an example taken from Coventry *see* footnote, p. 242-243 below.

A Case Study

The dominating position of Birmingham in the West Midlands is one which has often been noted but which is difficult to explain. It never enjoyed marked locational advantages and it is not on a coalfield; in the Middle Ages it was vastly overshadowed by Coventry and it was not even on an important routeway. Dury remarks aptly enough that "it did not take part in the earliest iron working of the West Midlands and its accession to the rank of regional capital is as difficult to explain as is the medieval success of Coventry." Closer examination suggests that it was the work of the local lords of the manor, the de Berminghams, which paved the way for future development. This energetic family purchased a market charter for the little settlement near the River Rea in 1166, and by energetic encouragement of the market persuaded many craftsmen to settle there. So the early textile and leather trades grew and prospered, and later came the metal interest which led to the vastly expanded industrial growth of the seventeenth and eighteenth centuries. The manufacture of swords gave place to the manufacture of firearms, while buttons, toys, jewellery and brassware streamed out in ever-increasing quantities from the Birmingham factories. The story is complex, but we should remember to give due importance to the early lords of the manor without whose enthusiasm and determination the original tiny settlement near the River Rea may never have developed into the undisputed industrial and administrative capital of the Midlands.

INDUSTRIAL INERTIA

It has often been observed that industrial development is apt to continue in any given region long after any original advantages for it have disappeared. This phenomenon is referred to as *inertia*, sometimes as *geographical inertia* and sometimes as *industrial inertia*. While the fact of inertia may at first sight seem surprising, a few considerations will show that there are, in fact, good reasons for it.

In the first place, no industry can operate without capital equipment (fixed capital) and such equipment is not infrequently highly immobile, while the period of its working life may be fairly lengthy. No firm can afford lightly to write off such equipment when it may have many years of useful life left in it, especially if it is an expensive piece of plant. It may be argued that rather than write off present equipment at a considerable loss and transfer activities to a new site it may be better to continue operations on the old site even if the continued use of the old equipment means slightly less economic working.

It should also be borne in mind that some of the most expensive pieces of equipment are the buildings and factories themselves, and the above argument applies here with even greater force. We earlier referred to the high costs usually involved in developing industries on greenfield sites, and here again it could be more economic to continue at an existing site rather than to move elsewhere. It is estimated that to construct a new steel plant on a new site costs about three times as much as adding a comparable extra output capacity to an existing works, and if the new site were a greenfield site the costs would be considerably greater. In this connection it is of interest to notice that no new steel mills have been built either in Pittsburg or in Youngstown since 1911, though in both cases a very considerable increase in steel output has occurred. The reason is, of course, that existing plants have been modernised and extended.

The question of labour has also a bearing on inertia. When an industry has been operating in any given area for a fair length of time the skills appropriate to that industry will have taken root and there will be no shortage of them. Transfer of such an industry to a new centre, however, can involve difficulties in securing labour at the new location, and sometimes this type of difficulty can be overcome only by importing skill from the original site. This happened, for example, when textile industries were established at Darlington and at Kaduna (Nigeria); in both cases skilled labour had to be "imported" from Lancashire into the new area.

Neither must the importance of managerial skill be overlooked in this connection, for, while the appropriate types of this form of skill are likely to exist at an established location, there may well be a dearth of it at a new situation. Similarly, banks and other credit institutions at an established centre are more likely to be knowledgeable, and therefore more accommodating, than at a new location.

The importance of the location of technology should be borne in mind in the same connection. Specialised knowledge and equipment are not infrequently localised in established areas and greater fluency in their use is encouraged throughout such areas by apprenticeship schemes, technical college courses and other training programmes. It is likely that the continued location of the aircraft engine and propeller industry of the northeast of the U.S.A. which we have previously commented upon owes much to this point. New devices to assist more efficient and more economic production are more likely to be thought up in established areas, which further adds to the favourable position enjoyed by these areas.

It will be noted that most of the foregoing are intangible factors, but intangible factors can affect industrial strength and progress very markedly and the reader will possibly understand why there is so often strong resistance to locational change in any industry of reasonable size. At the same time one must not suggest that such change never takes place, for any such suggestion would be patently absurd. For many years the location of the cotton textile industry in Lancashire was held up as an example of geographical inertia, but it is well known today that Lancashire is strongly feeling the pinch of competition from outside regions—so much so that, as pointed out by Dury, textile manufacture now employs no more than 11% of all workers in the south-east Lancashire conurbation and only 23% in the large towns of the Ribble Valley.

It is probably fair to argue that limited industrial localisation is a sign of industrial immaturity, and that it normally disappears with increasing maturity of industrial development. It typically arises because new discoveries and inventions are naturally first tried out in the areas where the discoverers and inventors live, and this fact can confer a considerable advantage upon the favoured areas. Thus, Lancashire benefited from the early textile inventions which originated there (pp. 9–10 above) and was able to reap the financial advantages of these inventions for many years, but with increasing development in the industrialisation of other regions this advantage was bound to disappear. Such financial advantages, while they last, have something of the appearance of the quasi-rent of the economist. If this analysis is correct we shall agree that inertia is a feature of the "short run" (which may be quite a long run in terms of years) and that it tends to disappear in the "long run."

HISTORICAL ACCIDENT

A final point which might be mentioned in this part of the chapter is that which may be termed "historical accident"; such a feature can play a very important part in helping to account for modern industrial development, and the facts that Ford happened to be born in Michigan and Morris near Oxford might be held up as examples of this. Stamp mentions the case of Coventry, a town which enjoyed no particular locational advantages in earlier times but which has grown into one of the leading industrial centres in the West Midlands. The town seems to have been interested in the manufacture of woollen textiles at an early stage in its growth and this interest was increased by the settlement of Flemish silk weavers, a circumstance which had a certain element of chance about it. Coventry, then, became well known for its textiles, including its silk ribbons, and alongside this industry there grew up another for the manufacture of the necessary machinery for the production of textile goods. This introduced the various skills required for the manufacture of high-quality machinery and these skills later encouraged the development of other types of machine production, including sewing machines and watches. Later came the manufacture of bicycles, motor-cycles, motor-cars, aero-engines and other industrial goods for the manufacture of which skilled precision is all-important. It is no doubt over-stating the case to argue that today Coventry manufactures aero-engines because

Flemish weavers settled there in the Middle Ages, but no one can deny that the links between the two facts are clearly visible.

THEORIES OF INDUSTRIAL LOCATION

It is clearly no easy matter for an entrepreneur to work out the optimum location for any proposed factory in view of the variable nature and the number of the factors involved. It would obviously be helpful if it were possible to devise some method of quantitative analysis which could be

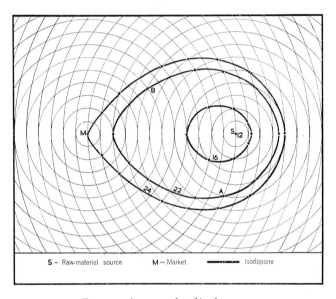

Fig. 24.—An example of isodapanes.

applied to industrial location; such a method could also be of very considerable value to the analyst who seeks to understand the patterns of industry which have developed and which are developing in the modern world.

In view of the importance of this whole subject it is not surprising that various workers in the field of industrial relations, notably von Thünen, Weber, Hoover and Isard (*see* references at the end of this chapter), as well as later writers, have attempted to formulate quantitative location theories. The whole subject of location theory, indeed, has developed into one of considerable complexity and magnitude, and it is not possible to deal with it fully in a book of this size and scope; the interested reader is referred for comprehensive information to detailed works on the subject. On the other hand, even a beginner in the study of industrial geography should realise that quantitative methods of analysis do exist, and in order to introduce the subject an elementary treatment is here given of Weber's concept of *isodapanes*—lines joining points of equal costs—and of his *locational triangle* device.

A possible isodapane diagram is set out in Fig. 24. For the sake of simplicity it has reference to the manufacture of a single commodity from one raw material for sale in a single market (M); the raw material is obtained from a single source (S). The concentric circles centred on M and S respectively are spaced to represent transport charges, and the spacing is the same in the two cases because the transport costs per ton-mile on raw materials is assumed to be the same as on the manufactured goods (if this assumption is not made the diagram becomes very complex). The spacing between any two concentric circles represents one unit of transport costs per ton of commodity moved; thus, the cost of transporting x tons of raw materials or of finished products between M and S amounts to $12x$ cost units. The further assumption is made that the raw material is gross* and that it loses 50% of its weight during the process of manufacture so that one weight unit of manufactured product leaves the factory for every two weight units of raw material consumed.

We are now in a position to examine Fig. 24 more closely, and the first point we might make is that, if the factory is located at M, every ton of manufactured product will have to bear 24 units of transport cost, since 2 tons of raw material have to be transported from S over 12 units of transport cost prior to manufacture. If the factory, however, is located at S only 12 cost units have to be borne.

But these are only two of the possibilities. Take, for instance, cases where the factory is located at A or B. In the first case 2 tons of raw material would have to bear 10 units of transport cost (over 5 transport cost units from S) and the finished product (one ton) 12 (to M), making a total of 22 in all; in the second case the corresponding amounts would be 16 and 6 units, against 22 in all. It is, in fact, possible to construct an *isodapane*, a line which links all points on the diagram at which transport costs total 22 units, and this has been done on Fig. 24; other isodapanes, similarly constructed, are also shown.

Isodapanes can be of value when it is desired to work out the effects on costs of some other variable such as labour cost. It is possible to examine on an isodapane diagram alternative factory locations and to work out

* Weber distinguished between raw materials which are *gross* and those which are *pure*; the former by definition lose weight during manufacture, the latter do not.

the comparative advantage in other costs which these locations must enjoy to make it profitable to establish the factory economically away from S, the point of lowest transport costs. For instance, a location on the isodapane of unit transport cost 16 can be economic only if at least 4 units of cost (16 — 12) can be saved on labour or on some other variable.

While the concept of isodapanes can here be only simply introduced, the techniques involved in the construction and use of this device can be refined to cover realistic cases. One necessary refinement, for example, must take note of the very uneven incidence of transport costs which we

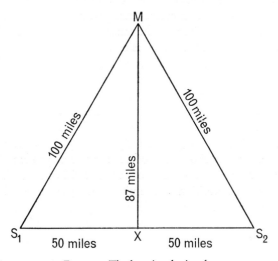

FIG. 25.—The locational triangle.

commented upon earlier. Meanwhile, we may pass on to study another of Weber's contributions to location theory, the locational triangle, of which a possible example is shown in Fig. 25. The assumptions behind the diagram are that two raw materials enter into the manufacture of a finished product (FP), one of which (R_1) is secured from source S_1 and the other (R_2) from source S_2; that both raw materials are gross and lose 50% of their original weight during manufacture; that 4000 tons of each raw material are required each year for the manufacture of 4000 tons of the finished product; and, finally, that the market is located at M on the diagram. To illustrate the working of the locational triangle we may examine three possible cases which might occur:

Case 1. If the factory is located at M the total transport costs involved will be based on the following figures:

4000 tons × 100 miles = 400,000 ton-miles to move R_1 to M
4000 tons × 100 miles = 400,000 ton-miles to move R_2 to M

TOTAL 800,000 ton-miles

Case 2. If the factory is located at S_1:

4000 tons × 100 miles = 400,000 ton-miles to move R_2 to S_1
4000 tons × 100 miles = 400,000 ton-miles to move FP to M

TOTAL 800,000 ton-miles

Case 3. If the factory is located at X:

4000 tons × 50 miles = 200,000 ton-miles to move R_1 to X
4000 tons × 50 miles = 200,000 ton-miles to move R_2 to X
4000 tons × 87 miles = 348,000 ton-miles to move FP to M

TOTAL 748,000 ton-miles

An interesting point, and one that is often overlooked, which emerges from a study of these figures is that the location which enjoys a comparative advantage in ton-miles is one located neither at a raw material source nor at the market, and it focusses attention upon the importance of such factors as weight-loss ratio and transport costs. While the example as given here is crude, the device can be adapted in use to cover actual examples.

These two examples taken from location cost theory are very simple and introductory but they may help to draw attention to the importance of this developing branch of economic geography. For more detailed information the reader is referred to other works such as those cited below.

SUGGESTIONS FOR FURTHER READING

Birmingham and its Regional Setting, British Association Handbook, 1950.
Cole, J. P. and King, C. A. M., *Quantitative Geography*, London, 1968.
Cole, M. M., "The Witwatersrand Conurbation: A Watershed Mining and Industrial Region," *Trans. Inst. Brit. Geog.*, 23, 1957.
Dury, G. H., *The British Isles*, London, 1961.
Estall, R. C., and Buchanan, R. O., *Industrial Activity and Economic Geography*, London, 1961.
Garwood, J. D., "An Analysis of Postwar Industrial Migration to Utah and Colorado," *Economic Geography*, 29, 1953.
Gibson, J. R., "Effluent Disposal and Industrial Geography," *Geography*, 43, 1958.
Hughes, M. E., and James, A. J., *Wales*, London, 1961, Chapter V.
Isard, W., *Location and Space-Economy*, New York, 1956.
Lewis, W. A., *Report on Industrialization and the Gold Coast*, Accra, 1953.

Miller, A. A., "The Use and Misuse of Climatic Resources," *Adv. of Science*, 13, 1956.
Mountjoy, A. B., *Industrialization and Under-developed Countries*, London, 1963.
Scargill, D. I., "Factors Affecting the Location of Industry: The Example of Halifax," *Geography*, 48, 1963.
Smith, J. Russell, and Phillips, M. O., *North America*, New York, 1942.
Stamp, L. D., *Applied Geography*, Pelican Books, 1963.
Weber, A., *Uber den Standort der Industrien*, Tübingen, 1909. Translated by Fridrich C. J., under the title *Alfred Weber's Theory of the Location of Industries*, Chicago, 1928.
White, C. Langdon, "Water—A Neglected Factor in the Geographical Literature of Iron and Steel," *Geographical Review*, 47, 1957.

Chapter VI

The Industrial Region

It has often been observed that by far the greater part of the world's total industrial output is produced in a few highly developed regions; this point was implied in Fig. 11 and on pages 47–48 above. Since the occurrence of these industrial regions is such a notable feature of contemporary world geography it is natural for the geographer to seek reasons for their development and location, and in this chapter we shall address ourselves to a broad consideration of the reasons for the emergence of the major industrial region.

THE DEVELOPMENT OF AN INDUSTRIAL REGION

Industrial development is normally initiated in an area which possesses certain inherent advantages for the industrialist. These advantages may take the form of comparatively easy access to raw materials, to energy resources or to markets, though as we recognise this we must be equally ready to recognise the importance of the human factor; we have previously seen, for instance, how the localisation and development of industry can owe a very great deal to the quality of the more progressive inhabitants of the area concerned.

Once industry is established in an area other factors come into the picture, factors which encourage further development. As workers are attracted by established industries, for example, the earning power, and therefore the purchasing power, of the area increases, and this may well attract more industries, especially those which produce consumer goods. Retail trade is likely to expand as a result, and this in turn attracts more workers—the process is, in fact, cumulative. And, alongside the development of industry and retail trades, tertiary occupations will almost certainly be increasing with regard both to actual numbers employed and to the range of services offered. So, for instance, we may well find that, while a single school is sufficient to cater for the needs of a village, a town of even moderate size will need several primary schools and one or two secondary schools as well. Even larger towns may well boast technical colleges at the apexes of local educational pyramids of considerable size and scope while some offer education at university level.*

We have taken the example of education from the whole field of tertiary occupations, but similar developments can be expected in regions of increasing populations in other tertiary fields of activity. For instance, there will almost certainly be a proliferation of offices, including government and local government offices, and these will require additional staff, while a likely improvement in public transport will produce further increases in population. The construction of new roads can be expected, together with additional facilities for the distribution of water, gas and electricity, and for the disposal of sewage. It is always possible that the provision of social services of this nature may act as an encouragement to the establishment of more industry, partly because of the increased accessibility to sources of energy (electricity and gas, for example) and the better transport facilities available, and partly because the increased population provides a better market and an expanded labour supply. Remember that labour normally exists in joint supply, and while a father works in a particular office, his wife, daughter or son may be offering their services in a broader labour market.

We should also bear in mind that, since industry itself can normally expect to secure very real benefits from large-scale organisation, there is therefore a constant urge to the industrialist to extend the scope of his activities. Such benefits arise from what are known to economists as internal and external economies. Internal economies arise within individual firms as a result of more efficient organisation and the better utilisation of capital and labour resources which can often accompany larger-scale activities, while external economies are economies in which many firms in a given area can share. Such economies can arise sometimes from intangible advantages such as the freer dissemination of new ideas and techniques, and sometimes from more tangible sources such as the presence of a larger and more mobile labour force, better roads and railways and increased energy supplies, and from association with related industries (*see* the section on "linkage" below). These possibilities all

* It so happened that while this text was being prepared (1966) the Vice-Chancellor of the University of Newcastle, N.S.W., was pointing out that the 280 members of staff of the university were "in a position to spend $1,500,000 in the area per year of salaries and wages earned." In addition there is spending by the University itself to be taken into account. Such an injection of spending power into the community is by no means negligible.

($ Aust. = 8*s*. sterling.)

stimulate a wider range of industrial development and encourage existing firms to increase their size.

We should not overlook in this respect the importance of the various financial institutions which are likely to establish themselves in a developing industrial region. All industrialists need capital, and at the same time capital needs industrial or other outlets if it is to be a profitable asset. The establishment of financial houses which have an intimate knowledge of the capital requirements of a region can prove of very great assistance to further economic development—a point which we have emphasised earlier (p. 102).

It is perhaps appropriate to mention that there is normally a limit to advantages which can be secured as a result of large-scale enterprise, though it is not always easy for those concerned accurately to determine when that limit is reached. Many factors may combine to produce "dis-economies," among them increasing competition for capital, land and labour as development forges ahead within a given area; such competition must at some point lead to increased costs of production. In addition, in established industrial regions, labour tends to become better organised and to secure for itself better advantages in the form of higher wages, shorter hours and additional fringe benefits, all of which must ultimately be paid for by the industries concerned, while the growth of urbanised regions above a certain size leads to dis-economies when the cost of living for the workers is sent up as a result of such features as high rents and the greater cost of travelling to work. Such rising costs are bound at some point to be reflected in higher costs to industry, and a tendency for industry to become more dispersed may then set in.

Linkage

In general, however, it remains true to say that the economies to be secured from concentration of industry are very real, and it is not surprising that industry shows a marked tendency to be attracted to established industrial regions. On the production side the factors of labour, technological skill and capital form powerful magnetic forces, while on the marketing side economies of transport, the existence of a wide market, and the prestige value conferred by association with an industrial region of proved reputation are no less powerful. We might at this point make a closer examination of a compulsive aggregating factor, one to which we have earlier made reference and which is known as *linkage*.

The point at issue here is the quite simple one that to a very large extent industries "feed" upon each other; the manufactured products of one industry form the raw materials of another. Often large-scale firms are able to secure substantial economies, as for example when certain manufacturing processes which result in the production of an accessory part are undertaken by self-contained firms. Thus, the manufacture of tyres and sparking plugs is normally undertaken by firms which sell these accessories to the automobile-manufacturing firms. Hiner tells us that the chemical industry of Humberside uses very large amounts of sulphuric acid (in mid 1961 about 13,000 tons a week) and that this demand has led to the establishment of the "greatest single concentration of sulphuric-acid manufacture in Britain" in the Grimsby area.

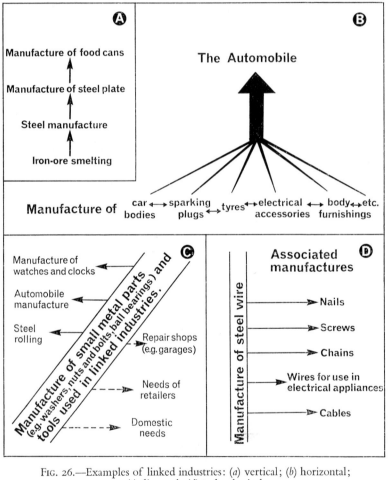

FIG. 26.—Examples of linked industries: (*a*) vertical; (*b*) horizontal; (*c*) diagonal; (*d*) technological.

Some typical ways in which linkage may bring together various industries into close association are indicated in Fig. 26, which distinguishes between "vertical," "horizontal," "diagonal" and "technological" linkage. Vertical linkage arises when a raw material undergoes a series of successive industrial processes, each of which constitutes an industry in itself, until a number of finally manufactured articles is produced. The example put forward comes from the iron and steel industry as iron ore undergoes processing with the purpose of eventually producing iron and steel goods for sale. Notice in the example given in Fig. 26 that the vertical linkage shown normally ceases with the manufacture of food cans, because the canning industry is located by quite different factors from those which affect the location of an iron and steel industry; questions of the availability of the fruit and of possible markets come into the picture.

Horizontal linkage occurs when a number of different industries produce goods which then are finally assembled as parts of a single larger product. The example chosen in Fig. 26 shows some of the various industries which help to make parts for an automobile. Diagonal linkage occurs when items are manufactured which are then used in a number of different industries. Washers, nuts and bolts, for example, are widely used in manufactured products, while tools of different kinds find wide usage in industry. Finally, technological linkage is a term used to denote the case when a single manufactured product is used as a raw material for a range of industries, just as steel wire may be manufactured into nails, screws, chains, electrical fuses, coated wiring and cables.

A Case Study

The foregoing examples represent only a small minority of the many industries which are closely linked together and which may normally be expected to find their most advantageous locations within easy reach of each other, and the reality of the forces which encourage the development of industries within the same general area (the industrial region) should be no longer in doubt. The case of Lancashire comes readily to mind in this respect, for we can observe in this industrial region an excellent illustration of the way in which industry "breeds" industry. We saw in Chapter I how the early inventions encouraged the development of the cotton textile industry in Lancashire, but the story did not end there by any means. The textile machinery itself had to be manufactured, and, while the early machines were made largely of wood, later ones were made of iron and steel, which would better withstand the strains and stresses imposed by fast-moving machinery. Hence it was natural that engineering interests developed in the region, in the first place to service the textile industry. Later on, however, these interests expanded in different ways. At one end, as it were, of the engineering thread, the mining of iron ore near Wigan developed, while at the other end the establishment of transport interests was a natural thing in a region which relies largely upon transport, both of raw materials and of finished products, for its very existence. So we find, for instance, the considerable manufacture of Leyland lorries at Preston which has expanded into an industry of world-wide significance. And, since large-scale industries of the kind we have been considering need considerable energy reserves, it was inevitable that coal mining should become established in Lancashire. A later period saw the setting up of thermal electric power stations.

But the textile industry itself makes considerable demands upon further resources; for instance, chemicals are needed in the manufacture of bleaches and dyes. It was a fortunate occurrence that the salt deposits of Cheshire were so conveniently placed to serve as a physical basis for the strongly developing chemical industry, but it is quite likely that such an industry would have developed in any case. And since chemicals are difficult things to handle and store, and since there is only one really suitable form of container which can safely be used for a wide range of chemicals (glass), it is again to be expected that a glass-manufacturing industry of the kind associated with St Helens should develop in association with the chemical industry.

Even here we have not come to the end of the story, for textiles need to be washed, and as we recall this obvious point we see that the establishment of the colossal soap industry of the region offers another example of linkage, while yet another connection is to be seen in the establishing of the artificial fibres industry near the Dee estuary. There is no doubt, as Moisley points out, that this perhaps unexpected location was chosen largely because of its proximity to the centres of the cotton textile industry, which is its largest single customer. The industrial region as a whole develops because of a closely woven network of industrial enterprise as one industry encourages advances in related industrial fields. It is a true organism and is not simply a region in which a number of disparate industries happen to have taken root.

Before we leave the concept of the industrial region there are three further points which we should make. These are as follows:

1. The industrial region is not static but dynamic.
2. The industrial region cannot develop unless conditions generally, human as well as physical, are favourable.
3. The industrial region is not an easy region to delimit in detail.

THE GROWTH OF THE INDUSTRIAL REGION

We made the point when we were dealing with in[dustry that] changing times are apt to bring changing circumst[ances, and] any given region was well suited to a particular type [of industry in the] past does not necessarily mean that it is equally well suited to [that] industry today. Regions have to adapt themselves to changing circumstances, and we find that sometimes industrial regions decline in importance, sometimes they retain their importance at the cost of modifying their industrial structure, while sometimes new industrial regions arise. Large-scale examples of industrial regions which have absolutely declined in importance are not commonly found, however, for a region which has been found suitable for one type of industry will in all probability be found suitable for others as well, and the forces which we have just been examining are likely to encourage a wide industrial development. The importance of such an industrial region will not easily decline in absolute terms, though comparatively, there may be a very real decline as other competitive industrial regions develop.

The "False Start"

While this is true, however, it is not difficult to recognise industrial "false starts," cases where industry was established but where it never really took root. It is significant, for example, that the earliest of the British coalfields to be developed was that of South Shropshire, an area which is by no means a leading industrial region today (see, for example, Stamp, L. D. and Beaver, S. H., *The British Isles*, fifth edition, London, 1963, Fig. 187). Yet it was on this coalfield, at Coalbrookdale, that coke was first used (in 1709) instead of charcoal to smelt iron ore, and it was here that it was first shown that iron could successfully be used to build boats and bridges and for general constructional purposes. Despite this early start, however, the region never developed into a large-scale manufacturing region, two main reasons for this being its comparative remoteness and the limitations imposed by the smallness of the coalfield. This example shows us that the "early start" so often invoked as a reason for later industrial importance cannot necessarily overcome inherent disadvantages of location and deficiencies of energy and raw materials.

Another good example of a false start comes from New South Wales, where in 1848 a group of pioneers built at Mittagong (Fig. 27) the first iron works in Australia. Local advantages included the occurrence of both coal and iron ore, while limestone was obtainable 15 miles away. Un-

A GEOGRAPHY OF MANUFACTURING

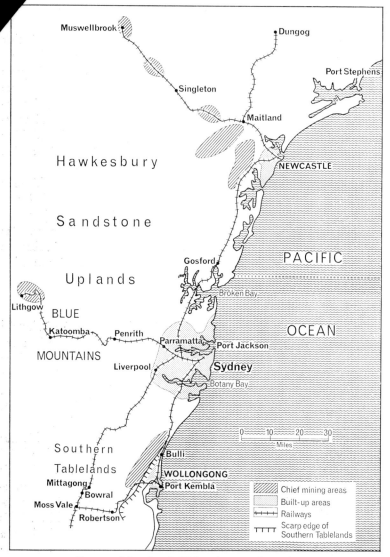

FIG. 27.—New South Wales: the central coastlands.

fortunately, however, both the coal and the iron ore proved to be of poor quality and limited in amount, though the factor which more than anything else caused the failure of the Mittagong enterprise was the remoteness of the small town from any large-scale market. The plant closed down in 1886. A similar fate overtook the second Australian attempt to establish an iron and steel industry with a location at Lithgow, which lies just over 90 miles inland from Sydney, and for similar reasons. It was not until iron works were established with a better appreciation of location requirements with regard to markets (first at Newcastle in 1915 and then at Port Kembla in 1928) that a real start was made in developing large-scale industry in New South Wales. (*See* also footnote, p. 196 below.)

A centre which provides a good example of inhibited industrial growth is that of Sydney, Nova Scotia. The Cape Breton Island area has many advantages for industrial development, particularly with regard to easy access to coal and raw materials. The coalfields of the island are very extensive and many of the seams extend for considerable distances beneath the sea; seventy seams of coal outcrop along the cliffs near Sydney, many being over 5 feet in thickness. Limestone occurs on the island and is also shipped from Newfoundland, while iron ore was produced in large amounts at Bell Island, Newfoundland, though the greater part of the output today is actually exported, mainly to the United Kingdom and Germany. The iron ore of Schefferville is also within comparatively easy reach of Sydney.

It is perhaps difficult to imagine a set of natural circumstances which are apparently more favourable than these for the establishment of an iron and steel industry,* an industry which might have been expected to act as the economic base for the development of an industrial region of some importance, but in fact the area has failed conspicuously to attract secondary industry. Early in this century a fair degree of prosperity was generated as the region specialised in the manufacture of rails, to keep pace with the expanding network of Canadian railways, especially in the Prairie Provinces, but after the First World War, the demand for rails sharply diminished. The tendency now is for the blooms, billets and rods manufactured in the Sydney area to be shipped for further processing to mills in the main manufacturing belt of southern Ontario and Quebec. The region, despite its apparent advantages, suffers from arrested development and has never fulfilled the hopes originally entertained for its expansion.

The reasons for this situation are not far to seek, and the following list is interesting as it touches on several points previously made:

1. Cape Breton coal contains a fairly high proportion of sulphur and is very volatile. The quality of coke made from it is much inferior to

* It is perhaps only at Birmingham, Alabama, that more favourable circumstances are found.

that made from imported American coal at centres such as Hamilton.

2. The Wabana ore from Bell Island is too rich in silica and phosphorus for the Sydney open-hearth furnaces, though it is excellent for European processing. Small amounts of ore from Schefferville and Brazil are mixed with Wabana ore at Sydney before treatment.

3. The whole region is remote and does not carry a large population. The local market is therefore very restricted, while the lengthy journey to the more densely populated parts of the St Lawrence lowlands and the Lake Peninsula makes it difficult for Sydney firms to compete in these comparatively distant markets.

4. The structure of freight charges works to the disadvantage of the Sydney region. Kerr notes, for example, that in 1955 freight charges on finished nails moving from Sydney to Montreal were $13.00 per ton, while corresponding charges to Toronto were $20.40 and to Windsor $23.20. On the other hand, shipment of the unfinished wire rods from which nails are made to each of these towns cost respectively $7.61, $10.26 and $11.32 per ton. This situation is one which offers a clear incentive for processing industries to establish themselves near the extensive St Lawrence-Lake Peninsula market rather than at Sydney.

5. Government policies must also be considered, for the U.S.–Canadian boundary effectively seals off Cape Breton from what would naturally be her largest customer for iron and steel goods—the Atlantic seaboard of the U.S.A. The possibility of opening up this market is, however, ruled out by the economic nationalism of the U.S.A. which manifests itself in prohibitive tariffs.

6. The lack of a substantial market near at hand robs Cape Breton of a valuable raw material—scrap iron. This is a serious disadvantage when other production areas can draw heavily on this comparatively cheap commodity which is so well suited to modern steel-making processes.

It is not perhaps surprising that the importance of Cape Breton as an iron and steel producer is declining, a fact which was emphasised by the closure in 1968 of the large Dominion Steel and Coal Corporation plant near Sydney. The location of the region is distinctly marginal, in an economic as well as in a geographical sense, and its continuance as a producer may be recognised as an example of industrial inertia. It provides an excellent example of an industrial false start.

A Dynamic Region: New England

Some industrial regions retain their importance by adapting their industrial structure to changing circumstances. Such adaptation requires considerable foresight and courage, for it often has to be carried out against a background of industrial depression engendered by competition suffered by established industries from newer, perhaps better-favoured, areas. The case of New England comes readily to mind in this connection for, like England itself, New England started early in the industrial race.* Estall reminds us that both areas at an early stage developed a high degree of dependence upon industries, particularly the textile industries, for the establishment of which they possessed no especial advantage except the existence in each case of a market; the textile industry was, however, bound to be one of the earliest to develop because the demand for clothing and cloth is one which has existed throughout historic time, and as it is widely based it was naturally one to attract attention from inventors and industrialists at an early stage of industrial development. Such was the early importance of this industry that it has been said that "the story of the New England cotton industry is the story of the industrialisation of America" (Ware). For further information *see* pp. 223–230 below.

This honeymoon period could not last for long. Other regions were bound to enter the textile-manufacturing industry, and New England had already passed through its period of industrial supremacy by the time of the First World War. Estall points out that between 1919 and 1939 the number of manufacturing firms decreased by 18% and the number of industrial workers by 26%. In the textile industry during the same period, employment fell dismally by 36%, while there was a notable fall in employment in the hitherto important leather industry. Even the metal-working industries (except those manufacturing electrical equipment) were depressed and increases in employment were recorded only in the clothing, printing and publishing industries.

Since 1939, however, this gloomy picture has completely changed, and an increase in industrial workers of about 25% between 1939 and 1954, restored the number of workers approximately to the 1919 figure. This metamorphosis has come about as a result of a large-scale change in the industrial structure of the region, and the following are the main points to bear in mind.

The manufacture of textiles still remains the most important of the non-durable goods industries, though only 16% of all textile mill operatives in the U.S.A. are now to be found in New England. The old supremacy in this industry has obviously gone, and this is especially true of the cotton textile industry, though the traditional woollen textile branch is also losing its former dominance. It is interesting to observe, however, that the finishing section of the industry has not suffered a comparable

* The following section draws heavily on Estall (*see* reference at end of chapter).

decline. It is also interesting to notice that the decline in the leather industry seems to have halted and that about 30% of all leather workers in the U.S.A. remain in New England, while the manufacture of rubber products claims nearly 11% of all U.S. workers in this field. The importance of these industries offers interesting examples of linkage, for both are closely connected with the concentration in the region of the manufacture of footwear, while the rubber industry makes not only tyres but also rubber accessories used in the electrical industries.

The manufacture of pulp, paper and associated products still shows a fair degree of concentration in New England, and the advantages for this industry include extensive reserves of pulpable hardwoods, the proximity of the enormous market of the north-eastern U.S.A. and the large resources of good-quality water. Unfortunately, energy costs are higher in New England than in competing areas, and this is a serious point, for the industry is a large consumer of energy.

With regard to the durable goods industries, the leading interests are concerned with the manufacture of electrical and non-electrical machinery, of transport equipment, and of instruments and related products. The importance of the electrical machinery group emphasises the strength of the emergence of new industries, as a result of research and industrial diversification. Facilities for research and training in electronics are well developed, especially in the Boston area, and a striking commentary on the newness of this leading industry is the estimate that, in 1955, at least a half of all employees in electronic enterprises were working on products developed as a result of research undertaken since 1950. Such products include electronic tubes, transistors, telephone and radio equipment, and other apparatus for use in the electrical industry. In 1968 95% of the world's production of computers was in the hands of American firms, many of which were located or based in New England. The non-electrical machinery group of industries produces a wide range of products, including machines for metal grinding and cutting, and also machinery designed for the textile, leather and paper industries. Here is another example of the strength of the urge for related industries to establish themselves in the same region; questions of market attraction and linkage are clearly involved. Neither should we forget that enterprise on the scale on which it is conducted in New England demands extensive office organisation, and so it should not be surprising that the manufacture of office equipment (typewriters, dictating machines, filing cabinets, etc.) is a thriving branch of industry in this region.

Transport equipment (including ships, locomotives and aircraft), is manufactured in New England, where, after a period of spectacular development, more than a quarter of all workers employed in the U.S.A. in the manufacture of aircraft engines are working. There seems little doubt that the large resources of skilled labour available have acted as a powerful attractive force in this industry, while the existence of numerous small firms able to undertake the manufacture of accessories is also advantageous—another example of linkage.

Shipbuilding is a notable industry in New England, and about 12% of the total number of workers in the U.S.A. employed in this industry are in New England shipyards, despite the increased shipbuilding developments in the north-west of the country which we commented upon earlier (p. 98). It is important to notice, however, that the motor-vehicle-manufacturing industry is poorly developed in this region. This may seem surprising since in the early years of this century, the main centres of automobile manufacture in the U.S.A. were New England and the Mid-West, but the reason seems to be that the New England industrialists manoeuvred themselves into an industrial cul-de-sac as they concentrated mainly upon the design and manufacture of steam-driven and electric vehicles; these later proved less popular than the petrol-driven models which the Mid-Western manufacturers had favoured.

The final category of durable goods industries mentioned above, is that of the manufacture of instruments and related products. Modern development is such that there is in constant demand, a whole new range of measuring devices, ranging from the requirements of aircraft and guided missiles to the computers needed over a wide range of industrial and educational enterprises. The long-established manufacture of watches and clocks has declined in recent years.

The point of the New England example is clear enough: an industrial region must be prepared when occasion demands to cut its losses in industries which lose their competitive cutting edge, and build up industries for which it has comparative advantages (*see* p. 85 above). Thus, New England was giving employment in 1919 to about $1\frac{1}{2}$ million workers, and to about the same number in 1956, but, whereas in 1919 40% of those workers were employed in textile and leather interests, the corresponding proportion in 1956 was less than 19%. This is evidence of a massive readjustment to changing conditions, which is further emphasised by the increase in the labour force in the main metal-working industries from 22% in 1919 to 36% in 1956.

We might finally note, however, that while the region as a whole has successfully made this adjustment, certain areas within New England have not. This ties up with the point we noticed earlier, that some fairly small industrial regions may suffer actual decline with changing circumstances.

The case which comes to mind is that of the textile group of industries sited near the Merrimack River, especially at Lowell and Lawrence, where industrial stagnation in the textile field has not been counterbalanced by advance in other directions. This may be accounted for partly because the old textile buildings are not suitable for or attractive to modern industrial requirements, while another point is that the Merrimack region is situated on the northern fringe of industrial New England and far removed from the main body of the U.S.A. The area is paying the price of a circumferential location.

Industrial Relocation: New South Wales and the U.S.A.

As a final comment upon the dynamic character of modern industry, we may note that it is inevitable that industry will relocate itself as opportunities arise, and this means that new industrial regions are likely to develop. We have previously observed that an industrial region is in process of developing along the seaboard of central New South Wales (Fig. 27), though this area is not rich in natural resources except coal. The seaboard location is, of course, a great asset, though with the exception of Port Jackson, and, to a lesser degree, the mouth of the Hunter River, useful harbours are not typical of this coast. The Hunter mouth is kept open for sea-going vessels only with the help of constant dredging. Port Stephens to the north and Broken Bay (Hawkesbury River) farther south may appear from the map to provide excellent harbourages, and so they do, but this is the only advantage in their favour. Port Stephens is located well to the north of the coalfields and suffers from its comparative remoteness, while the Hawkesbury ria system bites into the picturesque but desolate uplands based upon the unproductive Hawkesbury sandstone series. It is interesting to note that in the southern part of the region known as Illawarra, a completely artificial harbour has been built at Port Kembla to handle incoming iron ore and outgoing iron and steel goods; it is hoped that this new port may eventually develop into a general services port which could relieve some of the strain upon Sydney, but the difficulty so far has been that the coastal Illawarra area is backed by a notable escarpment, which is a real hindrance to the opening up of communications with the tablelands. The only railway running inland, the one to Moss Vale, is a single-track line which has experienced great difficulties because of steep gradients, expensive maintenance and periodic blocking of the track by earth slipping, but recent improvements hold out some hope that it may become a more effective link. Already, Port Kembla is making a bid to capture some of the wool export traffic from the Southern Tablelands which has traditionally passed through Sydney.

This, then, is the region with its three focal points of Newcastle, Sydney and Port Kembla–Wollongong which is developing into one of the chief industrial regions in the southern hemisphere. The motive forces behind this development include periodic shortages of manufactured goods in Australia in times past owing to European-based warfare, an increasing reluctance to continue to be dependent upon overseas sources of supply, which coincides with an increasing feeling of national pride, and an increasing realisation of the market potential available not only in Australia, but also in South-East Asia. Australian manufactured goods are even beginning to compete successfully in established world markets elsewhere. We may, therefore, expect industrial development and the further building-up of this industrial region to continue.

Other well-known examples of the relocation of industry come from the United Kingdom and the U.S.A. McLoughlin and Robock have examined the American case and have put forward a number of reasons why industry is moving southwards from the old-established industrial areas. The following appear to be the main points at issue:

1. In 45% of the cases studied, the attraction of expanding southern markets was a main reason for change.

2. In 30% of the cases the availability of raw materials and energy supplies was the compelling reason. Important factors here included the electrical energy derived from the Tennessee Valley Authority (T.V.A.), the mineral oil and natural gas of the Gulf region, and other assets such as the phosphorus reserves of the Gulf coastlands and the cotton from the Cotton Belt.

3. In 25% of the cases it was considerations of labour which governed the decision to move. Many factors are involved here, including alleged restrictive practices met with in the North (p. 104 above) and differential efficiency between the North and the South, but the main reason may well lie in the differing attitudes to factory employment in the two regions. In the North it is still something of a social stigma to work in the "mill," while in the South it is rather a social advancement. This change of attitude is a reflection of differing environments, for until recently the South has been a depressed area with few chances for social and economic advance; it has been a region with few large cities and with a low level of capital investment, and rates of return per unit of labour have traditionally been low, especially in agriculture (*see* p. 228 below). The result is that the new factories offer opportunities for increased earning power, which are to be welcomed and not viewed with something akin to contempt. Because of this favourable social climate, labour relations in the South are generally good and absenteeism is rare.

4. When once established it is observed that industry in the South benefits greatly from such features as a favourable climate (which means low heating costs), modern plant and, not infrequently, model housing and community schemes. The latter are in large measure advantages often enjoyed by new industrial regions as compared with older ones.

5. Industrial growth in the South has coincided with the entry of the government into economic affairs, a comparatively new departure in the U.S.A. The New Deal legislation of 1933 initiated a period of increased government interest and encouragement, the T.V.A. being the most spectacular manifestation of this changed social environment.

6. The outstanding "capture" of the South has been that of the cotton textile industry from New England, for the South has quite definite advantages in this respect. These advantages include lower wages (p. 81 above), closer proximity to cotton supplies, which are therefore cheaper than in New England, cheaper power, the advantage of newer and more efficient plant and mills, and lower taxes. (For further information about the cotton textile industry in the South, see Chapter IX, pp. 223–229 below.)

The United Kingdom

It does not seem surprising, when we bear all the relevant points in mind, that the South is developing into a major industrial region, while the correlated developments in older-established industrial regions include the modifications of the industrial structure of New England which we observed earlier in this chapter. The reader will not find it difficult to recognise that in the United Kingdom, there are comparable reasons for the southward movement of industry from the older-established coalfield areas. Such reasons include nearness to continental markets, raw materials and energy resources; the growth of the domestic market in the south, particularly the south-east; the mobility conferred on industry by a close network of roads and railways and by the comparative ease of distribution of electric, mineral oil and natural gas energy resources; the more genial climate and landscape which many people would agree characterise the south; the availability of advantageous sites such as that of Fords at Dagenham (p. 100 above); and the prestige which is commonly associated with a metropolitan location. In addition to these attractive forces we might add that a history of restrictive labour practices and poor labour relations in parts of the north has undoubtedly dealt industry in that region serious blows, while it is also the case that many of the old-established traditional industries of the north, such as the cotton textile industry, shipbuilding and coal-mining itself, have all been passing through difficult stages associated with the intensification of competition from newly developing regions overseas. The general impression often built up of the north is that it is a region of sub-standard terraced houses, labour queues and depressed industries, and this image, however unjustified it may be, has undoubtedly helped to intensify the drift to the south.

THE GROWTH OF THE INDUSTRIAL REGION (2)

The second major point which we made above (p. 126) regarding the industrial region, is that such a region cannot develop unless conditions generally, human no less than physical, are favourable. This principle has already been illustrated in the case of Sydney, Nova Scotia, while we might further take as an example the case of Spain. Spain has been until recently, one of the poorest territories of Europe (*see* below), but this does not mean that the country is unusually poorly endowed with natural resources; the coastlands of the north-west, for example, possess substantial reserves of iron ore and coal, iron ore mainly near Bilbao and coal near Oviedo. By the middle of the nineteenth century Spain was exporting iron ore from Bilbao, for the ore was easily mined and it is of high grade and non-phosphoric—an important point at that time (*see* p. 174 below). Capital reserves were gradually accumulated from the sale of this ore and in the 1880s an iron and steel plant was established near the location of the ore deposits. Export of ore continued, however, and Welsh coal was brought back as a return cargo. Great Britain and Germany were the chief customers though the comparative importance of non-phosphoric ore diminished after the invention of the basic steel making and the open-hearth processes.

Here, then, we might suppose that we have the right circumstances for the development of at least a minor industrial region. Raw materials either exist near by or can be imported, capital was available, while the initiative of the pioneers of the iron and steel enterprise which was established during the 1880s cannot be doubted; yet the attempts which were made to establish the industry were never really successful, and the main reason for this was simply that the country as a whole was not ready for industrial development on a large scale. The market for large amounts of manufactured goods just did not exist. The fact is that even today Spain is only emerging from the status of an agricultural territory, with comparatively low living standards. In 1962, for example, the national *per capita* income was only about £130, as compared with about £270 for the territories of the European Economic Community and £520 in Britain, while almost 40% of the total working population were dependent for their livelihood directly upon agriculture. Furthermore, until recently, agriculture has been backward and has not been able to afford the mechanised and other technical aids which industry could provide.

The narrowness of the domestic market meant that industry generally had to be organised on a modest scale; Quintana, writing in 1964, stated that 82% of Spanish firms at that time employed fewer than ten workmen. This in turn meant that the basic iron and steel industry had to be organised on a comparatively small basis because of the restricted market for its products, and this condemned the industry to small-scale uneconomic production. The *Summary of the Spanish Economic and Social Development Plan* (Madrid, 1964), for instance, states (p. 61) that the "ferric minerals extracting industry . . . is characterised by some out-of-date equipment and the large number of firms in it, which raises a difficulty in the way of reducing costs. The present production is obtained by 222 mining firms; 200 produce less than 50,000 tons annually and only 8 produce more than 200,000 tons, representing 55% of the whole production." It was inevitable that under these circumstances iron and steel goods could not be sold in foreign markets because they were too expensive, while domestic purchases were restricted for the same reason.

It is not surprising that the industry did not flourish until a change came over the scene in the period after the Second World War, largely as a result of increased government activities, for many of these activities (for example the expansion of airfields near Saragossa, Madrid and Seville, and the development of naval interests at Cartagena and Cadiz) opened up additional markets for iron and steel goods. In addition, financial measures initiated in 1959 stimulated the economy and helped to bring about a complementary stirring of economic activity in the field of independent industry, shown, for example, by the initiation of E.N.S.A.S.A. (Empresa Nacional de Autcamiones S.A.), a large firm with factories in Madrid and Barcelona which manufactures buses, motor coaches and lorries. This firm employs 8000 workers who produce a total output of about 7500 vehicles a year, about 15% of which are exported. This export figure is expected to increase. Results of these changing circumstances appeared in the post-war expansion of steel output, which rose from 1·6 million tons in 1956 to almost $2\frac{1}{2}$ million tons in 1961. Much of this increase was due to privately owned steel companies, but the state-owned integrated plant at Avilés came into production in 1959 with an annual output of almost 800,000 tons *per annum*.

In 1964 came the launching of the four-year plan for economic and social development, the aim of which was to deepen and broaden economic and social developments throughout the whole spectrum of life in Spain. Results were not slow in coming. In 1964 alone industrial production startlingly increased by almost 12% over the previous year, while the tertiary sector of the economy also grew by 12%. Gross national investment showed an increase of 8·5%, while long-term foreign investment amounted to almost £300 million. Developments under the four-year plan included the building of 15,000 classrooms for primary schools alone, with corresponding increases in post-primary education; a substantial improvement in the whole transport system of Spain which sparked off a demand, among other things, for very large amounts of steel for bridge construction; an increase in the output of motor-cars from 80,000 in 1963 to 260,000 in 1966; and a comprehensive range of developments in such activities as the canning industry, the chemical industry and the manufacture of telephones. It is estimated that about a million new employment vacancies were created in industry between 1964 and 1967, and many of these vacancies were filled by attracting workers away from agriculture. Despite this, however, agricultural production has greatly increased (by 2·3% between 1966 and 1967 alone) thanks to the introduction of new methods, increasing mechanisation* (which further stimulated demand for the products of industry) and a markedly increased use of fertilisers (also a product of industry).

It is difficult to avoid the impression that the Spanish four-year plan has been a very successful enterprise. In the five years 1963–67 inclusive the gross industrial product increased by 61% at constant prices; during the same period employment in industry rose by 20% and industrial productivity by 35%, while the *per capita* income increased to £237 in 1966, an increase of 178% over the 1962 figure (*v.s.*). This gives the staggering annual rate of growth of over 19%. In almost every sector of industry output increased, as is shown by some representative figures in the following table:

---

* It is not always realised to what an extent agriculture can now be mechanised, though many people have seen such equipment as combine harvesters and mechanical hedge trimmers in the fields. But mechanisation can today go much farther than that. The writer recalls seeing in Australia a progressive dairy farm in which cows are led to the milking parlour in relays. Manure is led away by mechanical means from beneath the floor of the parlour to containers some distance away. Foodstuffs (hay and cattle cake) are set before the animals as they are milked also by mechanical devices, while the cows are attached to what is virtually a pumping system, which extracts the milk from the udder and delivers it to a tank; from there it is again pumped to the container lorry which will transport it to the creamery. Very little labour is required to ensure the smooth operation of these processes; one man can supervise operations which needed twelve workers a quarter of a century ago.

	1963	1966
Coal ('000s metric tons)	15,547	15,624
Electricity (millions of kWh)	25,897	37,466
Steel ingot ('000s metric tons)	2,805	3,751
Sulphuric acid ('000s metric tons)	1,495	1,517
Cement ('000s metric tons)	7,152	11,832
Fertilisers ('000s metric tons)	504	646
Cotton textiles ('000s metric tons)	104	124
TV receivers (units)	315,460	525,856
Motor cars (units)	79,432	257,910
Tractors (units)	13,278	16,799
Ships launched (t. gr. reg.)	188,000	365,659
Dwellings (units)	206,700	287,160

Since the industrial worker's income is now almost 50% higher than that of his agricultural counterpart, it is likely that increasing numbers of workers will move from the land to the factories. Quintana believes that the 1960s will probably register "an unprecedented drop in the percentage of farm workers." Clearly, industry is on the move in Spain, but it has taken many years before the right combination of circumstances for large-scale industrial development finally emerged.

DELIMITING THE INDUSTRIAL REGION

While we may be fairly clear in broad outline what we mean by an "industrial region" we may find it necessary to agree, when we seek more carefully to delimit such a region, for instance on a map, that it is not easy to set bounds to it. Buchanan recognised a similar difficulty when he attempted to delimit an agricultural region, yet in some ways the agricultural region seems to be a more easily comprehended entity than the industrial. It is possible without much difficulty to visualise a region in which the dominant interest is agricultural—a region in which all but a minority of working inhabitants follow agricultural pursuits; the difficulties encountered by Buchanan were not so much difficulties of recognition of agricultural regions in general, as of recognising appropriate boundaries for major subdivisions such as the old Corn Belt of Baker.

Comparable difficulties arise in any attempt carefully to delimit an industrial region. On the one hand we are likely to encounter industrialised areas in which agricultural interests scarcely exist, while at the other extreme we may recognise the genuinely rural area. Between these two extremes lies a whole range of transition zones, and it is doubtful whether it is possible clearly to recognise exactly where an industralised urban region gives place to a rural one. This difficulty was encountered by Pounds when he tried to define "more specifically and with greater precision" the industrial region generally known as the Ruhr. He found that the density of urban development falls off gradually from the closely built-up city areas, through zones where agriculture is of some significance, to areas where agricultural interests are dominant, areas in which mining and industry are entirely subsidiary features in the landscape. His map of landscape types in the Ruhr (*see* Fig. 5 in the article cited at the end of the chapter) recognises four elements (closely built-up urban areas, industrial areas with subsidiary agriculture, agricultural areas with subsidiary industry, and agricultural areas) but he admits that the classification "is in some measure a subjective one." And the percipient reader will have noticed a further discrepancy—that urban areas are apparently equated with industrial ones.

This immediately raises the question of the appropriate criteria to use in any attempts to delimit an industrial region. Is it reasonable to equate urban and industrial regions when often even in built-up areas only a comparatively small proportion of the landscape is taken up with specifically industrial features? It is usual to find groups of factories or even single factories embedded in a matrix of non-industrial urban development, prominent among which will be residential areas for factory workers and others. Yet we may justifiably doubt whether it is reasonable to isolate for purposes of classification certain parts of an urbanised region in which factories happen to be concentrated, and in which most of the inhabitants are factory workers, from adjoining parts. Is it not true that a continuously urbanised region generally has some form of unity, a unity which is quite at variance with any conception of the isolation of residential and industrial cells? Is it not true that, while only a bare majority or even a minority of the workers in a given urban region may work in factories, the activities of many more of the fellows of these industrial workers directly assist them by helping to transport them to and from their work, by selling food, drink and other necessities to them, and in many other ways? And is it not true that even a tertiary service like education may be largely aimed in fact, though probably not in theory, at assisting industrial enterprise? This, indeed, will be the final effect as children are equipped with basic skills which enable them to take their places in office and in factory, in shop and in public service, while higher technical education may more consciously be directed towards providing essential skills needed in industry. Is it not true that all the foregoing activities form as much parts of the pattern of the industrial region as the activities which are met within the factories?

It begins to look as if a good case can be made for the inclusion of most large-scale urban areas in major industrial regions, and this may be some justification for Pounds's approach to the case of the Ruhr, but to leave the matter there would almost be equivalent to arguing that an industrial region is an urban region, and a map of industrial regions would be almost entirely a map showing the location of urban areas. This hardly seems reasonable, especially if we recognise that the questions which we raised in the last paragraph can sometimes quite appropriately be asked about non-urban areas too. Is it not true that in a highly industrialised region quite large numbers of industrial units will be located in the countryside? This may be particularly the case with respect to industries which process the products of agriculture, but it is becoming equally true of other industries. Brick works, cement works, gas works, breweries and flour mills, among others, have for long been typical of the rural scene and of small towns, while many manufacturers are now taking advantage of the possibilities of decentralisation which we referred to earlier (p. 123) to establish factories in the pleasant surroundings offered by villages and country towns. And is it not true that in an industrialised country many, probably the majority, of the agricultural workers are engaged in producing food and drink for the industrial workers, and thereby helping to make their way of life possible, just as much as the bus driver or the shopkeeper in the urban area help in their way? And is it not reasonable, therefore, that such agricultural areas should be recognised as forming part of the industrial regions which they help to sustain?

There is no easy answer to this question. The purist might be tempted to argue that if this line of thought is adopted we might just as reasonably nclude, say, the Canadian Prairies as part of an industrial region as the wheat of the prairies goes largely to feed the industrial populations of North-West Europe. While we might well feel that this is carrying the argument too far, it does remind us of the difficulty of fixing reasonable limits to industrial regions in cases where such a pronounced feature as a sea-coast does not provide us with a ready-made one.

Another line of approach to the whole question which is sometimes adopted, is to classify as an industrial region any area in which large numbers of employers are engaged in industrial pursuits. This kind of approach was adopted by Alexander in compiling the map shown in Fig. 28. On this map each separate circle represents either one county or one metropolitan area (which may include several counties), and from the given data and the pattern of circles it is simple enough to recognise in general terms the existence of industrial regions, *e.g.* those along the north-eastern coast, and to the south of the Great Lakes. We find, however, little guidance regarding the precise limits of such regions.

Yet another approach may be made through a more careful examination of the areal distribution of industrial activities. Figure 29, for example, represents a type of map commonly met with in textbooks and it is an attempt to delimit the main industrial regions of North-West Europe. (This map is taken from Dicken and Pitts: *see* reference at end of chapter.)

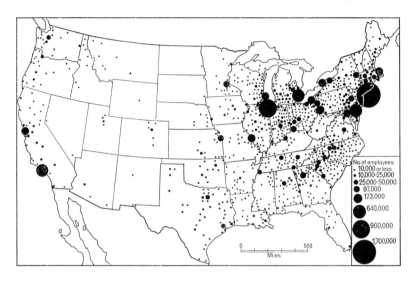

FIG. 28.—Manufacturing employment in the U.S.A.
[*Courtesy: Prentice Hall Inc*

The compiler of the map has recognised that the matter is one of some complexity as he has recognised the existence of "principal subdivisions" within the Western European manufacturing region; these are presumably regions of intensified industrial activity. No indication is given, however, of the criteria upon which this category of subdivision is based and one suspects that there is a strong subjective element about it.

It would also be helpful to know upon what data the boundaries of the general manufacturing region shown on Fig. 29 were drawn. Do these boundaries in fact correspond to any facts of landscape or of economic activity? Is there in the areas located just within the shaded side of the boundary a concern (direct or indirect) with an industrial economy which is lacking in adjacent areas in the region not shaded? From the fact that the boundaries as drawn are remarkably smooth and sweeping one is tempted to feel that they, too, have been drawn wholly, or almost wholly, subjectively and that they would be very hard to defend in detail.

What, then, is the exact significance of such a map? It cannot mean that the whole of the area shaded is submerged beneath a surface mantle of industrial activity; indeed, it is possible to travel over much of the area shown without setting eyes on a single factory or even on a town of any size. It seems that the compiler of the map may have attempted to use the general criteria suggested above and is arguing that the shaded areas are areas in which human enterprise is geared to industry, and that where agriculture appears (even where it is the dominant activity) it is concerned in some way with nourishing industry, either directly by providing

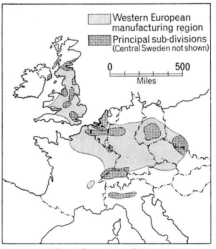

FIG. 29.—Industrial regions of Western Europe.
[*Courtesy: Blaisdell Publishing Co.*]

raw materials or indirectly by producing food and drink for industrial workers.

We can examine this point more closely with the help of Figs. 30, 31 and 32, which show in more detail part of the same region of North-West Europe. Figure 30 shows quite clearly that the concentration of industry in Belgium is in the main limited to a few areas and that over fairly considerable parts there is no industrial development at all. Figure 31 illustrates the point that the agricultural areas (which all fall within the Western European manufacturing region of Fig. 29) produce raw materials for use in industry as well as food and drink for industrial workers, and it therefore seems reasonable to include such areas within the general industrial region. What the map cannot show, however, is the propor-

THE INDUSTRIAL REGION 145

tion of rural activities which are so orientated; it is more than likely that some of the lightly populated areas of the south (Fig. 32) have in fact very slight industrial links and that they should be excluded from the industrial region shown on Fig. 29. The exact situation, however, could be resolved only after extensive field observations. The matter is obviously one of considerable complexity and all that can be attempted in a work of this scope, is to show why no clear and definite answer can normally be given to the question: "What is an industrial region and how can we recognise its boundaries?"

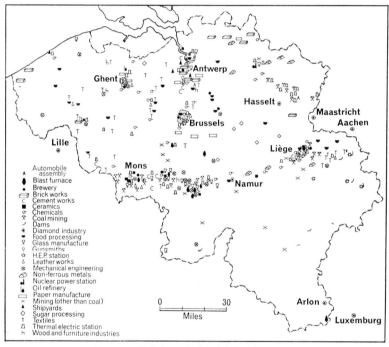

FIG. 30.—Belgium: industrial development.

In case this conclusion may seem too defeatist it is perhaps appropriate to point out that we are in this matter facing a problem which appears in other branches of geographical study. We may take, for instance, the example of the climatic region. No geographer imagines, when he studies a small- or medium-scale map which purports to show a pattern of climatic regions, that each separate region is climatically homogeneous throughout, neither does he argue (except in unusual cases) that the boundaries of the various regions as shown on the map can be defended in detail. Rather, it is true that, while the climate experienced within any given climatic region shows a fair degree of uniformity, local variations may be expected to occur; these local variations may be quite marked and on a detailed analysis may be regarded as giving rise to sub-regions. Similarly, in an industrial region, we may legitimately expect to find local variations, sometimes quite marked, in the general pattern of

Fig. 31.—Belgium: agriculture.

industrial development. It is again true that climatic regions generally merge almost imperceptibly into one another and the same may very well be true of the industrial region and its non-industrial neighbour (or its industrial neighbour if a distinction is being sought between different types of industrial regions). In so far as this is the case it follows that there is bound to be a certain measure of artificiality in any attempt precisely to delimit any industrial region.

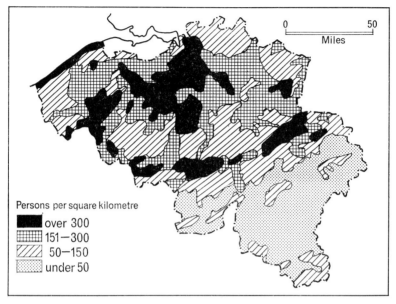

FIG. 32.—Belgium: population density.

SUGGESTIONS FOR FURTHER READING

Alexander, J. W., *Economic Geography*, New Jersey, 1963.
Buchanan, R. O., "Some Reflections on Agricultural Geography," *Geography*, January 1959.
Dicken, S. M., and Pitts, F. R., *Introduction to Human Geography*, New York, 1963.
Estall, R. C., "Changing Industrial Patterns of New England," *Geography*, April 1961.
"Focus on Spain," *Steel Review*, 34, 1964.
Hiner, O. S., "Industrial Development on South Humberside," *Geography*, November 1961.
Kerr, D., "The Geography of the Canadian Iron and Steel Industry," *Economic Geography*, 1959.
McLaughlin, G. E., and Robock, S., *Why Industry moves South*, New York, 1949.
Moisley, H. A., "Rayon Industry in Great Britain," *Geography*, June 1949.
Pounds, N. J. G., "Ruhr Area: A Problem in Definition," *Geography*, July 1951.
Quintana, E. F., "Spain: A Society in Full Development," *Progress*, 50, 1964.
Stamp, L. D., and Beaver, S. H., *The British Isles*, fifth edition, London, 1963.
The Times, Supplement on Spain, 5 June 1965.

PART TWO
SELECTED INDUSTRIES

Chapter VII

The Small-scale Industry

It will by now be apparent to the reader that in this book the term "industry" is used in the sense of "manufacturing industry," and that such activities as the perhaps unhappily named "tourist industry" are excluded from our survey. This point was explained in the Preface. We have in Part One of the book made a broad survey of the various factors which help in deciding the location of industrial enterprise and in Part Two we shall select one or two industries for special study; the space at our disposal does not permit more than this.

In this connection it is considered desirable to include first a short chapter dealing with some examples of small-scale industry in order to induce a sense of perspective. It is very easy for the passing observer in these days to conceive of that rather amorphous group of human activities known as "industry" solely in large-scale terms, and when so many industrial activities are in fact large-scale this is a very simple error to make. But the fact remains that much industry is not large-scale, even today (*see* the case of Spain above, p. 138), and any survey which implies by omission that it is must be incomplete. Not only is a considerable amount of industry carried on in small factories, but in some countries domestic industry is still of considerable importance.

We should perhaps note that there is no generally agreed criterion of what constitutes a small-scale industry, and this form of activity is variously defined in national statistics. One possible criterion, however, is that suggested by the United Nations Economic Commission for Asia and the Far East, that a small-scale industry is "one which is operated mainly with hired labour usually not exceeding 50 workers in any establishment or unit not using any motive power in any operation, or 20 workers in an establishment or unit using such power" (*Economic Survey of Asia and the Far East, 1958*, Bangkok, 1959, p. 100). While this may seem a reasonable assessment of the situation in an under-developed territory, however, it seems unduly restrictive when industrially developed territories are concerned (*see* the point made in the middle paragraph, p. 238 below) where questions of relative, as well as absolute, scales of operation would seem to be involved.

DOMESTIC INDUSTRY

It may be useful at this point to note that one very important type of industry which is still carried on in many countries on a domestic basis is the manufacture of textiles; Fryer remarks that even the flying shuttle, invented by Kay in 1733, is still unknown to many Indian handloom weavers. The industry, though broadly based, is inefficient and the output per worker could greatly be increased with the help of even simple technical appliances. In West Africa and in many other cotton-producing areas the raw cotton is ginned in the homes of the workers and the lint is then spun into thread with the help of very primitive equipment (*see* Fig. 47 and p. 206 below). Weaving is carried on in a crudely simple way. A series of very long threads is arranged side by side up to a breadth of about 18 inches and the threads are weighted at one end by being secured to a heavy stone. The free ends of the threads are passed through a wooden loom at which the weaver sits, and the weaver produces the cloth by passing a shuttle by hand to and fro between the threads, which are moved up and down by a hand-worked contrivance. A comb is used to make sure that the cross-threads are firmly pressed together. As the weaving proceeds the weaver winds the cloth round a spindle near his feet and as this happens the threads are dragged towards the loom but are kept taut by the heavy stone. In this way long but narrow bolts of cloth are woven and it is necessary to sew two or more of these bolts side by side to make broader cloth.

It is perhaps pertinent to point out that modern industry in the "developed" countries was not infrequently based in its initial stages on domestic industry. This point has been mentioned in Chapter I, where we saw that the change from the domestic system to the factory system was one of the more obvious features of the Industrial Revolution in England. Trevelyan has remarked that when George II began his reign in 1727 manufacture was still "a function of country life." The villages of England supplied themselves with such necessities as clothing, tools, bread and beer, and only the richer citizens could go elsewhere for the purchase of their best furniture, books, china and other manufactured goods. Many villages, in addition to supplying their own needs, specialised in some form of manufactured goods for the wider market, and the manufacture of woollen cloth and the rapidly growing cotton-manufacturing industry were both carried on in the homes of rural England.

The present author owns an eighteenth-century grandfather clock which still keeps excellent time and which was made, according to the inscription on the clock face in the small Worcestershire village of Pebworth by "W. Johſon." Things greatly changed in the manufacturing world during the course of the following century (*see* Chapter I).

LOCK MANUFACTURE AT ALIGARH

In addition to the manufacture of cloth, production of such necessities as pottery, leather goods, soap and simple metal goods (including pots, pans and ornaments as well as simple tools) is still widely carried on upon a domestic basis in many parts of the world. In the case of metal work, however, it is not unusual for a specialised class of metal workers to

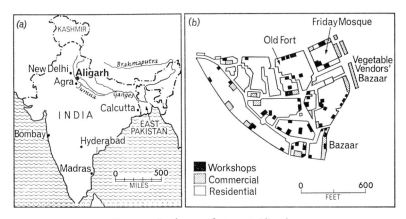

FIG. 33.—Lock manufacture at Aligarh.

develop, because the smelting of ores and the subsequent processing of metal require specialised equipment as well as skill. In some cases metal working is carried on in factories; such a factory may be very small and may comprise part of the dwelling of the manufacturer; a good example of such a handicraft industry is that of the manufacture of locks at Aligarh, in India (Fig. 33). (This section is based on Hirt, H. F., "Lock Making: A Handicraft Industry in Aligarh, India," *Focus on Geographic Activity*, Thoman R. S., and Patton, D. J. (eds), New York, 1964.) The origins of this industry can be traced back to 1897, when an Englishman established a workshop in Aligarh for the manufacture of metal stamps, locks and seals for the Post Office. This industry is no longer carried on and the workshop was converted in 1932 into a government printing plant, but from the manufacturing activities to which it gave rise sprang a nucleus of skilled and semi-skilled workmen. It was largely on the basis of this nucleus that the present lock industry was developed, and today the lock-manufacturing industry is the largest single employer of workers in Aligarh—in fact, the industry engages about half of the total number of employed workers in the town. Even so, the industry is run on a small-scale basis in the homes of the people; many workshops are run as family businesses, while some are rather more ambitious and employ a few workers to supplement the help given by members of the family.

This industrial structure almost inevitably gives rise to some of the undesirable features which were formerly associated, for example, with the early cottage textile industries of the north of England. Small-scale manufacturers such as those of Aligarh are not able economically to arrange for the purchase of raw materials or for sales of finished goods on their own behalf; they have not the knowledge of markets to do this, neither have they any bargaining power as small-scale producers, though the total consumption of raw materials by the industry as a whole is by no means negligible—according to Hirt, the total input of raw materials in 1956 included 4 million lb of brass, 1,600,000 lb of steel sheets and scrap, and as much steel again in the form of rounds and wire. It is almost inevitable under these circumstances that some form of bulk purchasing should develop and in a more advanced society this might well have come about under the auspices of a co-operative society, but in the unsophisticated circumstances which prevail at Aligarh the buying and the selling have passed under the control of equivocally termed "manufacturers" who are not manufacturers at all in the Western sense, but middlemen or dealers. These dealers purchase the essential raw materials (principally brass ingots and wires, steel sheets and scrap, steel wire—including spring wire—and zinc for galvanising) mainly from suppliers in large cities such as Bombay and Calcutta, though some materials such as tin and the zinc must be imported, while coke and some steel come from the iron mills and the mines of West Bengal and Bihar. The bulk purchases are then broken down by the dealers and sold to the actual producers at higher prices. In the case of the finished goods the dealers arrange for marketing since they have access to markets which the producers know nothing of; the middleman is thus able to purchase the locks comparatively cheaply and sell at a considerable profit.

The general result of this system is very much what might be expected —the "maufacturers" become rich while the actual producers remain very poor and can expect meagre wages if they are employed workers. One unfortunate result of this is that the producers are not able to bear the cost of installing efficient and modern machinery; this means in its turn that output remains low and further ensures that earnings are depressed. We might note that the employment of women and children acts as a further depressant to wages.

The homes in which the industry is carried on consist for the most part of small mud or brick buildings with enclosed courtyards; quite often the workshop opens directly on to a narrow street, though this is not always the case since any available space in the house is likely to be used as a working place. Manufacture is carried on mainly with the help of hand tools, lathes and grinding tools which are hand-driven, though larger operators may possess old presses, shears, lathes, grinders and polishers driven by small petrol or electric engines.

Despite these unfavourable working conditions the 1300 workshops of Aligarh produce each year over 20 million locks of varying kinds, and this output comprises almost 70% of the total annual output of locks in the whole of India. When we bear this in mind and when we further remember that handicraft industries are carried on in hundreds of small towns all over the country, we shall realise that the total output of such small-scale industries must be very considerable indeed. A further important point is that, because these small-scale industries attempt to supply the more humble type of manufactured goods, the large-scale firms are left free to concentrate upon the production of large-scale equipment. It also seems likely that unskilled workers (who are probably uneducated) are able to secure employment more readily in these small-scale enterprises than on the floors of large factories. These workers thereby receive some training and this means that technical skills (admittedly of a comparatively low order) are more widely disseminated than would be the case without the contribution made by the small-scale enterprise.

SOYA BEAN PROCESSING IN THE U.S.A.

It is important to bear in mind, however, that small-scale industrial enterprise need not be conducted along such inefficient lines and under such depressing conditions as is the case at Aligarh, and our next example is selected as an illustration of a small-scale but modern and efficient enterprise. It also serves to elucidate an important point—that industrial enterprises engaged in the processing of agricultural produce often need to be located at fairly closely spaced intervals in the appropriate agricultural region. Thus, we typically find that creameries are normally located in dairy-farming regions, while sugar mills are usually met with at fairly close intervals in sugar-producing regions. This is because the raw material, whether it is sugar cane or sugar beet, carries only a comparatively small proportion of its own weight as actual sugar, and the weight-losing-factor (p. 22 above) comes strongly into play. About seven tons of cane may be required to produce one ton of sugar. For a case study see Duncan, C., "The Trevor Farm: A Case Study of Cane Sugar Production in the Queensland Coast," in Thoman and Patton, op. cit.

The example selected for present study is that of the processing of the soya bean in the U.S.A. Much of what follows is based upon "The Soybean, a Versatile Product of American Farms," in Highsmith, R. M., *Case Studies in World Geography*, Englewood Cliffs, N. J., 1961. The soya (or soy) bean came originally from the Far East, where it is used in many different ways. As the bean has a high protein and fat content, for example, it forms a useful substitute for meat in many territories where stock rearing may not be practicable; in China it is used after fermentation to produce the various sauces which form the base of much food; it yields cooking oil; the young plant can be used as a vegetable, while the bean itself forms a very palatable vegetable when it is picked before ripening; the ripe bean can be ground to provide flour, or it can be soaked, ground and mixed with water to produce bean milk. Such a wide variety of uses renders the soya bean a most valuable crop.

Figure 34 shows the chief areas of production of the soya bean in the U.S.A., a country where, despite its great value, the crop did not become really important until the 1930s; since the commencement of the Second World War its importance has considerably increased and it now ranks as the fifth most important cash crop in the country. Illinois is the leading producer, followed by Minnesota, Iowa, Indiana and Missouri. The general disposition of these six states which between them account for 64% of the soya bean acreage in the U.S.A. gives point to a recent suggestion that the "Corn Belt" of Baker should now be known as the "Corn and Soya Bean Belt." In the South and East the crop is grown largely as a green manure and for forage but in the Mid-West it is grown almost entirely for the bean.

The main uses to which soya bean oil is put in the U.S.A. include the manufacture of margarine, cooking fat (these are easily the two most important uses), soap, salad cream and mayonnaise. The oil is also used as a raw material in the manufacture of paints and varnishes, rubber substitutes, insecticides, fertilisers, and even in the production of linoleum and oilcloth. The main centres engaged in the initial processing of the soya bean are indicated on Fig. 34, and it can easily be seen that processing is widely carried out in the general producing region.

The following is a method of processing commonly adopted. The beans are first dried and cooled before being cracked between rollers, after which the flakes are thoroughly soaked in a chemical solvent (hexane) which dissolves out the oil. The solvent is later removed in evaporators and the crude oil is sold to refiners in the Mid-West or on the West Coast, after which it is used in one of the ways already mentioned. The residue left after the extraction of the oil is ground into meal, which also finds a variety of uses. The most important use is as a stock feed but its role as a source of industrial protein is increasing. For example, a leading user of this protein is the paper industry, which uses it in the manufacture of high-grade printing paper, while it is now being used in the manufacture of glue. It is, in fact, used in many instances where a binder, size or adhesive is required.

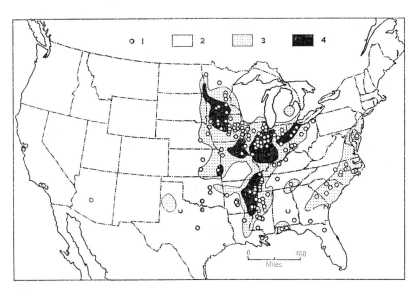

FIG. 34.—The processing of the soya bean in the U.S.A.: (1) processing centre; (2) areas in which few or no soya beans are grown; (3) areas of moderate production of soya beans; (4) areas of greatest production.

We might further note that the whole bean after hulling is sometimes ground to produce soya flour, commonly used as a food supplement. It is mixed, for example, with wheaten flour in the baking trade and may also be added to processed meat to increase its bulk without lowering its quality, while its low starch content means that it is a valuable raw material in the production of dietary foods.

FRUIT CANNING IN NEW SOUTH WALES

It is perhaps desirable to enter a precautionary word at this point. It is not always the case that the processing of agricultural produce is most economically carried out in the producing area; and the reasons for this are probably best seen from an actual example. In the Murrumbidgee Irrigation Area (the M.I.A.) of New South Wales a considerable amount of fruit is grown, and much of this fruit is canned, particularly at the local centres of Griffith and Leeton. This seems to be a reasonable situation and one to be expected. What is perhaps more surprising is that a great deal of fruit from the M.I.A. is sent for canning to Melbourne, Sydney and even to Brisbane, and the reason for this appears to lie in the transport costs involved. A breakdown of costs involved in the fixing of a retail price of 2s. 10d. (A) for a tin of peaches canned in Leeton works out as follows:

Peaches		6d.
Sugar		6d.*
Tin	1s.	0d.†
Transport		3d.†
TOTAL	2s.	3d.
Retail margin		7d.

* Cost in Leeton after transport from producing or manufacturing area.
† Transport of packed tin to markets—average figure.

The figures in the table were supplied to the author by a local producer in August 1965. The currency given was Australian, but since 1965 Australia has adopted a decimal form of currency.

Now Leeton is located at considerable distances from the sources of supply both of the sugar (which comes from Queensland) and of the cans (which come from Port Kembla, Fig. 27), while it is also far removed from the main markets which lie along the comparatively densely populated coastlands between Melbourne and Brisbane. Notable transport costs can be saved, therefore, if the fresh peaches are taken in bulk rather than in the canned form to the market areas for processing. Transport costs are saved not only on the packed tins but also on the sugar and empty cans; in Brisbane, of course, the main sugar-producing zone lies not far distant, while the sugar can be transported comparatively cheaply by sea to Sydney and Melbourne, and the lengthy, expensive overland haul to Leeton is avoided. The empty cans can also be despatched from Port Kembla much more cheaply by sea to coastal centres. As a result of these savings M.I.A. fruit can be canned as far away as Brisbane more cheaply than it can be canned in the producing area! This example should warn us against facile generalisations which argue that food processing is necessarily best carried on in the food-producing areas.

This chapter is a short one, and advisedly so. There is no suggestion that the importance of the output of small-scale industry in any way rivals that of the large-scale concerns, but it is important that we should realise that small-scale enterprises continue to render valuable service in developing and developed territories alike. We have not considered in this chapter the small-scale firm which produces a component part of a large manufactured unit such as a motor-car. Such a firm sells its output to a larger company and is essentially an integral part of a large-scale enterprise, while such firms, though classed sometimes as small-scale, are quite often of substantial size: it is only in a comparative sense that they are small. Mention of such firms, and the reasons for their persistence in the motor-car industry, is made below in Chapter X.

SUGGESTIONS FOR FURTHER READING

Fryer, D. W., *World Economic Development*, New York, 1965.
Highsmith, R. M., *Case Studies in World Geography*, Englewood Cliffs, New Jersey, 1961.
Thoman, R. S., and Patton, D. J. (eds), *Focus on Geographic Activity*, New York, 1964.
Trevelyan, G. M., *Illustrated English Social History*, Pelican Books, 1964, vol. 3.

Chapter VIII

The Iron and Steel Industry

STEEL is a commodity which is indispensable to the modern way of life. It enters into almost every phase of present-day activity in one form or another; there can be scarcely a single room in all the houses in the towns and cities in which we live which does not contain at least one object fashioned from steel, whether it is an ornament, an household article, or simply a nail helping to keep some part of the room (or a picture) in place. On a larger scale steel enters widely into constructional work of all kinds, into the manufacture of machinery and all types of tools, and into the construction of transport equipment such as motor-cars, locomotives and rails. Even agriculture is today heavily dependent upon steel implements and appliances which make possible the large-scale crop production of our own day. Annual *per capita* consumption of steel varies widely from territory to territory between such extremes as the 20 lb of China and India and the 1230 lb of the U.S.A. It is very largely thanks to the part played by iron and steel that our modern economy and way of life have been developed, and it is no coincidence that the world's big producers of iron and steel are the world's most powerful states. Because iron and steel are today in such wide demand they act as a barometer of economic conditions, for demand soars when business is good and falls when depression strikes. No other metal possesses such strength, toughness and malleability, and these qualities can be produced to almost any required degree with the help of alloys and suitable treatment. It is indeed fortunate that supplies of iron ore, the basic raw material from which steel is made, are widespread and abundant, while new techniques and processes permit comparatively cheap manufacture.

HISTORICAL DEVELOPMENT OF THE INDUSTRY

While the points made above are true, however, we should not forget that the history of steel manufacture has been a very lengthy and chequered affair, and a brief review of some of the main points will serve to remind us that we have arrived at our present-day competency only after a long-drawn-out series of discoveries and inventions. Such a review will also help to emphasise some of the main points regarding manufacturing processes which should be borne in mind. It is perhaps permissible to point out, for the sake of those beginning the study of industrial geography and who may be tempted to wonder if much of what follows is "really geography," that some knowledge of industrial processes is essential for a proper understanding of the geography involved. The situation is comparable to that of the geomorphologist, who must to some degree be a geologist, or to that of the geographer interested in climate, who must acquaint himself with the processes of meteorology; the geographer interested in industry must similarly be something of an industrialist, and a certain knowledge of metallurgy is an essential prerequisite for the proper understanding of the geography of the iron and steel industry.

We may never know just how the smelting of iron ore was first undertaken by man, but Pounds makes the suggestion that primitive man must many times have kindled his fires on a hearth of iron-bearing rock, and it is therefore likely that sooner or later he would come to realise that an impure form of metal could be extracted from rock of a certain type. The metal, impure as it was, could be fashioned into useful objects; the Hittites, for example, are known to have possessed iron swords which gave them a considerable advantage in battle over their neighbours. The presence of beams composed of a very high quality of wrought iron in early Indian temples suggests that iron of unusual quality was produced in India as early as 2300 B.C. The famous *wootz* steel, for example, was manufactured in Hyderabad and so excellent was its quality that it was marketed throughout most of the known world; it provided the raw material for the famous sword-blades of Damascus and Toledo which could be bent into a circle and yet which could take a cutting edge rarely, if ever, surpassed, even by modern methods of production. Such excellence in production was, however, rare.

Early iron smelting was not a complicated process. Broken pieces of ore were heated on an open hearth and the temperature was raised by the use of a draught. With increasing experience better techniques of handling the draught came into use; for example, at first the naturally moving air was simply funnelled on to the hearth, while later a bellows was used. The bellows were operated initially by muscle power (man or animal), though later on power was supplied by a wheel turned by running water. Heat was supplied by charcoal fuel and, as the charcoal burned, carbon combined with oxygen present in the ore to form carbon monoxide or carbon dioxide; these impurities were discharged into the air and lost.

Unfortunately other impurities are normally present in the metal and these impurities were not easily got rid of, because temperatures in the primitive furnace were not sufficiently high; pure iron melts at the temperature of 2799° F (1537° C) though at temperatures considerably below this, the metal takes on the form of a pasty mass which can be kneaded. The molten metal can dissolve carbon and this action lowers the melting point of the iron to a minimum of 2066° F (1130° C), but temperatures of this order could not be reached in the early furnaces. The result was that the iron finally produced even in the improved "Catalan forges" of the Middle Ages was not a homogeneous metal; it was uneven in texture and it varied considerably in quality throughout its bulk. It absorbed little carbon and was therefore generally soft and malleable. Sometimes, however, a locality became famous for the quality of its metal, and this could in some instances be attributed to a superior furnace which developed higher temperatures than were usual. Such a development was often made unknowingly, but it made possible the absorption of a small amount of carbon by the iron, and this had the effect of producing a metal somewhat resembling steel; sometimes an unsuspected constituent, manganese, was present in the ore, and this could result in the production of a metal with some of the qualities of manganese steel. Some regions such as Styria, the Siegerland and Sweden were renowned for the quality of their steel, but it is likely that the iron-masters did not really know how the superior quality was achieved.

In later years, more was discovered about the actual processes of manufacture involved and it was realised that the addition of small amounts of carbon to iron improved its quality—it imparts a hardness and strength that "pure" iron (which is comparatively soft and malleable) does not possess. It was later discovered that the carbon content could be safely raised to about 1·4%, but that if it were raised above that figure the steel became brittle and was inclined to snap without warning; ordinary cast iron has a carbon content of up to 4% and it is useful for work in which toughness is required, but its liability to snap makes it unsuitable for constructional purposes. It is necessary to add that with the help of modern techniques the proportion of carbon can now be increased considerably beyond 1·4% without the danger of producing brittleness, and special high strength cast irons may today contain up to 4% carbon.

To return, however, to our historical survey, we may note that even during the Middle Ages certain regions possessed advantages for the manufacture of iron, and that some regional specialisation therefore developed.* Iron ore was comparatively widespread in the limited quantities which were needed in those days, but forested areas possessed a strong advantage because they provided raw material for the supply of charcoal. Swiftly flowing watercourses offered the further advantage of providing the power necessary to drive the bellows. Forested regions, therefore, which possessed running water and ore supplies were among the most favoured areas, examples being the Weald of Kent and Sussex, the Rhineland and Styria.

This level of development was completely changed by the series of discoveries and experiments which finally led to the emergence of the blast furnace; this piece of apparatus developed in a series of stages from the old hearths, the first stage being the erection of a wall to enclose the hearth. The early retaining walls were low, but when they were built higher the forced draught became more effective and temperatures within the furnace rose substantially higher than ever before; this had the effect of causing the iron actually to melt and to absorb carbon before "flowing" to the bottom of the furnace, where it solidified into an ingot, the useless slag resting on the top.

Because of the retaining walls round the furnace the ingot of iron could not directly be retrieved and for a very long time it was customary to break down the walls before lifting out the ingot. This was clearly wasteful, and the next stage was to bore a hole near the bottom of the furnace, which could be plugged at will with clay and through which, when open, the molten metal could flow on to a sand floor. Here the metal solidified into "pigs." Molten slag could be released through a notch higher up the retaining wall. European ironmasters discovered how to cast iron in this way in the middle of the fourteenth century, and the process was in fairly common use early in the fifteenth century. The advantage of this method was that when once in operation a furnace could be kept going for a long time, until its lining was destroyed by the intense heat; in a modern furnace lined with refractory bricks this can be several years, though the fireclay used in earlier days would not last anything like as long as that.

The first true blast furnace was constructed in the fifteenth century, probably in the Rhineland, and this innovation rapidly changed the character of the iron-producing industry. Not only were greater amounts of

* Regional specialisation was, in fact, limited more than might be expected because of the difficulty of transporting heavy pig iron over any considerable distance in those days. This fact made it desirable that iron manufacture should be carried on as widely as possible. Pack-horses were generally used, but these animals could carry only limited amounts of pig, and they could travel only slowly over the poor roads of the time.

water power necessary to produce the stronger blast which was essential to the successful working of the furnace, but the larger scale of working meant that considerably larger supplies of charcoal and ore were needed than for the old hearths of "bloomeries." These more exacting requirements led to the emergence of specialised iron-producing regions on a scale previously unknown, and such areas as the Weald, the Forest of Dean, Burgundy, the Siegerland, Central Sweden and Styria became large-scale centres of supply. Another important consideration was that iron production became a highly capitalised industry, for the blast furnace was (and remains) an expensive piece of equipment, and this fact ensured that control of the industry became concentrated in the hands of a limited number of large-scale producers.

A concomitant development alongside that of the blast furnace was that of the iron refinery or foundry, a development made necessary because in the high temperatures reached in the furnace the metal became thoroughly molten and absorbed a fair amount of carbon. The 4% of carbon usually attained in pig iron renders the iron unsuitable for many purposes, and it became necessary therefore, to refine the iron to make it more malleable and less brittle. Sometimes the foundry grew up near the blast furnace, but Pounds points out that such an arrangement placed a very heavy burden upon local supplies of charcoal. It was therefore, not uncommon for the pigs to be transported by water or even by road to foundries which were located more advantageously with respect to markets; thus, pig iron from the Forest of Dean and also from the Welsh Marches found its way to the Black Country for refining.

The constantly increasing scale of development of the iron industry inevitably placed a very heavy burden upon the wood and charcoal resources of the iron-producing regions, and to an ever-increasing extent the threat of fuel shortages imperilled the existence of the industry. It is not surprising, therefore, that increasingly urgent efforts were made to discover an alternative form of fuel to that which was fast denuding the countryside of its forests (to the mounting alarm of shipbuilders and those concerned with national safety in England), and the alternative was finally discovered in the form of coke. Coal itself is normally of no use since the volatile constituents driven off during combustion would combine with the iron to give iron compounds, and the production of iron from these compounds would be a prohibitively expensive process. It is not always realised, however, that the qualities needed in blast-furnace coke are many: for instance, there must be a minimum of impurities to affect the quality of the iron (some coals contain sulphur, which rendered them almost useless for coking in the early days, though in the modern blast furnace this is not a serious problem); the coke must be structurally strong to withstand the tremendous pressures from material loaded above it in the blast furnace, while still permitting the free movement of circulating air and gas; and it must produce a minimum of ash and clinker. There is also the point that coke burns less readily than charcoal, and in the early, comparatively small, furnaces the temperatures attained were not sufficiently high to permit combustion of this fuel.

When we bear in mind this rigorous list of requirements, we may be almost surprised that it was ever found possible to use coke as a blast-furnace fuel, and it does seem that Abraham Darby, who first successfully used the new fuel in 1709 (*see* p. 127 above), succeeded as much by good luck as by good judgement. Darby used a furnace which was above the average in size for those days and which was fed by a more powerful bellows than was usual, and under these conditions it proved possible for the coke to continue burning and effect the smelting. It was also fortunate that coke of a suitable quality was used, although the various requirements for metallurgical coke were not at that time understood.

Even after Darby's success the use of coke in the blast furnace developed very slowly,* and we can understand this if we remember that such a happy series of accidents as Darby experienced was most unlikely often to recur. More had to be known about the precise conditions required for the successful use of the new fuel. One very troublesome feature was the difficulty caused by the sulphur content of so many coals, for some of the sulphur always remained in the coke and affected the quality of the smelted iron. This difficulty was not surmounted until larger furnaces were used and until a hot blast of air replaced the cold. These innovations produced much higher temperatures within the furnace, and the higher temperatures encourage chemical decomposition of the iron sulphides initially formed, so that pure iron remains and sulphur compounds are swept away harmlessly as gases. It was not, however, until 1829 that Nielson at the Clyde iron works showed the advantages to be derived from the use of a hot blast, and by the middle of the nineteenth century coke had in fact become the fuel generally used. During this period the iron-smelting industry forsook the forested areas and was relocated on or near the coalfields.

The other raw material which is necessary for blast-furnace operation is limestone, which acts as a "flux," combining with most of the minerals present in the country rock (especially alumina and manganese) except phosphorus to form a slag, so leaving the metallic iron in a free state.

* It is interesting to note that charcoal smelting in the U.S.A. continued until 1830, when the use of anthracite began in eastern Pennsylvania.

Limestone is of comparatively wide occurrence over the earth's surface.

Blast furnaces have continued to develop in size and efficiency over the past century. This increase in efficiency is graphically illustrated by the fact that, whereas in 1873 683 furnaces in Britain swallowed between them 16,820,000 tons of ore and 16,719,000 tons of coke to produce 6,566,000 tons of pig iron and ferro alloys, in 1963 64 furnaces took in 28,678,000 tons of ore and 10,489,000 tons of coke to produce 14,591,000 tons of metal.

THE MODERN BLAST FURNACE

At this point it may be useful if we examine in some detail modern techniques of iron smelting, as we have now noted many of the difficulties which had to be overcome in earlier years. The apparatus in which smelting is carried on is still the blast furnace, but a modern blast furnace is a monster which may tower upwards to a height of 120 feet or more and have a hearth diameter of over 30 feet. It comprises a tapered steel shell thickly lined with fire-resisting bricks; it works continuously until the refractory bricks are burned out (a matter of several years; in a modern furnace more than $2\frac{1}{2}$ million tons of pig iron can be produced from a single lining), and the larger furnaces can produce more than 2000 tons of pig iron a day. A new furnace may cost up to £10 million while ancillary equipment such as a battery of coke ovens may cost several million pounds more. It immediately becomes apparent why the production of iron has become limited to a few large-scale firms and to comparatively few territories—at present about 90% of the total steel-producing capacity of the world is concentrated among half a dozen countries.

The operation of the furnace is illustrated in Fig. 35. The coke, iron ore and limestone are loaded into the furnace at the top of the tower, the actual feeding into the furnace being regularised by means of two bells which also perform the function of preventing the loss of gases into the atmosphere. While the temperature near the top of the furnace is a comparatively modest 400° F (205° C), near the bottom it is likely to be over 3000° F (1650° C), and perhaps 3500° F (1930° C). Under these conditions complete smelting is achieved, the limestone fusing with the country rock of the ore to produce a molten slag while the coke provides carbon, some of which is absorbed by the iron and some of which combines with the oxygen in the ore and in the blast to form carbon dioxide or carbon monoxide before being swept out of the furnace. It is important to notice, therefore, that the coke acts as a raw material as well as a fuel. The molten iron falls to the bottom of the furnace before being drawn off as molten pig iron.

The whole process of combustion and consequent smelting, however, is made possible with the help of the hot blast, the working of which is indicated on Fig. 35. A battery of (usually) four stoves (one of which is shown in the diagram) stands near the furnace; three of the stoves are normally being heated and through the fourth (which has previously been heated) cold air is drawn from the atmosphere. After the air has completed its passage through the stove it has been heated to a temperature

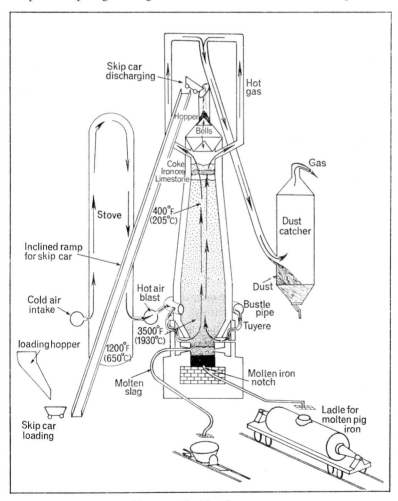

FIG. 35.—The blast furnace.

of 1200° F (650° C) and it is then drawn into the "bustle pipe," a large pipe which encircles the lower part of the blast furnace. As the speed of the blast is very great, comparatively little heat is lost during the rapid passage of the air from the stove to the furnace. From the bustle pipe the hot air is injected into the furnace.

The greater part of the combustion takes place in the bottom third of the furnace, and the resultant hot gases rise and work their way through the descending materials, combining with oxygen from the ore as they do so. One important result of these processes is that a large amount of combustible gas is thrown off at the top of the furnace and this is drawn through a series of pipes as shown in Fig. 35, to pass first of all through a dust catcher which traps about 85% of the considerable amount of dust and ore particles blown out with the gas. After further cleaning and cooling, about a quarter of this gas is used as fuel in the hot-blast stoves while the rest is used for the production of thermal power and for general purposes throughout the iron works.

RAW MATERIALS

These, then, are the main features of the operation of the modern blast furnace, but before leaving the topic we ought to say something about the raw materials used—iron ore, limestone and coke—while the importance of a large water supply is one which should not be overlooked. Although it is true that iron ore is comparatively plentiful within the earth's crust, it is also true that ore bodies commercially exploitable are of fairly uncommon occurrence. The main reason for this is that modern operations are carried on on such a large scale that, to be economically workable, ore bodies need to be very large indeed; we are concerned with a mining industry which utilises electric shovels capable of handling up to 12 tons of ore *at a single bite*, and in which an annual output of one million tons of ore, or ore concentrate, is not uncommon.*

The chief ores of iron are magnetite (Fe_3O_4), haematite (Fe_2O_3) and limonite ($2Fe_2O_3.3H_2O$). Magnetite is the richest but it is comparatively rare; in a pure form it may yield over 72% by weight of metal but unfortunately it not infrequently contains impurities such as phosphorus, sulphur and country rock, which may limit the value and reduce the iron content to 59% (as in Utah) or even less. Haematite is the most widespread type of ore and when pure may yield between 60% and 70% by weight of metal, or 50% in an impure state, while limonite yields considerably

* Reference might be made back to Chapter II for mention of some of the problems connected with the mining and processing of ores.

less. Treated limonite commonly yields up to 60% iron by weight but the naturally occurring ore from such areas as the Midlands and east of England contain only about 25%. A less common ore, siderite, (FeCO$_3$) may contain up to 48% iron but this figure may drop to little more than 30% in the case of impure ore. The quality of steel obtained from siderite, however, is high, and it was this fact which helped to give England and

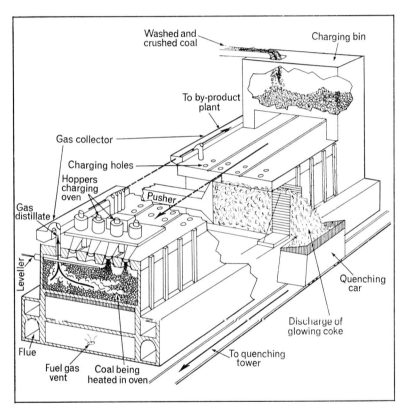

FIG. 36.—Diagram of a battery of coke ovens.

Sweden such high reputations in the manufacture of cutting instruments.

Limestone, the raw material used as a flux in the blast furnace, is of comparatively wide occurrence. The problems to be faced in the case of coke, however, are more numerous, and some of them have been referred to earlier (p. 164 above). The process of coking is in theory fairly simple; coal is distilled, the volatile constituents being driven off to leave a residue, coke, which in composition comprises at least 80% carbon. Originally, coking was carried on in wasteful beehive ovens which allowed the gases liberated to escape into the atmosphere, but today coking is normally carried on in the by-product oven.

The coal is first washed and crushed, and impurities such as shale or rock are removed, after which it is pulverised and blended before being charged into the coke ovens. These ovens each consist of a compartment about 43 feet long, 13 feet high and 16 inches wide, and they are grouped side by side to form a battery (Fig. 36). Coal is fed into them from the top and after about fifteen hours' intense heating this is converted into coke.* Both (narrow) ends of the oven are then opened and a "pusher" is forced in at one end to thrust the glowing coke out at the other, into a quenching car. As the glowing coke comes into contact with the air it bursts into flame, so the quenching car is quickly moved under a quenching tower, where a douche of water quenches the coke; at the Port Kembla works in New South Wales, over eight tons of water go up in steam from the quenching tower every six minutes—an interesting comment on the enormous amounts of water which are required in a modern iron and steel works. The water must be of good quality otherwise impurities are introduced into the coke and thence into the blast furnace.

The volatile parts of the coal driven off during the coking process are drawn off, and after treatment yield a wide range of commodities. Chief among these is coal gas, which can be put to good use,† while other products include tar, naphtha, sulphate of ammonia, benzol, toluol and others which enter into the production of dyes, insecticides, paints, plastics, drugs, explosives, fertilisers, cosmetics and many others. At the B.H.P. iron and steel works in Newcastle, New South Wales, the four batteries of coke ovens between them can coke up to 40,500 tons of coal a week: the average yield from this input of raw material is as follows:

Coke	23,900 tons
Breeze and nuts	2,800 tons
Coal gas	480 million cubic feet
Sulphate of ammonia	440 tons
Benzol, toluol and solvent naphtha	110,000 gallons
Coal tar	300,000 gallons

* An oven charge of $15\frac{1}{4}$ tons produces about $9\frac{1}{2}$ tons of blast furnace coke, though these figures can vary slightly according to the quality of the coal.

† Most of the gas is normally used for heating the hotblast stoves and the open-hearth furnaces (see below), while the remainder is mainly used as boiler fuel and for general purposes.

MODERN DEVELOPMENTS IN IRON PRODUCTION

Before we go on to examine steel manufacture, however, we should take note of the fact that in recent years a great deal of attention has been paid to possible ways in which the efficiency of the blast furnace might be improved, and since the Second World War the productivity of the furnace has, in fact, been greatly increased.* The following are the chief ways in which this has come about:

1. *Beneficiation.* Most ores as they are won from the ground contain much earthy or clayey material which is entirely useless, and which might, in fact, make the transporting of the ore over considerable distances impossible on economic grounds; examples of this kind of situation have been given in Chapter II. Under these conditions it is desirable to treat the ore before transport, and to the extent that this means that a higher quality ore is fed into the blast furnace this produces a more efficient performance. Very considerable fuel costs are saved in this way. Beneficiation is of course normally carried on near the mines to secure maximum economies in transport.

2. *Sintering and pelletising.* Some ores are unsuitable for direct feeding into the blast furnace even after beneficiation, a common reason being that the ore is too fine. A powdery ore will restrict the flow of gases through the furnace and severely impair efficiency, while a considerable amount of any fine ore will be blown from the top of the furnace in the outgoing blast. The process known as sintering is designed to overcome these difficulties and it comprises essentially a fusing of the fine ore, perhaps with crushed coke and lime, in the form of agglomerated lumps of fairly uniform size and quality which can be fed into the furnace. Sinter plants are normally located near the blast furnace. Pelletising is a process which subjects fine ores to wetting and moulding into pellets about the size of a small egg ($\frac{3}{4}$ in. to $1\frac{1}{2}$ in. across). The pellets are then dried, heated and hardened, and after cooling they can be stock-piled

* The following table gives an indication of the tremendous economies in fuel consumption in the manufacture of pig iron over the past two centuries:

Amounts of coal required to produce one ton of pig iron (tons)

Mid 18th century	8–10
Mid 19th century	4
1900	2
1939	1·3
1965	0·8

Based partly on figures given in the *B.H.P. Review*, February 1966, p. 13. See also p. 9 above.

without fear of damage. Pellets are more durable than sinter and they can be made from very fine ores which in earlier years would have been too fine for blast furnace use.

3. *Pressurised tops.* This involves a technique difficult to describe whereby gases are maintained in a state of pressure at the top of the furnace but at the same time the throughput of gas is increased. Because of the pressurised top, however, fewer ore particles are swept out from the furnace. This technique is still in the experimental stage, but where it has been employed considerable economies in fuel consumption have been reported.

4. *The oxygen blast.* This is simply the technique of using a blast of hot oxygen-enriched air instead of ordinary atmospheric air, and it has been found that this refinement greatly speeds up the changes inside the furnace so that substantial fuel economies are made. It is found that the coke is more completely used with the help of the oxygen-enriched blast so that a greater tonnage of pig iron per ton of coke is obtained. Even greater efficiency is secured if steam is injected with the oxygen.

5. *Fuel-oil injection.* This is effected by the injection of fuel oil into the bustle pipe, whence it is drawn with the blast into the furnace. Part of the carbon requirements of the smelting process is therefore met in the form of hydro-carbons; if convenient in any particular location injection of natural gas can produce the same result. It is interesting that these refinements to blast furnace operation in recent years have resulted in Australia in a direct saving of up to 15% in coke consumption, linked with a substantial proportional increase in the output of pig iron. Experiments are continuing with the injection of fine coal, flue dust and powdered lime with the oil.

It might be appropriate to mention at this point that various attempts have been made and are still being made to find a substitute for the expensive blast furnace. Such methods include the low-shaft furnace, which is not as tall as the blast furnace proper, which is cheaper to install, and in which the pressures are reduced so that a very hard coke is not necessary (even coke produced from lignite can be used); the charcoal blast furnace, which is especially useful in heavily forested countries such as Brazil; electric reduction furnaces; and the Krupp-Renn rotary furnace. Such plants all suffer from two main drawbacks, however: they are comparatively small so that output is much reduced, while the continuous operation which is such a valuable feature of the modern blast furnace is sacrificed. It seems likely that the position of the blast furnace in the iron and steel industry will be virtually unchallenged for a long time to come.

These, then, are the main features of the manufacture of pig iron from iron ore, and even from this cursory survey, it will be clear that the process is one of great complexity in which a whole series of related industries is concerned. It is a sobering thought that we have so far dealt only with the production of pig iron and that the subsequent manufacture of steel entails yet another series of complex processes.

DEVELOPMENTS IN STEEL MANUFACTURE

We should now turn our attention to the processing needed by the pig iron before it can be used as a raw material. Earlier in this chapter we made the point that pig iron from the blast furnace contains about 4% of carbon, and this, while it renders the metal harder, also makes it brittle; it can be cast in a molten form but it is not malleable. It was realised soon after the development of the blast furnace that some economic means would have to be found of treating pig iron and removing the carbon, because although in the early stages much iron was used in the form of castings, the shortcomings of pig iron were apparent enough. The first big advance was made in 1781 when Henry Cort patented the puddling process which Pounds describes as the heaviest form of labour ever regularly undertaken by man.

The purpose of the puddling process was to remove the carbon by oxidation. This was achieved by placing the pig iron in the shallow pan of the puddling furnace and by drawing over it flames from a coal fire; there was no direct contact between the fuel and the iron or impurities would have been introduced into the metal. High temperatures were attained in the furnace and the oxidising process was made possible by the addition of haematite ore (Fe_2O_3). The oxygen liberated from the ore combined with the carbon and was swept away as carbon monoxide or carbon dioxide. The chemical reaction was diffused throughout the mass of metal by the puddler, who stirred the molten mass with a long iron rod introduced into the furnace through a hole in the side, and as the carbon was lost the metal became more viscous so that finally the puddler could "ball up" the iron into a roughly round mass of wrought iron. The ball could then be rolled into bars. Wrought iron made in this way was fibrous though tough and it was much used for the manufacture of railway lines until superseded by steel (*see* pp. 17–18 above).

It will, however, probably be clear to the reader that the puddling process was highly extravagant of both fuel and human effort, and that it could only be an interim process as it was not suitable for large-scale working. Many workers therefore continued to conduct experiments designed to discover alternative methods of refining pig iron and the first really notable success was scored by Henry Bessemer in 1856. His invention was a "converter," a container in which air is forced from the bottom through the molten pig iron above (Fig. 37). The capacity of a Bessemer converter varies between 15 and 60 tons and processing takes only up to twenty minutes; impurities such as carbon, silicon and manganese are burned out with the help of the oxygen in the blast. The reaction itself generates enough heat for the process to be self-sustaining, and no fuel is required.

It is not surprising that the invention of the converter was immediately hailed as a resounding success, but sad disappointment quickly followed as it was realised that comparatively few kinds of pig iron could be refined in this way. The main difficulty arose because the widespread phosphoric ores were the ones most frequently used, and pig iron smelted from such ores retained an appreciable amount of phosphorus; this phosphorus remained in the metal even throughout the "conversion" process, with the result that the final product was brittle and almost useless. It was a fortunate accident that Bessemer had used a non-phosphoric type of pig iron in his original experiments* (*see* also p. 178 below).

It was natural that until an alternative form of processing could be devised non-phosphoric (acidic) ores should be greatly in demand, a point which was mentioned with reference to Spain (p. 138 above). Meanwhile, a competitor to the Bessemer process appeared on the scene—the open-hearth furnace, developed by Martin and Siemens in 1862. The open-hearth process is still widely used; about 90% of the steel produced in the U.S.A., for example, is made in open-hearth furnaces while only about 2% is manufactured with the help of the Bessemer process. In Britain, about 75% of steel produced comes from open-hearth furnaces.

The layout of the open-hearth furnace is illustrated in Fig. 38, which shows that the central part of the apparatus consists of a shallow container for the metal. At each end of this container are chambers encasing a

---

* A distinction is normally drawn between "acidic" and "basic" iron. Acidic iron, which is made from haematite, contains little phosphorus but a comparatively high proportion of silicon, while basic iron contains less silicon and more phosphorus. The following table illustrates the point:

Typical Analysis of Steel-making Irons

	Basic iron (%)	Acidic iron (%)
Carbon	3·5	3·75
Silicon	0·85	2·0
Sulphur	0·08	0·04
Phosphorus	1·6	0·045
Manganese	1·0	0·5

Source: The British Iron and Steel Federation.

THE IRON AND STEEL INDUSTRY

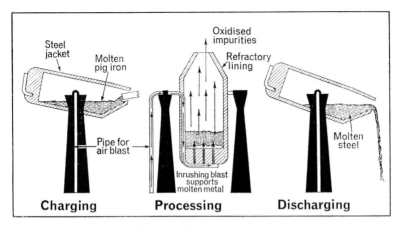

FIG. 37.—The Bessemer converter.

complicated arrangement of brick checkerwork through which air and coke-oven gas are drawn. Operations are commenced by the charging of the furnace with scrap and limestone to which iron and more scrap is added. This charge is melted in about three hours by the ignition of the coke-oven gas, drawn in from one end of the furnace along with the air which makes combustion possible. The flames are drawn right over the charge as in the old puddling process, and the hot gases following combustion are swept out through the checkerwork at the other end of the furnace and dispersed into the open air via a damper through a tall chimney. As the escaping gases pass through the checkerwork they heat it up very considerably and after a period of working, the whole direction of flow of the gas and air is reversed so that the incoming gases are pre-heated by

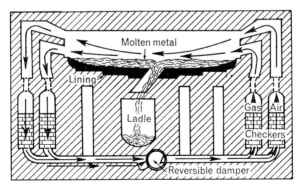

FIG. 38.—The open-hearth furnace.

G

passing through the hot checkerwork. This periodic reversal and the consequent utilisation of residual heat, known as "regeneration," make possible the attaining of very high temperatures (over 3000° F (1650° C)) inside the furnace.

After the initial charge has melted, molten pig iron is poured into the furnace and the complete processing of the metal may take between eight and fifteen hours, depending largely on the size of the charge. Although this is much longer than the corresponding Bessemer process, up to 600 tons of metal can be treated at the same time, and there is the further advantage that since the process is slower, a closer watch can be kept on the changes which are taking place inside the furnace. Samples are taken periodically of the charge and any required additions (*see* below) can be made. There is the further considerable advantage that the open hearth will accept scrap whereas the converter will not, and this has led to very substantial economies of operation. As a result of these new processes, steel production rose considerably between 1856 and 1870 though the price was halved.

The early open hearths, however, suffered from the same disability as the converter—they could accept only charges of non-phosphoric pig iron—but this difficulty was surmounted by the Gilchrist–Thomas technique of introducing limestone into the charge; this renders the slag alkaline, and this in turn causes the phosphorus to be oxidised to phosphorus pentoxide which then passes into the slag. This is the basic slag widely used as a fertiliser and the whole process is known as the "basic" process for producing steel. Unfortunately, the story is not quite as simple as has so far been implied, for at the high temperatures prevailing inside the furnace, basic slag will react with an acid silica-brick lining and destroy it.★ The main contribution of the Gilchrist–Thomas process was to manufacture a refractory brick with an alkaline reaction by using dolomite, and such a brick is unaffected by the chemical changes we have described. Final success was achieved by this means in 1879, and the way was then open to steel manufacturers to make full use of the widespread phosphoric ores. It was not long after this that steel superseded wrought iron as the leading product of the iron industry.†

★ The early furnaces were lined with refractory silica bricks.

† *See*, for example, the footnote on p. 18 above. It may be of interest to note that the construction of a single open-hearth furnace together with its extensive system of checkerwork may require the use of more than one million refractory bricks of different kinds.

MODERN PROCESSES OF STEEL MANUFACTURE

Despite the tremendous success of the open-hearth furnace for steel manufacture, experiments are still being conducted with a view to perfecting more efficient methods of production. For example, the operation of the open hearth has in recent years been rendered much more efficient by the enriching of the coke oven gas injected as fuel with atomised tar or oil or with oxygen. Completely new processes for steel

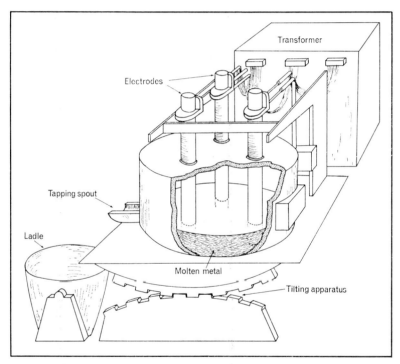

FIG. 39.—The electric furnace.

production, however, have now been devised, and two modern methods which we might mention are the use of the electric furnace and the basic oxygen process. Figure 39 shows in diagrammatic form the chief features of the electric furnace, a large container for the metal into which protrude three giant electrodes that conduct the electric charge to the molten-metal content. The furnace can process pig iron in from three to five hours and scrap iron in from five to seven hours. These furnaces are very efficient in the use of scrap and since they can be very exactly controlled they are excellent for the manufacture of specialised alloy steels.* Because it is becoming increasingly possible to closely regulate the operation of the open-hearth furnace, however, competition between the two methods is keen. In the U.S.A. electric furnaces account for about 9% of steel made, mostly of high-grade quality, while the corresponding figure in Britain is 10%.

We have already seen how the increasing use of oxygen has improved the efficiency of the blast furnace; in a somewhat comparable situation it is probably true to say that, since the development of tonnage oxygen has made large supplies of oxygen available to manufacturers, an increasing use of oxygen in steel making has been the most important innovation since the discoveries of Gilchrist and Thomas. The use of an oxygen blast in the "basic oxygen" process (which is essentially similar to that of Bessemer) obviates one of the chief disadvantages of the Bessemer technique—absorption of nitrogen by the metal—and produces a steel comparable in quality with open-hearth steel. A further advantage is that the additional heat generated by the use of oxygen means that a higher proportion of scrap can be used in the charge; the Bessemer converter can barely take 5% scrap while the basic oxygen process can deal with a charge containing up to 30%—even up to 50% in the rotating Kaldo furnace.

The main features of two basic oxygen processes are shown in Fig. 40. In the LD process† a jet of oxygen is directed from the top of the converter into the charge but this method can deal only with ores of very low phosphoric content. Later developments have included the blowing in of powdered lime with the oxygen blast; the adding of lime to the molten metal during discharge; the two-stage blow with an intermediate tipping of slag; and the Swedish Kaldo‡ rotating converter which permits production of a wider range of steels. Phosphoric charges can now be processed even with a single-stage blow. The vacuum-treatment technique (the molten steel is subjected to a vacuum) is found to reduce the

---

* The manufacture of alloys and special steels now comprises complex and highly specialised processes. The three main groups of alloy steels are (a) stainless, (b) manganese and (c) tool steels. Chromium and nickel are the two metals chiefly used in the manufacture of stainless steels, while the addition of manganese produces a very hard steel which becomes harder under constant hammering and which is best known as the steel used in railway points. Tungsten, vanadium and molybdenum are used in the manufacture of tool steels.

† Named after two Austrian steel-manufacturing centres (Linz and Donawitz) where the process was first tried during the Second World War.

‡ So called because the inventor, Professor *Kal*ling carried out his practical research at the Swedish steelworks at *Do*mnarvet.

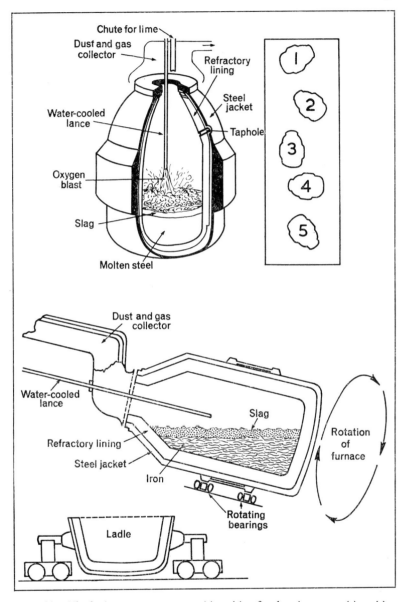

Fig. 40(a).—The basic oxygen converter: (1) position for charging scrap; (2) position for charging pig iron; (3) position for blow; (4) position for tapping steel; (5) position for emptying slag.

Fig. 40(b).—The Kaldo furnace. The converter is swung into a more upright position for charging and swung farther down for discharging.

gaseous content of the finished metal and to produce a steel of superior quality.* Since the quality of steel manufactured by the basic oxygen process is good and since a comparatively modest degree of capital investment is needed for the installation of basic oxygen converters, it is likely that this process will be used on an increasing scale. About 14% of British-made steel is now produced in converters.

It might be useful to bear in mind that carbon steels (as opposed to "alloy steels"—footnote, p. 178 above) are of differing qualities according to the uses to which they are to be put. The distinction between differing grades rests upon the amount of contained carbon shown in the following table:

Grade	%age of carbon
Low carbon steel	Less than 0·15 ("soft" steel)
Mild steel	0·15 – 0·25
Medium carbon steel	0·2 – 0·5
High carbon steel	0·5 – 1·5 or higher ("hard" steel)

Source: The British Iron and Steel Federation.

It is not desirable in a work of this nature to describe the whole series of processes which go into the manufacture of steel goods; the description of the various processes as far as the manufacture of raw steel has been given at some length, because an understanding of the requirements of the industry (especially the requirements of raw materials) is essential if its geographical setting is to be appreciated. It is perhaps useful at this stage simply to say that after manufacture the steel ingots are brought throughout to a uniform temperature of 2400° F (1300° C) in a soaking pit before they are rolled. In the various rolling processes the ingots are passed between rollers which successively press the steel into different shapes—plates, bars, rails, plates and so on. These commodities form the raw materials of a whole series of metal-working and constructional industries (Fig. 41). It is interesting to notice that at some of Britain's largest steelworks continuous casting plants now turn out 6000 tons of steel castings per week. This new technique by-passes the ingot stage of production as the molten steel is led directly into the casting plant to produce the billets, blooms and other items. The process is described at greater length on page 201 below. Further developments in steel production are to be expected; for instance, production control computer systems are being introduced in some steelworks. These systems will eventually

* A revolutionary spray process for the manufacture of steel developed in Britain has not so far had the impact on the industry which was anticipated in some quarters. The firm which pioneered the new process at Millom, Cumberland, has now closed down.

cover entire integrated works, from sinter plants and coke ovens to blast furnaces and steel output. To a degree greater than ever before the production of iron and steel is becoming a capital intensive industry.

THE INTEGRATED IRON AND STEEL WORKS

We have so far dealt with the production of pig iron and the subsequent manufacture and processing of steel, as though these are distinct industries, and so in a sense they are and in certain cases they are carried on

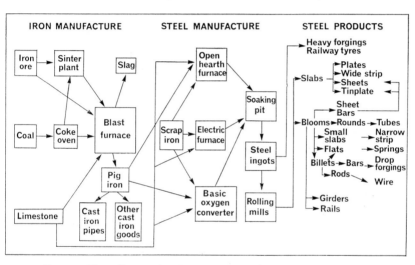

FIG. 41 —Flow diagram showing conversion of raw materials into iron and steel products.

at different locations. In fairly recent years, however, it has become increasingly more usual for the whole series of processes leading from the initial processing of raw materials and fuel to the output of finished steel parts to be conducted on a single site, and this has given rise to the development of the integrated iron and steel works, a vast type of enterprise needing up to £100 million of capital investment. Despite the massive scale of investment required, however, the economies to be secured from integration are so great that economic considerations weigh heavily on the side of the integrated works. It is not difficult to see, for example, that very substantial fuel costs can be saved if molten pig iron can be transported rapidly over a short distance to the steel works for treatment rather than if the metal is allowed to cool down; if this happens the ingots must be reheated at a later stage and this will introduce heavy additional fuel demands. In addition, we have already mentioned that gas secured from the coking ovens and from the blast furnace can be used as fuel in the hot-blast stoves, in the open-hearth furnaces and in the soaking pits; what in effect we have is a series of linkages between the different stages of the industry, and these linkages inevitably tend to draw the various stages of manufacture more closely together.

LOCATION OF THE INDUSTRY

We are now in a position to attempt an assessment of the relative importance of the various forces which affect the geographical location of the iron and steel industry, and it is probably true to say that six main points should be considered:

1. Raw materials.
2. Markets.
3. Site requirements.
4. Water requirements.
5. Capital availability.
6. Government policy.

Raw Materials

The importance of raw materials in affecting location will need very little emphasising in view of what we have already said, for it will be clear enough that the materials needed are both bulky and heavy; the industry cannot afford to neglect a careful consideration of the location of its raw materials. Figure 42 illustrates the pattern of raw materials which go into the manufacture of one short ton of ingot steel, and it can be seen that it is the iron ore and the coal which form the most significant of these. The ideal location, of course, would be that at which coal, iron ore and even the market all coincide, but such a combination of circumstances is rare. It did exist, in fact, in earlier days in the Midlands of England and in the Central Lowlands of Scotland, and it still exists at Birmingham, Alabama. More usually, however, a choice of location has to be made and it would appear from Fig. 42 that the stronger pull must be towards the ore fields because of the greater weight of ore required for the production of any unit weight of steel, but if we consider the volumetric situation we find that the coalfields have a much stronger pull. Because iron ore, bulk for bulk, is considerably heavier than coal, the requirements of coal on a volumetric basis for the blast furnace were until recently of the order of 5·3 as opposed to 1·5 units of iron ore. For this reason it has been the

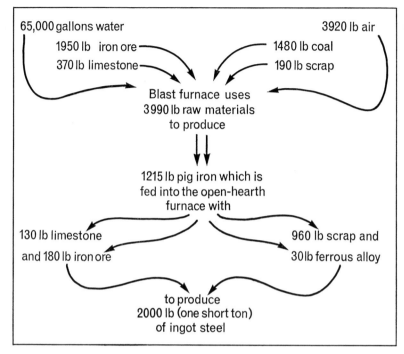

FIG. 42.—Diagram to show the raw materials needed to produce one short ton of ingot steel. Based on figures issued by the American Iron and Steel Institute.

coalfields which have traditionally attracted the iron and steel industry, the case of the Ruhr, which draws its ore from a very wide area, coming immediately to mind. In Britain, the Black Country, the West Riding and Lanarkshire coalfields have attracted the industry (Fig. 43).

The situation, however, is far from clear-cut. The series of improvements in technique mentioned earlier have had the effect of economising notably on coke requirements, a fact emphasised in the footnote on page 171. Because of these substantial economies the pull of the coalfields is today weaker than it has ever been, and modern iron and steel works are not infrequently located away from the coalfields (*see* the cases of Sparrows Point and Fairless Hills, p. 186 below). In Britain, to take another example, other locational factors come into play. Estall and Buchanan have pointed out that after about 1860 the need for foreign (*e.g.* Spanish) nonphosphoric ores for use in the acid Bessemer and open-hearth processes, drew part of the steel-making plant to tide water; the case of Tees-side comes to mind in this connection. Another consideration is that British ores are often lean, and progressively poorer ores are being worked; in

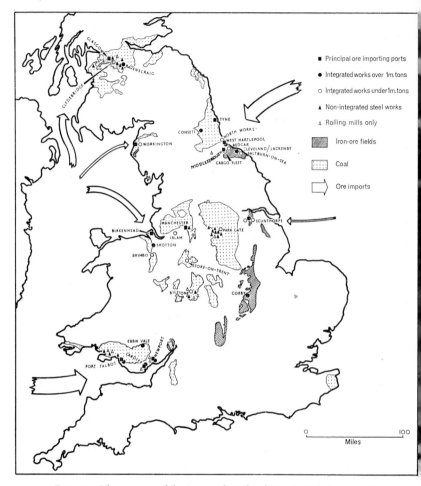

FIG. 43.—The pattern of the iron and steel industry in Britain in 1966.

the low-grade Jurassic ore belt, for example, the average iron content of worked ore fell from 30% in 1948 to 27% in 1964,* while the average content of imported ores rose from 55% to 59% during the same period. In order to allow for these differences in iron content, delivered prices should be expressed in terms of the cost per unit of iron rather than in terms of the cost per ton of ore; such an adjustment roughly halves the cost advantage of domestic ores over imported if this advantage is expressed crudely as cost of transport per ton of ore.

* *Steel Review*, No. 38 (April 1965), p. 11.

One result of the comparatively poor quality of British ore is that a substantially greater amount of capital investment is needed to make possible any given level of pig-iron production if domestic ore is to be used rather than foreign. The former British Iron and Steel Federation (in *Steel Review*, April 1965) estimate that at a new sinter and blast furnace installation, the domestic ore user needs to invest roughly half as much capital again as the user of imported ores. This imposes a heavy additional burden of capital—an important point in times like these when interest rate are high.

Under the circumstances described above, there is a pull towards the lean ore fields on the one hand* (exemplified by plants such as that of Appleby-Frodingham, Scunthorpe) and towards tide-water location (exemplified by Tees-side and by the Abbey Margam works near Port Talbot) on the other, especially since the iron content of imported ores is expected to rise to somewhere near 62% in the foreseeable future. The result of this situation is portrayed in Fig. 43.† An interesting feature which has developed as a result of this dichotomy is that each group of works shows a marked preference for its own ore supplies; there is little or no marginal shift from home ores to imported or vice versa to reflect changing circumstances of supply, demand or technology. One main reason for this appears to be the very high freight rates now charged by British Rail on ore traffic; the standard ton-mile freight rate for home ore rose by no less than 61% between 1954 and 1964.

Of the other raw materials required, limestone is needed in much smaller amounts and since it is fairly widespread in distribution it is not normally considered as a locating factor, while the alloys in normal use are needed in comparatively small amounts only. The alloy is normally added to the molten steel immediately after the Bessemer or open-hearth processing as the metal is flowing from the converter or the furnace, and it is commonly added in the form of *spiegeleisen*, an alloy of iron containing up to 25% manganese. Sometimes ferromanganese which has a higher proportion of manganese is used, and sometimes crushed coke is added to bring up the carbon content to the required amount.

We might usefully observe that the attractions of a tide-water location such as we have observed in Britain had parallels in other parts of the world, notably in the U.S.A., where the middle Atlantic seaboard has become a leading producer of iron and steel. We have previously

---

* See p. 29 above.
† This paragraph is not intended to be taken as a systematic examination of Fig. 43, for which the reader is referred to regional studies. The situation in South Wales, however, is examined in Chapter XII, pp. 288-294 below.

mentioned the enormous Sparrows Point plant, while other large plants are in operation at Fairless Hills (on the Delaware River opposite Trenton) and at Bethlehem, slightly farther inland. One reason for this choice of location is the increasing dependence of the U.S.A. (following the depletion of the Lake Superior ores) upon imported iron ore, especially from Venezuela, Canada, Chile and West Africa. Another reason is that very large-scale markets are easily accessible from this seaboard; such markets include the heavily populated coastlands themselves and the developing Gulf Coastlands with their mineral oil developments and their expanding industries. A third point is that the middle Atlantic location is well placed to receive supplies of scrap, now a very important raw material.

Markets

The importance of the market is a point which we have previously emphasised in a general way in Chapter IV, and it is a very relevant point in connection with the iron and steel industry. The products of this industry are heavy and bulky and transfer costs can be very burdensome if they have to be sent to distant markets; this is the sort of situation we examined in the case of Sydney, Nova Scotia (p. 129 above). Admittedly, the bulk and weight of steel products are not as great as in the case of the original raw materials (Fig. 42), but neither can they be handled in bulk as the raw materials can and handling costs increase sharply in consequence. Freight tariff rating is apt notably to increase the cost of despatching finished products (*see* the examples quoted in Chapter IV, pp. 86–87 above). The contrasting cases of Duluth and Houston, Texas, come to mind in connection with market location. The iron and steel industry at Duluth was established to avoid a threatened state-government tax on export ore (p. 104 above) and it has the advantages of close proximity to large-scale iron-ore deposits, and of a situation on Lake Superior which makes reception of coking coal from Pennsylvania a comparatively simple matter. Limestone can also be despatched via Lakes Huron and Superior while the waterside location of the plant ensures that adequate water supplies are available. It would therefore seem that conditions are favourable for the development of a major iron and steel industry yet the enterprise has never been successful, and the main reason for this is the lack of a substantial near-by market for its products.*

* The fairly long and very cold winter has also proved to be a notable disadvantage, for measures had to be taken to protect all vulnerable points exposed to the weather against the low winter temperatures. For instance, all pipes carrying slowly moving water were laid seven feet below ground level; the coke quencher is self-draining; a steam-heated appliance is available for thawing frozen ore; and five stoves are provided to feed each blast furnace instead of the customary four. These points may

Houston, on the other hand, is well placed with regard to the market offered by the productive mineral-oil and natural-gas fields of the south-west U.S.A. Prospecting for and developing these oil fields has sparked off a demand for very considerable amounts of steel needed in the construction of drilling equipment and other appliances, and the importance of this demand as a locating factor is underlined by the fact that a new integrated works has recently been set up at Dangerfield in north-eastern Texas, in a location which is well placed with regard to the same market provided by the oil fields. It is also well placed with reference to raw materials (coking coal and dolomite from south-eastern Oklahoma and iron ore from near-by deposits in Texas). The establishment of the Fontana plant near Los Angeles (p. 189 below) was primarily due to market attraction.

Site Requirements

An iron and steel works is a large-scale enterprise and it needs an extensive site on which to develop. The Abbey Margam works in South Wales, for example, occupies about 4 square miles (Fig. 63) and the Sparrows Point plant $6\frac{1}{4}$ square miles. The establishment of the iron and steel industry of Chicago and Gary was greatly assisted by the existence of the sand-dune country fringing the southern shore of Lake Michigan which was almost entirely unused. This feature brings out an equally important point—that the land must be cheap, while it is a further advantage if the site faces tidal water (or at least water which can carry heavy mineral-carrying vessels). Opportunity costs for such a site as that offered by the sand-dune country south of Lake Michigan must be limited and the cost of securing the land will therefore be low.

Water Requirements

The water requirements of the iron and steel industry are very great indeed; we have already mentioned, for example, that at the Port Kembla works in New South Wales, eight tons of good-quality water are used for coke quenching alone every six minutes, and the Port Kembla enterprise is a very modest one judging by world standards. The world's largest steel producer, the plant of the Bethlehem Steel Company at Sparrows Point, Maryland, is located on the brackish Patapsco River near the point where it flows into Chesapeake Bay, and in the earlier years of operation this plant drew brackish water for cooling purposes from the river and fresh water from wells. So much water was drawn from the wells, how-

---

serve to remind us that the climatic factor is by no means an unimportant one as regards the location of an iron and steel works.

ever, that they became contaminated by an inflow of impure water, and since 1955 the iron and steel plant has processed for use almost the entire flow of sewage of the city of Baltimore, 150 million gallons daily, in addition to the consumption of 540 million gallons of brackish water. It is quite impracticable to transport water on this scale over any distance, and it seems inevitable that the location of future iron and steel works will be more and more closely governed by water availability.

After being used in the iron and steel works the water is normally flushed away as waste, and this leads us to make mention of another advantage of the tide-water location. Iron and steel works excrete very large amounts of fouled waste water, a fact which is brought out by the table on page 284 below. A tide-water location means that such wastes have not far to travel before they can be ejected into the sea, and this is a very helpful feature for the manufacturer.

Capital Availability

The modern iron and steel industry is one which requires very large amounts of capital; it is *capital intensive*. While this in itself is not a geographical point it does have geographical repercussions. Developing territories, for example, do not find it easy to raise sufficient capital to permit them to establish iron and steel industries, and the further difficulty arises that unless a large enterprise (with an annual output approximating to about one million tons per annum) is established the small scale of operations condemns the enterprise to uneconomic working (pp. 109 and 181 above).

Government Policy

Government action, as we have seen in the case of Egypt, can initiate the development of an iron and steel industry even in the face of apparently unfavourable factors, but only at the cost of economic workings, and in some cases even government backing proves insufficient to ensure success. This has been illustrated by the failure of the efforts of the Communist Chinese government in the 1950s to stimulate pig-iron production in primitive furnaces operated locally. It was realised from the outset that such furnaces could turn out only limited amounts of pig iron, but it was anticipated that the large number of such enterprises would between them make a substantial total contribution towards the efforts being made to promote the national industrialisation programme. The scheme failed on two main grounds—economic and technological. Despite the immense overall effort, the millions of workers involved with their rudimentary equipment were able to produce less than half the iron output of the limited number of blast furnaces which did exist, while the poor-quality pig iron produced was found to be unsuitable for use in any steel works and was almost useless. By way of contrast we may note that the now prosperous Japanese iron and steel industry owes its origin to the establishment at Yawata in 1901 of a government undertaking, built primarily to provide iron and steel for military needs.

The case of Ebbw Vale, where a major iron and steel plant was located as a result of government intervention, was mentioned in Chapter V (and *see* p. 292 below). Pounds argues that this project, "however justifiable from the social point of view, was economically a retrograde step," and certainly the Ebbw Vale works, located on the floor of a narrow winding valley which leaves little scope for expansion, and under the necessity of securing its iron ore from Northamptonshire after a lengthy rail haul, has experienced great difficulties in operation. On the other hand, an example of a local government action which has had very beneficial effects comes from Canada, where the city of Hamilton is now the largest producer of iron and steel in the country. The industry was first attracted to Hamilton because in 1895 the city was willing to offer tax concessions to the manufacturing company concerned.

In some instances strategic considerations have played an important part in locational decisions; the cases of the establishment of the Hermann Goering Works at Salzgitter, in Nazi Germany, in what was then considered an area "safe" from likely enemy attack, comes to mind (p. 108 above). The location of the Fontana plant shows an appreciation of market advantages, but strategic considerations led to its location at Fontana, 50 miles from Los Angeles, rather than on the coast where the manufacturing company wished to develop it.

IRON AND STEEL IN COMMUNIST CHINA

In view of the unusual character of and the interest attaching to the efforts made by Communist China to develop her iron and steel industry, it might be useful to say a little more about this topic.

The "Backyard Furnace Experiment"

The Communist government was determined to accelerate the rate of industrial expansion after the conclusion of the first Five Year Plan, and one method of bringing this about was the initiation in 1958 of a very large number of small industrial enterprises which were to be operated by part-time workers and by peasants mobilised under the commune system. An essential feature of this plan was to be a large increase in iron and steel production and fabrication based upon the establishment of a large number of small-scale iron-smelting ("backyard") furnaces; some reports mention as many as two million of these appliances. According to available data, the production of native pig iron increased enormously as a result of these measures from 4·2 million metric tons in 1958 to 11·0 million metric tons in 1959, while "native" ingot steel production over the same period rose from 3·1 to 4·7 million metric tons. At the same time we might note that "non-native" production in modern-type conventional iron and steel mills in the same years remained fairly constant (pig iron, 9·0–9·5 million metric tons; ingot steel, 8·0–8·6 million metric tons). The most productive modern-type mills are located as follows (the figures show the estimated capacity output of each mill in 1960 in thousand metric tons):

	Centre	Province	Capacity
(a) Pig iron			
	An-shan	Liaoning	4170
	Ma-an-shan	Anhwei	2200
	Hsiang-t'an	Hunan	1800
	Liu-chou	Kwangsi	1750
	Shih-ching-shan	Hopeh	1600
	Wu-han	Hupeh	1467
	Hai-nan	Kwangtung ⎫	
	Shanghai	Kiangsu ⎬	1000
	Ta-yeh	Hupeh ⎭	
(b) Ingot steel			
	An-shan	Liaoning	5570
	Shanghai	Kiangsu	4000
	T'ai-yüan	Shansi	2267*
	Ma-an-shan	Anhwei	1500
	Wu-han	Hupeh	1470
	Ta-yeh	Hupeh	1260
	Liu-chou	Kwangsi	1250
	Chungking	Szechuan ⎫	
	T'ang-shan	Hopeh ⎬	1000
	T'ung-hua	Kirin ⎭	
(c) Finished steel			
	An-shan	Liaoning	3300
	Shanghai	Kiangsu	2680
	Wu-han	Hupeh	1100

Source: Yuan-Li Wu.

* Two mills, one producing 2,167,000 metric tons, the other 100,000 metric tons.

It was expected that many benefits would accrue from the backyard furnace experiment. Not only would the furnaces provide the large-scale mills with additional raw iron, but they would expand the areal base of the industry and provide an impetus to the economic development of a large number of districts; they would greatly increase the scale of economic development in the country; and they would make it possible for the large mills to experiment more widely than before with new techniques and with more refined and sophisticated products. Unfortunately, however, the bold experiment yielded most disappointing results, and in 1959 and 1960 many of the small furnaces were either abandoned or replaced by rather larger furnaces of improved design. It was admitted as early as 1959 that the pig iron produced in the small furnaces was of poor quality, while the costs of production turned out to be considerably higher than was expected. Yuan-Li Wu considers that the failure must be blamed partly upon an over-optimistic estimate of the possibilities of using labour-intensive methods in iron and steel manufacture, partly upon an under-estimating of the costs involved, and partly upon an undue emphasis on quantitative production without due regard for quality. After a careful examination of the available data he comes to the conclusion that "the contribution of the backyard furnace movement to national income was . . . worse than nil"—in other words, the movement proved economically a burden rather than a positive contribution to the national effort.

It is not fair, however, to write off the experiment as a total failure. It is true that some visitors to China have stressed the allegedly worthless nature of backyard pig iron and have implied that it was completely useless for further fabrication ("useful only for ballast"), but Yuan-Li Wu, who cannot be accused of any pro-Communist bias, emphatically disagrees with this appraisal. He admits that it is almost certainly true that only a very few of the backyard furnaces met standard requirements in the quality of their product in as much as 90% of their output and that a figure of 80% was probably well above the average, while in the case of the semi-modern furnaces the comparable figure was probably under 70%. It cannot be denied that this performance represents technically a very low level of efficiency and one which could not be tolerated in a free society, but at the same time it cannot be classed as a total failure.

There are, moreover, further points which are worthy of consideration. Professor Keith Buchanan (in a personal letter to the author) argues that the significance of the 1958 experiment should be measured not simply according to the amount of iron or steel produced, but also in terms of the value of the campaign in introducing the Chinese peasantry to the techniques of industrialisation. Particularly important is the blow at the widely held belief that state aid and large "injections" of technicians and technical skills are essential before the stage of economic take-off (*see* p. 304 below) can be attained. The technical skills which the peasants acquired during 1958 and in succeeding years, were mobilised subsequently in the semi-modern furnaces and in the small-scale workshops which are now attached to many communes and which can undertake a surprising range of metal-working tasks; the success of this mobilisation has undoubtedly exercised a markedly buoyant effect upon the morale of workers and administrators alike. "My own feeling, and this is reinforced by my visits in 1964 and again in the middle part of this year [1966], is that this psychological aspect has been greatly underestimated by the West" (Buchanan).

Modern Developments

In common with their counterparts in other countries, Chinese technologists are experimenting with modern methods of iron and steel production. Some of these methods have been noted above, but we will say a little more about some of the most notable.

1. The use of self-fluxing sinter has proved very successful. This sinter has a higher basicity ratio (CaO/SiO_2) of 1·0 instead of 0·5. This technique has been tried out at An-shan with the following results:

   (a) the consumption of limestone in the blast furnace has been reduced by 60–70 kilograms per ton of pig iron produced;
   (b) there has been a 4–5% reduction in the coke ratio;
   (c) there has been a 5–7% increase in the overall efficiency of the blast furnace;
   (d) "lumping" of slag has almost disappeared.

2. The use of controlled moisture in the blast has notably improved efficiency. By adding 20–25 grams of steam to each cubic metre of hot blast (which was pre-heated to 760–810° C) the productivity of the furnace was raised by 5%.

3. The temperature of the blast has been raised. When the temperature of the blast at No. 2 furnace at Shih-ching-shan was increased from 680 to 720° C, productivity went up by 2·5% while the coke ratio fell by 2·17%.

4. Productivity has also improved from the use of pressurised tops. At An-shan every increase in top pressure of 0·1 kilograms improved output by 3%.

In the face of facts such as these it is clearly misleading to imply that Chinese iron and steel technology has been bounded by the backyard movement. It is rather the case that we are here dealing with an industry which is very much alive to modern developments and possibilities, and a future of considerable promise lies ahead of it.

THE AUSTRALIAN IRON AND STEEL INDUSTRY

To conclude this chapter, a brief note is added on the organisation and setting of the iron and steel industry in Australia. This example is chosen, partly because the writer has some first-hand knowledge of Australia and partly because this particular example is rarely dealt with in any detail in textbooks. On the other hand, many writers have dealt with the iron and steel industry in Great Britain, North America, Europe and Japan (*see*, for example, references to Stamp, Dury, Jones and Bryan, and Yamamoto at the end of the chapter).

The Australian iron and steel industry is largely concentrated in the hands of two major producers—the Broken Hill Proprietary Co., Ltd. (B.H.P.), and its main subsidiary company, the Australian Iron and Steel Proprietary Ltd. (A.I.S.). The B.H.P. own coal mines (at Newcastle), iron-ore mines (Iron Knob and Iron Baron near Iron Monarch),* limestone quarries (at Rapid Bay), dolomite quarries (at Ardrossan) and magnesite mines (at Fifield)—magnesite is required for the manufacture of refractory bricks. The A.I.S. own coal mines (near Port Kembla) and iron-ore mines (at Yampi Sound). The B.H.P. operates integrated iron and steel works at Newcastle and Whyalla together with the rolling mill at Kwinana, while the A.I.S. operates an integrated plant at Port Kembla. Associated engineering industries are located particularly at Newcastle, Port Kembla, Whyalla, Sydney, Melbourne, Geelong and Hobart. Figure 44 illustrates the geographical setting of the industry.

It may be of interest to note that the B.H.P. took its name from the well-known mining town of Broken Hill, near the western border of New South Wales, where in 1883 a rich silver–lead–zinc lode was discovered. The Broken Hill Proprietary Co., Ltd. was formed in 1885 to exploit these ores, and from the ensuing activities developed the town of Broken Hill and the associated lead smelters at Port Pirie in South Australia. When the open-cut section of the lode was worked out in 1939 the mine was closed (the company no longer conducts operations at Broken Hill: "Today the only link with Broken Hill is in the name"— pamphlet published by B.H.P.), but foresight within the company had ensured that the firm had developed other interests to take the place of the Broken Hill activities. These other interests stemmed from the fact that ironstone was needed in the Port Pirie smelters for use as a flux, and the consequent prospecting for iron ore led to the discovery of the enormous Iron Monarch reserves which were mined, at first on a modest scale, for

* B.H.P. also hold a substantial interest in the Mount Newman iron-ore mines. Most of the ore, however, is railed to Port Hedland and is exported to Japan (Fig. 44).

despatch to Port Pirie. When it became clear, however, in the early years of this century that the Broken Hill deposits were showing signs of exhaustion, the company cast about for another field of interest at first to

FIG. 44.—The pattern of the iron and steel industry in Australia in 1966. The iron-ore deposits shown are those now being exploited for the export of ore mainly to Japan.

supplement and later to replace the original enterprise, and an obvious line of development appeared to be to exploit further the Iron Monarch iron ore and to enter the iron and steel industry. After a close study of the problems involved the company decided to proceed with such a venture at Newcastle, the main reasons for the choice of location being as follows:

1. The company already held a title to land on the south bank of the Hunter River (Fig. 45).

2. This site lay adjacent to tide water and could be reached by ocean-going vessels bringing raw materials (Fig. 44) and carrying manufactured goods.

3. The site was unused as it was swampland, and while this meant that heavy initial costs were involved in drainage and in providing foundations

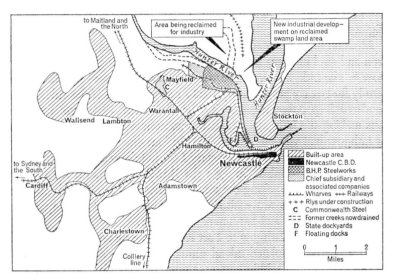

Fig. 45.—Newcastle, N.S.W., and the iron and steel industry. Almost the whole of the area shown on the map lying south of the Hunter River is underlain by coal measures. The chief mining areas lie west of Wallsend, west and south-west of Cardiff and south of Charlestown.

to carry heavy plant, it also meant that there was no competing use for the land.

4. The site was well placed to accept coal from the coalfields of Newcastle and the lower Hunter region.

5. Newcastle was already a fair-sized city which could offer a reasonably adequate labour supply.

The first foundation pile was driven into the swamps in 1913 while the first tapping of the original blast furnace (which had an output capacity of 350 tons a day) took place in 1915, and development has gone ahead, with temporary setbacks due to economic causes, since that modest beginning. A fully integrated plant is now in operation with four blast furnaces and associated coke ovens, sinter plant, basic oxygen equipment, soaking pits, bloomeries, rolling mills, skelp and rod mills. Associated works on adjacent or near-by sites produce a wide range of products including steel wire and wire products, rock bolts (for the support of mine roofing), tubes, drills, lathes and a wide range of mining and agricultural equipment. In the suburb of Waratah a subsidiary company, the Commonwealth Steel Company, manufactures special steels needed in engineering work, including stainless steel, and produces wheels and axles for rolling-stock and for heavy forges (*see* table on p. 199 below and Fig. 45).

Mention has been made (p. 127) of the early attempts to establish an iron and steel industry in Australia, first at Mittagong and then at Lithgow.* The Lithgow enterprise was owned and operated by the Hoskins family for twenty years, but by 1928 a reorganisation of the industry was rendered imperative because the success of the Newcastle enterprise was making the Lithgow production uneconomic. By 1929 production at Lithgow had virtually ceased, the Hoskins interests passed to the new Australian Iron and Steel Proprietary Ltd. and the scene of activities was moved to Port Kembla. In 1935 the B.H.P. acquired all the ordinary shares of A.I.S., so making the smaller company the principal subsidiary of the B.H.P. This move also made possible planned development as between Newcastle and her newer competitor.

The move from Lithgow to Port Kembla can be seen as a move away from a relatively isolated inland centre of very modest size towards a tide-water location within very easy reach of a coalfield (Fig. 27). Coal outcrops along the scarp slope overlooking Port Kembla and coal adits are driven into the hillside; it is a comparatively simple matter to transport the coal the few miles from the openings of the adits along a route which is predominantly downhill to Port Kembla. Limestone comes from Marulan (Fig. 44) by rail, and iron ore from Iron Monarch and Yampi Sound by sea, and, although there was no natural harbour at Port Kembla comparable to that provided by the River Hunter at Newcastle, a new "Inner Harbour" has now been constructed which provides a safe anchorage for the large ore carriers at present in operation. As in the case of Newcastle, an extensive site adjacent to the tide water was available for

---

* There was also an attempt to establish an iron industry between 1873 and 1884 at Lal Lal, fifteen miles from Ballarat in Victoria, but the attempt failed largely because pig iron could not be produced to compete in price with imports from England. Pig iron was exported from England in the form of ballast for the return journeys of the clippers, and freight costs were therefore low. The ruined blast furnace is still in existence at Lal Lal. Oddly enough, peak production (800 tons of pig iron) was achieved in the last year of operations, 1884.

development as it was originally swampy low-value land for which there was no competing use; the plant now occupies a site of $3\frac{1}{4}$ square miles in extent.

Activities at Port Kembla are comparable with those at Newcastle though on a rather larger scale. Four batteries of by-product coke ovens and an ore-screening and sinter plant feed four blast furnaces, while open-hearth furnaces, soaking pits, a bloomery and continuous rolling mills prepare raw steel in strip, slab, plate, billet and girder form for industrial and constructional use. Slabs, blooms and billets are square or rectangular in section, though the width of a slab is considerably greater than its thickness. Plates are produced from slabs and are relatively thin, sometimes no more than one-eighth of an inch. Strip is supplied in coils; if the steel is 18 inches or less in width it is known as strip, but if it is wider it is referred to as sheet (*see* Fig. 41). Associated engineering industries are concerned with the manufacture of such products as tin plate, cast-iron pipes, steel plate, rails and girders. Billets (which have a fairly small cross-section of not more than $4\frac{1}{2}$ inches) are produced on a large scale in continuous mills★ and are further processed in merchant mills into a wide range of shapes and small items† for use by manufacturers in the production of component parts of machinery, motor-cars, tractors and other equipment.

An interesting feature of the Port Kembla organisation is that, of the approximate total of 720,000 tons (173 million gallons) of water used *daily*, no less than 700,000 tons (almost 160 million gallons) is salt water pumped directly from the sea, the remainder being fresh. It is interesting to note that this 173 million gallons used each day, on a site of about $3\frac{1}{4}$ square miles is not so very far short of the total daily consumption (about 200 million gallons) of the entire city of Sydney, less than 100 miles distant. At Newcastle, comparable figures of daily water use are 160 million gallons of salt water and $6\frac{1}{2}$ million gallons of fresh.

It is instructive to compare the relative performances of the iron and steel industries at Newcastle and Port Kembla and to enquire into the reasons for any differences noted. The clearest point is that, while the Newcastle enterprise had the advantage of being the first to be established, in recent years output at Port Kembla has overtaken that at Newcastle. In 1946, for example, Newcastle produced 63% of Australia's pig iron,

---

★ So called because the steel which is being processed passes through continuously in one direction instead of being passed and repassed through the rollers.

† Such small items were in earlier years distributed to manufacturers by steel merchants and came thus to be known as merchant bars, while the plant which produces them became known as a merchant mill.

Port Kembla 33% and Whyalla 4%, but in 1960 the corresponding figures were 31, 61 and 8% respectively. This change occurred because a 40% increase in output at Newcastle was matched by a more than fourfold increase at Port Kembla. In 1967 Newcastle produced 32% of Australia's pig iron, Port Kembla 50% and Whyalla 16%, while corresponding figures for the output of steel ingots for the same year were 29, 57 and 10%. The following table illustrates the pattern of change clearly:

Output of Pig Iron in Australia (Thousand Long Tons)

Year ending May	Newcastle	Port Kembla	Whyalla	Total
1946	563	290	38	830
1947	559	367	218	1143
1948	635	395	207	1237
1949	565	335	154	1053
1950	520	450	106	1076
1951	597	567	142	1303
1952	659	575	185	1419
1953	712	773	193	1678
1954	714	911	193	1817
1955	646	1030	181	1857
1956	687	1063	148	1898
1957	740	1112	232	2084
1958	793	1237	237	2267
1959	803	1237	227	2237
1960	801	1598	209	2608
1961	1104*	1857		2961
1962	1289*	2309		3328
1963	1145*	2195		3340
1964	1639*	2079		3718
1965	n.a.	n.a.		3986
1966	n.a.	n.a.		4335
1967	1594	2465	792	4851

* Newcastle and Whyalla.
Source: Figures supplied by the B.H.P.

A somewhat similar change of pattern in the annual production of steel ingots is noticeable, as shown in the table opposite.

It is clear from these figures that, while the output of pig iron and steel ingots is markedly on the increase, the increase in production at Port Kembla has considerably outstripped that at Newcastle. Perhaps three main reasons may be advanced to account for this development:

1. While it is true that both Newcastle and Port Kembla have comparatively easy access to adequate reserves of coking coal (B.H.P. collieries produced an average of 10,000 tons of coal per day in Newcastle

Output of Steel Ingots in Australia (Thousand Long Tons)

Year ending May	Newcastle	Port Kembla	Whyalla	C.S.C.*	Total
1946	703	333		19	1055
1947	801	476		34	1311
1948	805	505		33	1178
1949	711	432	0·2	34	1178
1950	683	496	2	36	1217
1951	756	642	0·7	43	1442
1952	818	653	4	44	1420
1953	959	808	1	35	1802
1954	1021	1041	5	52	2119
1955	970	1167	1	61	2199
1956	995	1272	2	45	2315
1957	1146	1552	3	68	2770
1958	1206	1756	3	64	3029
1959	1240	1883	4	66	3193
1960	1261	2162	6	78	3507
1961	1311	2332	11	n.a.	3736
1962	1328	2666	8	n.a.	4061
1963	1118	3063	n.a.	n.a.	4257
1964	1550	3122	n.a.	n.a.	4746
1965	n.a.	n.a.	n.a.	n.a.	5018
1966	n.a.	n.a.	n.a.	n.a.	5549
1967	1793	3554	634	n.a.	6047

* Commonwealth Steel Co., Newcastle. See p. 196 above.
Source: As above.

and 17,000 tons per day in Illawarra in 1967), the composition of the coal slightly favours Port Kembla. This is brought out by the following table:

	Newcastle	Port Kembla
Volatile constituents	35%	27%
Fixed carbon	55	63
Ash	10	10
TOTAL	100%	100%

The higher production of by-products from the Newcastle coal is valuable, for it not only provides the steel works with gas from the coke ovens but forms the basis for the manufacture of a wide range of products (p. 170 above). But the Port Kembla coal, with its higher proportion of fixed carbon, yields a coke which is denser and stronger than that of Newcastle and which can therefore withstand greater pressures within the blast furnace. This is the main reason why it has been possible to construct larger blast furnaces at Port Kembla than at Newcastle, and this in turn means greater output and greater economy of operation.

2. Since the Second World War the Australian demand for flat products (steel plates, bars, etc.) has increased enormously. This increase in

demand was not unexpected, and an anticipatory decision was taken by the B.H.P. to establish a merchant bar, rod and strip mill using the continuous-rolling technique. Such a mill must be housed in a very lengthy building to accommodate the plant necessary to take in the steel slabs for rolling at one end and rolling them to produce, for example, thin strip; to produce the strip the slabs pass through a series of rollers which successively reduce their width and thickness, and, although in the early stages of continuous rolling the reduced slabs, about one inch in thickness, pass fairly slowly through the rollers, the progressive reductions in dimension without any reduction in volume mean that the strip must move faster and faster after successive rollings until, near the end of the process, it is travelling at 25 miles an hour before it is rapidly wound into coils. The Port Kembla strip mill is about one mile in length. Such a mill must also be fairly broad because not all the slabs are required for strip but some are needed for the manufacture of plates, bars, etc.; such slabs are removed *laterally* from the strip production line before they enter the continuous-rolling plant. This means that quite distinct processing operations are going on side by side.

This may seem a lengthy digression from our main line of enquiry but the purpose of it is to emphasise the simple fact that a very extensive site is necessary for the establishing of a continuous-rolling mill and flat-products plant, and the situation was that such a site could be developed on the unused swamplands of Port Kembla, while land was not at that time available so easily at Newcastle (much additional land has since that time been reclaimed at Newcastle (Fig. 45)). Hence it was primarily the factor of site requirements which led to the further development of Port Kembla, and when once the rolling mill and the flat-products plant were in operation they attracted other industries, a notable example being the manufacture of tin plate and cans for the expanding fruit-canning industry. Questions of linkage are clearly involved. Additional blast furnace output was in turn needed to feed the continuous-rolling mill and the flat-products plant, and this led to the blowing in of number 4 furnace, the largest in the southern hemisphere.

3. Although the establishment of industry on a large scale in Newcastle is comparatively recent, it is unfortunately true that (rightly or wrongly) the city has acquired a poor reputation in the field of labour relations. This may go back to the rough-and-ready early days of coal mining. There is probably some justification for the view that the iron and steel interests were encouraged to develop the new Port Kembla site because it was hoped to avoid labour troubles in the newly developing industry. The experiment seems to have had considerable success, and one factor

which has undoubtedly helped in this respect is that a high proportion of workers are migrants from areas such as southern Italy. To such workers the opportunities for work and the wages received appear magnificent and there has been as yet little incentive for labour unrest to take firm root (compare the case of New England in the second half of last century (p. 225 below)).

We should not suppose, however, that efforts are not being made in Newcastle to secure an increasing share of the Australian market for iron and steel products. In recent years additional land has been made available for industry through swamp drainage and reclamation, while a fourth blast furnace was blown in during 1963. Three developments, however, are of more than usual interest:

1. In December 1965, after continuous operation for more than fifty years, the open-hearth department of the B.H.P. was finally closed down,

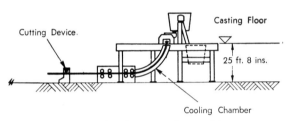

Fig. 46.—Continuous casting machine.
[*Reproduced by courtesy of B.H.P. Newcastle, N.S.W.*]

and steel manufacture is now carried on almost entirely by means of three basic oxygen furnaces (in 1967 all but 46,000 tons of the 1,792,762 tons of ingot steel produced at Newcastle came from the basic oxygen plant). Two of the furnaces each produce about 200 tons of steel in between 60 and 70 minutes, and the third about 50 tons in 35 minutes.

2. In 1967 a continuous-casting machine came into operation at the B.H.P. works. This is a very modern type of development and the working of the machine is illustrated in Fig. 46. Molten steel is poured from the melting furnace ladle into the tundish from which water-cooled nozzles feed the still molten steel into copper moulds along curved leaders; further cooling completes the solidification of the metal. The withdrawal is effected by rollers; as the strip of metal passes through these rollers its temperature is about 1650° F (900° C) and immediately afterwards the steel is cut to the required lengths by hydraulic shears.

3. The Commonwealth Steel Company also closed down its open hearth activities early in 1966 following the installation of a 50-ton

electric arc furnace. All steel produced by the company in future will be made in electric furnaces, and there are five arc furnaces now (1968) in operation and two high-frequency plants.

There should by now be no doubt in the mind of the reader as to the suitability of the Newcastle and Illawarra areas for the establishment of heavy industry, and the factors which have led to industrial concentration in these areas have been emphasised. In recent years, however, changing factors have led to the establishment of the iron and steel industry, or at least branches of it, in other parts of Australia which do not appear to offer any special advantages. The outstanding examples of this are the integrated works and shipbuilding yards at Whyalla, in South Australia, and the rolling mill at Kwinana in Western Australia. Kwinana (an aboriginal word meaning "pretty maiden") lies some miles south of Perth and was so named after a small freighter which was driven aground there in 1922. Before we take note specifically of these two examples, however, we might examine the changing social, economic and technological conditions in Australia which have helped to bring about these new developments.

1. *Political and social considerations.* Increasing disquiet was felt in many quarters in Australia, especially during the uneasy peace of the 1930s, regarding the concentration of heavy industry along the eastern seaboard —and along such a restricted stretch of the seaboard too. The strategic weakness of such a situation was fully realised and careful thought was given to the possibility of establishing heavy industry on a less exposed location. This urge towards decentralisation of industry has received a sharp impetus in more recent years as federal states, notably South Australia, are anxious to develop their own industries and seek to be less reliant on the older establishments of New South Wales and Victoria; political issues are therefore involved as well as strategic.

2. *Economic considerations.* Perhaps the most important point here is that the populations of Southern and Western Australia are sharing in the general population increase of the continent and that markets in these states are correspondingly increasing in size. In view of the importance of the market as a locational factor today it is only natural that industries should increasingly be attracted to these developing states.

3. *Technological considerations.* These are of considerable importance. We have noted reasons why heavy industry was attracted to the coalfields years ago and we might further note that in earlier days it was a matter of some urgency that coking coal be coked and used as soon as possible after mining, since prolonged exposure to the atmosphere results in deteriora-

tion and crumbling, probably because of drying out and oxidation though the actual causes are imperfectly understood. If blast furnaces were not on coalfield sites, however, considerable stockpiling of fuel was necessary at pit-head and furnace to ensure continuous supply, and this made for difficulties in view of the point just made. There was also the danger of spontaneous combustion in the fuel heaps, a danger particularly acute in the dry, hot Australian summers.

Today, however, the picture is completely changed. One reason for this is the development of the sintering and pelletising processes which have been of the greatest benefit to the Australian iron and steel industry since many domestic ores are very fine. These can now be rendered suitable for the blast furnace, while a great deal of coke is used in the sintering process, which eases the fuel situation referred to in the previous paragraph. Perhaps the most important point, however, is that, because of the various ways now available of economising on fuel, the amount of coke required per unit output of pig iron is much reduced (footnote, p. 171 above) and it is now practicable to locate an iron and steel works away from a coalfield (compare the situation in Britain and in the U.S.A. (pp. 185 and 186 above)).

Work at the Whyalla site was well under way at the outbreak of the Second World War, and the artificial deep-water harbour was completed in 1940. Meanwhile, the B.H.P. had been approached by the naval authorities, who wanted the firm to establish shipbuilding yards as a contribution to national security, and Whyalla was chosen as the constructional location; this was a tremendously important point as from the outset it assured the iron and steel works of a substantial local market for its products. The shipyard was opened in 1940 when keels were laid for two naval patrol vessels. In the following year, 1941, the blast furnace was blown in and soon afterwards came the electric steel-melting shop and the forge shop. With the opening in 1965 of a new £50 million steel works Whyalla has developed a fully integrated iron and steel industry with basic oxygen equipment of the most advanced type. A second blast furnace was blown in also in 1965.

The rapid development of a town of over 20,000 inhabitants located in a semi-arid region receiving an annual average rainfall of nine inches is a remarkable achievement, especially when we recall that about 32% of the residents are migrants from overseas. Most of the fresh water essential for the life of the town and for industry has been pumped through a pipeline from Morgan, on the River Murray, a distance of 223 miles! This pipeline was completed in 1944, and it can deliver 1200 million gallons of water daily as well as delivering a further 900 million gallons to other places along the route. Work was completed in 1967, however, on the installation of a second pipe, 33 inches in diameter, which runs roughly along the line of the original one but which cuts across the head of Spencer Gulf instead of going around the head through Port Augusta. Work on storage and pumping plant continues and the new system (which was designed to have a throughput of 14,000 million gallons a year) is not expected to be working at full capacity until 1980. The new pipe, together with the existing one, will relieve Whyalla of problems of water supply for many years to come.

The reasons for the success of the Whyalla enterprise can easily be summarised. Strategically, the location is well placed in a comparatively remote area well removed from other industrial centres, while the establishing of the shipbuilding yards has been a tremendous asset because of the market they provide for steel products. Among physical advantages, the very extensive iron-ore reserves of Iron Monarch are well placed in relation to Whyalla, while the dolomite and limestone deposits of Ardrossan and Rapid Bay respectively (Fig. 44) are easily accessible by sea. The warm dry climate is an undoubted asset in at least one respect, since much of the work connected with shipbuilding must be carried on in the open, while the availability of land on an ample scale has been most advantageous.

The Kwinana steel-rolling mill was completed in 1956 on the deep-water harbour of Cockburn Sound, and billets shipped from the rolling mills of New South Wales are unloaded directly on to the site. This merchant-bar mill represents the first stage towards the development of a fully integrated works. The next stage towards this development will be the completion of a sinter plant and a blast furnace in 1968.

The example chosen above of the iron and steel industry of Australia clearly demonstrates the changing factors which are at work today. Social (including strategic and political), economic and technological factors are combining to produce new locational patterns, patterns which may well undergo further modifications in the future now that a high degree of emancipation from coalfield sites has been achieved. The example also illustrates the dynamic character of modern industry and gives some indication of the manner in which new problems are being overcome.

THE IRON AND STEEL INDUSTRY

SUGGESTIONS FOR FURTHER READING

Britton, J. N. H., "The Development of Port Kembla, N.S.W.," *Geography*, July 1961.
B.H.P. Ltd., *Seventy-five Years of B.H.P.*, Australia, 1960.
Estall, R. C., and Buchanan, R. O., *Industrial Activity and Economic Geography*, London, 1961.
Jones, L. R., and Bryan, P. W., *North America*, tenth edition, London, 1957.
Kerr, D., "The Geography of the Canadian Iron and Steel Industry," *Economic Geography*, 1959.
Pounds, N. J. G., *The Geography of Iron and Steel*, London, 1959.
Rodgers, A., "Industrial Inertia: A Major Factor in the Location of the Iron and Steel Industry of the U.S.A.," *Geographical Review*, 1962.
Stamp, L. D., and Beaver, S. H., *The British Isles*, fifth edition, London, 1963.
Dury, G. H., *The British Isles*, London, 1961.
Taylor, G., *Australia*, London, 1951.
The Times, Supplement on Australia, 1963.
White, C. Langdon, "Water—Neglected Factor in the Geographical Literature of Iron and Steel," *Geographical Review*, 1957.
White, C. Langdon, and Primmer, G., "The Iron and Steel Industry of Duluth: A Study in Locational Maladjustment," *Geographical Review*, 27, 1937.
Wills, N. R., "The Growth of the Australian Iron and Steel Industry," *Geographical Journal*, 1950.
Yamamoto, T., "The Steel Industry in Japan," *Oriental Economist*, 26, 1958.
Yamamoto, T., Australian Iron Ore: Great Prosperity and Some Problems," *Steel Review*, 41, January 1966.
Yuan-Li Wu, *The Steel Industry in Communist China*, New York and London, 1965.

Chapter IX

The Cotton Textile Industry

THE production of textiles and textile goods must be one of the oldest industries known to man, and it has been traditionally a handicraft and a domestic enterprise. Man has needed some form of clothing from the earliest times, and the production of cloth from which garments could be made naturally developed as soon as he realised the possibilities of natural fibres for this purpose. The most commonly used fibre in earlier years was wool and it was this which accounts for the great importance traditionally placed upon sheep rearing in many lands, while silk and linen were in some cases available in limited quantities. Use of these fabrics, however, was restricted to the wealthy.

THE COTTON INDUSTRY AND DEVELOPING TERRITORIES

This situation was modified, at first mildly and later substantially, by the entry into the textile field of a new competitor—cotton. The reader might at this point turn back to Chapter I (pp. 9–11) for some of the main reasons for this. We shall give in this chapter further consideration to the cotton textile industry, which is widespread in the world generally and which is commonly one of the first industries to be established in developing territories that are seeking to establish domestic industry. The following are the more important reasons for this:

1. The demand for cotton cloth, both for garments and for other purposes, is widespread, even in comparatively poor countries; indeed, we might call the need for clothing a prime necessity, as important in its way as the need for food. There are clear advantages if cotton goods can be produced within national boundaries because home industries are stimulated and foreign currency saved.

2. The various processes involved in the machine production of cotton cloth are comparatively simple, especially in the case of the coarser

fabrics, and the necessary skill can therefore be acquired fairly rapidly even by workers without previous industrial experience.

3. The capital requirements of the industry are comparatively modest. This is particularly the case since the economic advantages of large-scale production are not very great and a small-scale enterprise is quite often at no marked disadvantage.

4. The raw cotton required by the industry, though fairly bulky, is easily handled and transported; furthermore, it carries a fairly high value per unit weight and can therefore bear comparatively heavy costs of transport (p. 23 above). These points mean that the industry is not particularly tied to a raw-material source location.

CONTRASTS WITH THE IRON AND STEEL INDUSTRY

The cotton textile industry has been chosen for consideration in this chapter partly because of its very wide importance and partly because it offers such a marked contrast to the iron and steel industry. We might at this point say a little about this contrast since it is fundamental in explaining the markedly different patterns of organisation and location between the two industries. The following are the main points to consider:

1. The amount of wastage of material incurred during the various processes of manufacture in the cotton textile industry is minimal, and there is nothing like the unavoidable wastage which occurs, for example, when iron is extracted from iron ore in the blast furnace. In other words, little weight loss is incurred during manufacture.

2. It is not possible to secure anything like such large economies of production as are possible in the iron and steel industry by employing large-scale plant. One reason for this is that bulky by-products of one process are not used as the raw materials of another. It is true that short fibres which are removed from the raw cotton during the early cleaning processes are used for the manufacture of cotton lint but there is nothing comparable in scale to the use made of the enormous amounts of gas produced in by-product coke ovens and in the blast furnaces in other parts of an iron and steel works. Neither is there anything comparable to the heating up of the raw materials in the iron and steel works and the economies which are possible if pig iron and raw steel can pass directly through succeeding stages of manufacture with a minimal loss of heat.

3. Because of these factors the various branches of the cotton textile industry can be located independently of each other, and in practice this has in many instances led to regional specialisation within the industry.

H

A notable case is that of the concentration of weaving in the towns north of Rossendale in Lancashire while the towns to the south (those of the Manchester region) are more interested in spinning. Another interesting case is that of the comparatively early migration of spinning mills from New England to the Appalachian Piedmont in order to take advantage of low wage levels. The weaving branches of the industry were much slower to make the change, while the "finishing" firms have hardly moved at all (p. 131 above). Regional specialisation of this nature, which is also noticeable in Japan (p. 221 below), stands in marked contrast to the present tendency towards large-scale integration in the iron and steel industry.

4. Another factor which encourages the development of comparatively small firms in the cotton textile industry has to do with the very wide range of yarns and fabrics which are now produced. Few firms could hope to manufacture such a wide range of textiles as is now in quite general demand, and it is found as a matter of experience that the small-scale specialist is at a considerable advantage. This point has more reference to the weaving and finishing branches of the industry than to the spinning and it is another reason for the economic and geographical cleavage which often exists between the two major branches of the industry as a whole. This situation is less typical of the iron and steel industry, the nearest approach to it being found in the small-scale firms which manufacture special kinds of steel.

5. No branch of the cotton textile industry needs such large-scale and expensive equipment as the blast furnaces or the open hearths (to take just two possible examples) of the iron and steel industry; capital requirements are comparatively small. This further encourages the development of the small-scale enterprise. On the other hand, the industry is labour intensive and labour costs form the second most important element in its cost structure (raw cotton is the first), and this encourages the industry to seek low wage rate locations. We have mentioned the case of the Appalachian Piedmont in this connection, and this same advantage is commonly enjoyed by developing territories as opposed to established industrial ones. There is an element of irony, for example, in the fact that, while an important reason for the success of the Japanese textile industry has been the low wage rates payable in the past, the Japanese industry itself is now being threatened by industries such as those of China and India where wages are lower than in present-day Japan. This illustrates a general principle; while a country can in its early stages of development benefit substantially from low wage rates, the very success of industrial enterprise will normally result in a powerful move for wage increases so that early comparative advantages may be wiped out.

We have now said enough to show why it is that the cotton textile industry has commonly developed a small-scale, cellular structure which stands in marked contrast to the giant iron and steel corporations of the present day. It is also true that there is tremendous competition within the industry, and as in other instances where this happens we find that the industry suffers from a low scale of expenditure on research into possible new techniques and materials. There is a marked contrast here between the cotton textile industry and the man-made fibres industry.*

THE ORGANISATION OF THE INDUSTRY

We should now be in a position to understand the underlying reasons for the widespread distribution of the cotton textile industry, and it might be useful at this stage to move on to a brief consideration of the various processes of manufacture. We may distinguish the following chief stages of production:

1. *Ginning.* This process involves the separation of the cotton seeds from the lint.
2. *Spinning.* The raw cotton as received by the spinner comes in the form of a tightly packed bale of fibres, and a preliminary process loosens these fibres and removes fairly coarse impurities. A final cleaning process removes not only dirt but also the shortest fibres, while the cotton is then arranged in ropes (known as "slivers") of clean fibres held together simply by the natural cohesion of the cotton. This process is called "carding."

Further preliminary processes ensure that the individual fibres lie parallel to each other, while the ropes are drawn progressively finer before the final twisting; this is carried out by the spinning machine, which will be either a mule or a ring frame. Mule frames handle the thread more gently and produce a high-quality yarn but they need a considerable amount of skilled attention; production is therefore comparatively slow. The number of mule spindles has decreased notably in recent years; in fact, they have almost disappeared except in the United Kingdom, where about 90% of the world's mules are now in use. The ring spindle is faster in production and requires less skilled attention.

After the initial spinning the thread is further treated to ensure that the

---

* It is not possible within the compass of this book to make a study of the man-made fibres industry, but we might note that this comparatively new activity, which is in some respects a specialised branch of the chemical industry, benefits very considerably from economies of large-scale production, and that as a result giant production combines have become a dominant feature of this type of enterprise.

final yarn is of an even quality and thickness. Individual threads are spun together as part of this operation to give a stronger yarn.

3. *Weaving.* The finished yarn is woven into cloth on the loom. Looms vary considerably in type, in the speed at which they operate, in the width of cloth they turn out, and with regard to the number of operatives required. Hand-operated looms are still found in areas where the domestic production of cloth is carried on, for example in the Hebrides and in many countries which have barely begun the process of industrialisation. Non-automatic looms persist in some countries such as the United Kingdom which made an early start with the factory system of production, while the automatic loom is typical in countries such as the U.S.A. and Japan which made a later entry into the industrial world. A single weaver can attend to about five looms of the non-automatic type but, with some assistance, he can watch over sixty automatic looms. The greater efficiency of the latter type of loom makes it possible to produce low-cost cotton fabrics on a large scale and has been a factor in the decline of the cotton textile industry of Lancashire.

4. *Finishing.* The "grey" cloth produced as a result of the above series of processes is not an acceptable article of commerce until it has undergone further treatment designed to improve its appearance and possibly to impart particular qualities now made possible with the help of synthetic resins and other materials. The main stages in the finishing process are those of bleaching, dyeing and printing, while later stages may include calendering and mercerising. The purpose of the bleaching, dyeing and printing will be clear; calendering imparts a smooth and shiny finish to the fabric while mercerising produces a lustre comparable to that of silk. In addition, fabrics are now produced which may be crease resistant or fire resistant, which do not shrink, or which are waterproof.

STRUCTURE AND LOCATION OF THE INDUSTRY

After this preliminary survey we are in a better position to examine the geographical location of the industry, and it will immediately be apparent from what we have already said that no simple generalisation regarding location can possibly suffice. The various stages of the industry have their own particular requirements, and it is not unusual to find that different stages of production are carried on in different areas.

Ginning

Ginning the raw cotton is normally carried out in the cotton-growing areas (Fig. 47). There are good reasons for this. The cotton seeds carry

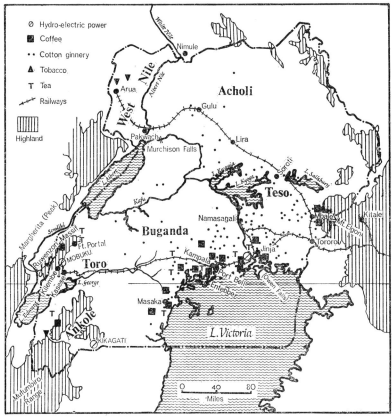

FIG. 47.—Uganda. Physical and economic features including cotton ginning. It is easy to overlook the contribution made to the economies of cotton producing countries by the cotton textile industry. Not only is the growing of the raw material important (raw cotton provides 26% by value of the total exports of Uganda) but the ginning introduces a simple industrial process to the inhabitants of the country concerned.

about two-thirds of the weight of the harvested crop so that ginning very substantially reduces the weight of the baled cotton, while it is clearly useful for the cotton-growing areas to retain seed for future sowing. Cotton seed is also used in various other ways (*e.g.* for the extraction of oil and the manufacture of fertilisers and cattle cake) which have no relation to the textile industry.

Spinning and Weaving

The spinning and weaving processes together constitute the major part of the textile industry, but while they may be located in the same general region this is not always the case; we have already mentioned some reasons why this should be so. We might at this point approach the whole matter more systematically by examining the physical, historical, structural and economic factors concerned:

1. *Physical factors.* The physical factors necessary for the successful establishment of a cotton textile industry exist in many areas, and the industry therefore has a very wide choice of location in this respect. One of the most pressing requirements is that of power availability. In the very early stages of the industry, when human effort provided the necessary power, a very high degree of mobility was possible, and this was an important reason why the manufacture of textiles was so widely distributed, but with the introduction of water power much of this mobility was lost; a location adjacent to strongly running water became essential as the siting of cotton mills in many Pennine valleys testifies. During the nineteenth century coal gradually replaced water as the energy source and a coalfield or near-coalfield location then became the most advantageous. The Appalachian Piedmont enjoys a clear advantage over New England in this respect since it lies considerably nearer to coalfields (West Virginia and Alabama), while with increasing use of electricity the Piedmont further gains over New England. Lancashire benefited enormously because no major locational change was needed with the change from water power to newer forms of energy.

The presence of soft water is a very helpful feature for the cotton textile industry for three main reasons: soft water lathers easily, it produces a minimum of scale in boilers and in pipes, and it is essential for use with bleaches and dyes. Other physical advantages which may be mentioned include a humid atmosphere and a location well placed with regard to transport routes. The former advantage is less important today than once it was since artificial humidification of factory air can competently be achieved, but efficient transport is vital to an industry which generally needs to import its raw material from a distance and to export most of its finished products to markets more or less distant.

2. *Historical factors.* The importance of historical factors will, of course, vary from one region to another and we can here take only two examples as case studies. In the case of Lancashire an important point was that the various inventions affecting the spinning and weaving branches of the industry did not keep in step with each other, and this played an important part in affecting location. The spinning side of the industry was mechanised long before the weaving. The early inventions (*see* Chapter I, p. 10 above) were mainly concerned with improvements in the spinning pro-

cess, because weavers for long experienced great difficulties in obtaining enough yarn, especially after the invention of Kay's flying shuttle. Guest, in his *Compendious History of the Cotton Manufacture*, has stated that a weaver not infrequently had to walk three or four miles in a single morning to call on five or six spinners before he could collect enough weft to see him through the rest of the day; during the summer things were even more difficult as the spinners often left their spindles for agricultural work in the fields. The new inventions swung the balance the other way as the spinning industry established itself in the Manchester region, but hand-loom weaving still persisted to the north of Rossendale; even as late as 1840, when spinning had been completely mechanised, there were only two weaving mills in Lancashire as against 17,000 hand-looms. Separation of the spinning and weaving interests was perhaps natural under such conditions and this separation persisted when powered weaving more firmly established itself, because the weaving mills were set up mainly north of Rossendale, where skilled labour could be recruited from the hand-loom weavers of the district.

In the U.S.A. the cotton textile industry established itself in New England, not because of any great comparative physical advantages, but because the region was at that time the most advanced in development generally and because it had close access to the largest market in the country. This situation came about simply because the modern phase of the historical development of the U.S.A. began along the eastern seaboard, especially in the north-east. This example is examined at some length later in this chapter.

3. *Structural factors.* We have observed in previous pages of this chapter that the structure of the cotton textile industry tends to be cellular, and we have examined some of the reasons for this. Given this form of structure it was probably inevitable that a geographical line of division should at times be drawn between the spinning and weaving interests. A case in point is that of Japan, where the spinning firms tend to be located near the large ports which import the raw cotton, while the weaving firms are more widespread in response to the pull of the market (p. 221 below). This example shows that the sets of factors which affect the operation of the spinning and weaving interests respectively are not always coincident.

4. *Economic factors.* The first point to make in this connection is that a very wide range of locations lies open to the spinner and the weaver. We have seen that this is true on physical grounds and it is none the less true with regard to the human situation. The demand for cotton fabrics is widespread, even in the developing territories, which is another way of saying that the market is wide—almost world-wide. When we link this

circumstance with the facts that the raw material is widely available and that the capital requirements are modest, and when we recall that a domestic cotton industry is valuable because it develops local industrial skills and helps to save foreign exchange, we shall realise that there are very strong reasons for the establishing of cotton textile mills in many countries, not least in developing territories.

Labour requirements can also be met over extensive areas. The fact that the basic skills are not difficult to acquire is a great help to those countries which have little or no industrial tradition. An under-employed rural population can often constitute a source of labour, and the introduction of a textile industry into a developing territory can provide a useful entry into the broader industrial field, as the case of Japan illustrates (*see* also the case of New England, below). It can also assist in building up domestic supplies of capital as well as associated skills (*e.g.* in the building and constructional trades) which will greatly benefit the country which aspires to further industrialisation.* In a developed country the textile industry can provide openings for female labour and can therefore complement the group of heavy industries with their predominant demand for male workers* (*see* the somewhat analogous case of Newcastle, N.S.W., p. 80 above).

A further point might very well be made at this stage. The economist is quick to stress the importance of comparative costs, which arise from the simple fact that it is very rare indeed that a given situation, geographical or economic, is suitable for the development of a single form of economic activity only. Almost any region offers opportunities (hence the term "opportunity costs" which is sometimes used) for a wide range of economic activities and it is not unusual for the most efficient use of the opportunities available to be discovered only after a period of trial and error. In the present connection, therefore, we may argue that a region may be well suited to the development of a cotton textile industry, but according to the theory of comparative costs such an industry is likely to develop *and successfully maintain itself* only if it has clear advantages over possible competitor industries.

A case in point is that of the region of the lower Clyde, in Scotland. This case has been examined by Beaver, to whose work the interested reader might refer for more detailed information (see Stamp and Beaver: reference at end of chapter). Physically, with its moist climate, its water power, its soft water and its coal, the region seems just as well suited to the manufacture of cotton textiles as does Lancashire, while labour and

---

* Australia offers an interesting example of an industrialised country which has never developed a textile industry of any magnitude. This phenomenon is rare.

capital were available from quite early times. Furthermore, the American War of Independence caused great dislocation in the established interests of the region during the period 1775–80, especially in the tobacco trade, and much capital, labour and managerial ability were therefore looking for new outlets at that time. This was the great period of the early inventions in the cotton industry and it was natural that this industry should take root in the region—which it did. It is an extraordinary tribute to the suitability of the region for the iron and steel industry and for associated industries such as shipbuilding that after this initial development of textile interests the heavy industries so gained in strength that they slowly smothered the cotton industry (with certain exceptions) during the second half of last century, though it must be recalled that the Scottish cotton industry never rested upon so secure a foundation as did that of Lancashire. Lancashire derived much of its prosperity from pedestrian but secure and widely based markets for fairly coarse-quality cloths, particularly in India, China and West Africa, while the Clyde developed an interest in fine-quality goods such as the famous Paisley shawls, fancy muslins and cloths which proved very susceptible to changes of fashion; such products could not enter into the wider but low-price markets supplied by Lancashire, and they could scarcely maintain existing markets in the teeth of economic depression. The total result of these circumstances was that with the exception of certain specialities such as sewing thread the cotton industry has almost disappeared from the Clyde region.*

An interesting and very important point arises from the foregoing discussion. Since cotton textile industries can be introduced into many different areas, and since such industries are particularly suitable for introducing into developing territories, many countries have indeed made a start in industrial development by setting up textile mills. One result of this is that textile industries in old-established countries such as the United Kingdom have severely felt the pinch of foreign competition and have had to reorganise their textile industries. During the world depression of the early 1930s, for example, British textile exports diminished to about 35% of those of the prosperous period before the First World

* An interesting point related to the above discussion concerns the relations between physical and economic geography. It is well recognised that physical factors may affect economic developments but it is less often realised that economic factors may at times affect physical circumstances. It is tempting to look at a map and to argue that the presence of a navigable waterway (the Clyde) must have helped the development of industry in the region under discussion, but in fact the Clyde, originally a poor stream encumbered with sandbanks and rocky protuberances, was *made* navigable because of the need of industry for the watercourse. It has been said that the lower Clyde is as much an artificial waterway as the Suez Canal, and that, while "the Clyde made Glasgow, Glasgow made the Clyde."

War; in 1880 Britain possessed more than one-half of all the cotton spindles in operation in the world, but this amazing proportion was reduced to one-third by 1913 and to one-fifth by 1960. The situation in New England is comparable to that in England though in this case the competition arises from another region within the same country (p. 227 below).

We might further notice that since the early part of this century the world output of cotton textiles has increased by about 50% but the amount entering directly into international trade has *decreased* by 40%; the main reason for this is the increasing establishment of domestic textile industries and it means that exporters are finding it more and more difficult to find export markets for their goods.

It is perhaps understandable under these circumstances that there should be a considerable outcry from the older producers, many of whom see their livelihood threatened, and there is a strong demand in these countries for government policies designed to safeguard established textile interests by subsidy, by tariff manipulation, or by any other means. Yet it is possible to argue that in the long run this is a mistaken policy; that it would be far better to encourage the establishment of textile industries in developing territories as part of a general aid programme to poorer countries who are trying to escape from their present precarious dependence upon a small group of primary products (in some cases even upon a single such product): for examples of such cases see Jarrett, *Africa*, pp. 487–8. Developed countries such as Britain appear today to suffer from a chronic shortage of labour and it would seem that here lies at hand a possible way of alleviating this difficulty—the release of labour from an industry which today can well be carried on in other lands. Territories with considerable industrial experience could, in the unlikely event of such a programme being adopted, devote themselves to the production of goods which require advanced techniques in production in a manner comparable to that in which New England has succeeded in hauling herself out of her inter-war industrial depression (p. 131 above). The topic of what might be termed "industrial sequence" receives further attention in Chapter XII, (pp. 287–294 below).

It is sometimes argued by producers in industrialised countries that the undercutting of their goods by products made in developing territories with the help of low wage rates is unfair. In all fairness, however, we must bear in mind that though wages paid in newer territories may be low in comparison with those paid in the developed countries they are not low in comparison with wages in the developing countries themselves— indeed, they are often well above the average for those countries. Such industries are, in fact, helping these territories to climb out of the low wage-limited market rut, and that such a climb can successfully be undertaken has been triumphantly demonstrated by Japan. It is possible to argue that the developed territories may in this regard have a moral duty to the rest of the world, and that such a form of aid as was suggested in the preceding paragraph would be far more dignified and valuable than aid made available in the form of gifts, for the giving and receiving of gifts on too large a scale is apt to degrade both the donor and the recipient.

Finishing

When we turn to the finishing branches of the cotton textile industry we notice a marked change in the general pattern of location. Experience seems to suggest that the finishing trades are by no means so ready to leave established locations as are the spinning and, to a lesser degree, the weaving interests. The main reasons for this may be that the finishing processes are far more closely linked (*a*) with the market, and (*b*) with other industries, than are the earlier stages of manufacture. Fashion changes rapidly in the textile markets and finishers must be keenly aware of such changes. A linkage also exists with the clothing industry, which is strongly attracted by the market (p. 89 above), while another linkage exists with the chemical group of industries which supply many of the necessary raw materials used in the finishing processes; the attractive force of this linkage is very powerful. It is significant that despite the migration of large segments of the cotton textile industry from New England the finishing branches have moved hardly at all (p. 131 above), while a comparable case is noticeable in the United Kingdom. It is a striking fact that *imports* of cotton textiles into Britain have markedly increased during the past two decades, and one reason for this is that much "grey" cloth has been imported for final processing, after which much of it is re-exported. Clearly, we are here dealing with a branch of the industry which must take keen account of the vagaries of the market and which makes considerable demands upon skill and upon the products of other industries; the fact is that the finishing stages of the industry have their own locational requirements which do not necessarily coincide with those of the earlier stages of cotton manufacture, and it may be necessary to add a rider to the point which was made earlier regarding the possibility of countries such as the United Kingdom giving up the manufacture of cotton textiles. The rider simply is that the argument as set out earlier may well not apply with the same force to the finishing industry as to the spinning and weaving stages, for finishing firms in the developed territories have a contribution to make to world markets for a long time yet.

WORLD PRODUCTION OF COTTON TEXTILES

Figure 48 shows the main features of the world supply of raw cotton and the production of cotton yarn. The largest textile producers are the U.S.A., China, the U.S.S.R., India, Japan, Western Germany, France and the United Kingdom. The limits set by this book do not permit us to

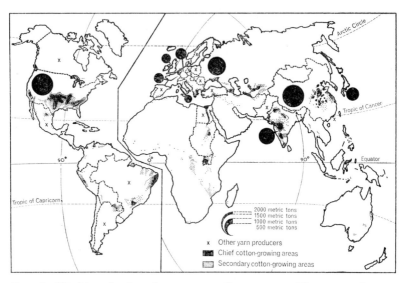

FIG. 48.—World production of raw cotton and cotton yarn. The output of cotton yarn (in thousands of metric tons) of the chief producing countries is shown by shaded circles. Other producers are indicated by a cross.

make a comprehensive study of each of these territories, and we shall limit ourselves to making a fairly detailed study of the cotton textile industry in the U.S.A., but before doing that we shall first make a few general points about the situation in the other countries mentioned.

China

Cotton spinning and weaving were traditional handicrafts in China, but the modern cotton textile industry began only in the 1890s. At the end of the Sino-Japanese War in 1894 foreigners were granted the right to establish factories in the Treaty Ports, and cotton mills were set up in Southern Manchuria, Tientsin, Tsingtao, Shanghai and Canton.

Great Britain, Germany and the U.S.A. each set up a factory in Shanghai and the importance of that city as a cotton-producing centre was therefore established.

Until some time after the First World War, however, the development of the industry was slow. The reasons for this included widespread social unrest and even warfare; competition from overseas, in the first place from Britain and later from Japan; inefficiency of labour and management; lack of equipment, both in the textile industry itself and in essential services such as transport; and the limited size of the domestic market. Under these conditions it was probably inevitable that Shanghai should increase its overall lead in the industry, partly because of its position near the mouth of the Yangtze Valley, an enormous region which carries about one-third of the total population of the country and which therefore constitutes the largest market area, partly because it was a peaceful oasis in a troubled country, partly because of the concentration in the city of labour, capital and managerial ability, and partly because it was comparatively well placed for transport facilities. At the end of the First World War Shanghai possessed 80% of the total cotton spindles of China.

Since that time the story has very largely been one of greater decentralisation of the industry as a whole and of considerable growth. The Communist régime places great emphasis on this industry and has encouraged the establishment of hundreds of small factories or workshops equipped with modern machinery which give employment to millions of textile wokers, and which make a very significant contribution to total production. It is interesting to notice the contrast here with the attempted production of pig iron in small-scale blast furnaces which proved such a failure (p. 189 above). Many large cotton mills have also been established; nearly one-half of the existing centres of production lie within the Yangtze Basin, which underlines the point made earlier regarding the importance of this region as a market area, and these centres account for about three-quarters of the total output of yarn and cloth, but there is increasing development in the cotton-growing areas of the basin of the Hwang Ho and the Great Plain of northern China. The total yarn output of the country is now little behind that of the U.SA.; most of the domestic demand (admittedly modest judging by Western standards) is met and there is an increasing export trade.

The U.S.S.R.

The story of the manufacture of cotton textiles in the U.S.S.R. began in the village of Ivanovo with the establishment of a calico-printing factory in the middle of the eighteenth century, and from that modest beginning the cotton-printing industry spread to adjoining areas in the Moscow region. The necessary fabric was imported from England. Actual weaving began a few years later, using yarn imported from England, while spinning developed in the late eighteenth and early nineteenth centuries. The central (Moscow) region became easily the leading area in Russia for the manufacture of cotton goods though the industry was also introduced into the Leningrad region. The availability of water power, labour and markets acted together to encourage this concentration of the industry within a comparatively small region, and even today four-fifths of the cotton-manufacturing centres of the U.S.S.R. lie to the west of the Urals.

In recent years output has increased sharply, and consumption of raw cotton is about two and a quarter times as great as in 1939 though it is still low by Western standards. Great efforts have been made to stimulate cotton growing and the Soviet Union now supplies all its own needs of the basic raw material, while the policy of industrial decentralisation pursued since the Second World War has resulted in the setting up of new mills nearer to the cotton-growing areas at such centres as Gori (in the Caucasus), Tashkent and Barnaul. Even so, about 80% of the mills are still located in the Moscow region, and Ivanovo remains the "Manchester" of the U.S.S.R., though there is some tendency for this area to specialise in better-quality goods as production becomes more widespread. The U.S.S.R. will not for some time be in a position to enter the competition for export markets.

India

India is now the second-largest cotton textile producer outside the Communist countries. Although the production of cotton goods goes back far into the past in India, the first powered mill was established only in 1854, and the subsequent expansion of the industry was very low, mainly because of competition from Lancashire. It was not until the First World War, when exports from Britain were cut off, that a chance really came for India to expand her production, and, while she was able in some measure to protect her growing industry in the inter-war period by fiscal controls, competition from Japan at that time proved severe. Since independence, however, the cotton textile industry has grown rapidly.

The first centre of development was Bombay, and this city has remained the leading centre of production though there has been a considerable extention of the industry in recent years. Bombay is close to the main cotton-growing area of India, and this fact conferred an early advantage. Machinery was in the early stages of development imported from England so that a port location which faced towards Suez was a helpful feature, while the city was also the centre of a transport system which provided links with the northern and southern coastlands of peninsular India and with the interior. Labour, capital and managerial skill were available in the city (the case is clearly comparable in many respects with that of Shanghai), while the humid climate conferred an advantage over drier centres, for instance Karachi. Even today about one-half of the cotton textile production of India comes from Bombay, though other producing centres include Ahmedabad, Baroda, Cawnpore, Nagpur, Delhi, Coimbatore and Madras.

Japan

The story of the dramatic rise of industry in Japan is well known, and only the outline need be sketched in here. The manufacture of cotton textiles formed the first stage in the development of the Japanese Industrial Revolution, the first cotton mill being opened in 1868. Like India, however, Japan was not able to make very big advances until the First World War, when she was able to break strongly into markets in China, India and Africa. This advance was maintained during the inter-war period and Japan became the world's leading exporter of cotton goods in 1933.

The industry suffered great damage in the Second World War, but within eight years she had rebuilt her factories, equipping them with the most modern types of machinery, and she had again become the world's leading exporter of cotton goods. Indeed, output is now such that she finds some difficulty in marketing her full production.

The location of the cotton textile industry has largely been determined by the fact that Japan grows no cotton of her own on a commercial basis and she therefore has to import all her requirements of this raw material. There is therefore, a natural tendency for cotton mills to be set up at the larger ports, the points of entry of the raw cotton, and in fact we find that most spinning mills are so located. The chief production areas are centred on Osaka, Nagoya, Tokyo and Yokohama, while mills are also operating in some smaller cities on the northern shore of the Inland Sea. It is interesting to notice, however, that weaving mills are much more widely scattered, and they are not controlled by a few large corporations as are the spinning mills. While most weaving mills are located in the same general areas which produce the yarn, many are situated in small cities and in satellite towns; this bears out the general principle which we noticed earlier, that there is no necessary congruence between the spinning and weaving interests in the cotton industry. It is interesting to note that in recent years the number of weaving firms has greatly increased, and the reason for this is a change in the nature of the demand for cotton goods. In the inter-war period the export market was content with standardised articles, and this demand could be well satisfied by large weaving firms, but in recent years the broadening of the market and the greater affluence in many export markets have led to a strong demand for a wider variety of cotton fabrics, and the small specialised weaver has so far been more successful than the large-scale producer in meeting this form of demand. Much of the success of the small-scale specialist weaver has come about through successful experiments in the combining of cotton yarn with man-made fibres.

The great success of Japan in the cotton textile markets of the world has been the subject of much study, and the main reasons for it are well known. The reasons include the almost continuous operation of machinery to secure the maximum efficiency of operation; low-cost labour, including child labour; the early introduction of machinery suitable for spinning short-staple Indian cotton, which was not in great demand elsewhere and which was therefore cheap; and the later introduction of the automatic Toyoda loom, which increased the output per worker fivefold. Powerful ancillary reasons include the fact that interest rates were low, as were building and transport costs.

Today, Japan exports about 35% of her total production of cotton fabrics, but she is now faced with the problem whether she will be able to retain her markets against producers in China and India. She has already lost her markets in China and is finding increasing difficulty in maintaining her Indian ones, while exports from China are already threatening markets elsewhere. There is a certain irony in the fact that the country which was able to capture so many of the world's markets largely because of cheap labour may be threatened because of undercutting from other territories where labour is low-cost. This illustrates a general principle, that while wage rates in any given area may be low for a time they tend in due course to establish some sort of equilibrium with corresponding wage rates elsewhere.

Western Germany

The cotton textile industry is of wide occurrence in Western Germany, but its greatest concentration is in the Rhine–Westphalian region where it forms an important part of the general industrial structure. The dominating localising factor has been the Ruhr coal, partly because the coal provides power for the mills and partly because by-products obtained from the coking ovens are treated to produce, among other things, aniline dyes; before the First World War the Ruhr was the world's leading producer of aniline dyes.* The question of linkage is clearly important here. Another important point was that the textile industry offers employment to women in a region largely devoted to heavy industry which demands a predominantly male labour force. The greatest concentrations of mills are to be found in the Wupper Valley, particularly at Barmen and Elberfeld (Wuppertal), a location that was favoured because it was slightly removed from the iron and steel centres, which are large consumers of water; the Wupper provided the only large source of water (for power and processing) not being used by other industries. A domestic market of considerable size was a powerful encouragement to the textile industry.

France

The leading cotton-manufacturing region in France lies in the northeast, Lille being the chief centre. This area, together with adjacent areas in Belgium and the Netherlands, has a long tradition of textile production. The industry has been modernised since the Second World War.

The United Kingdom

Reference has been made to the industry in the United Kindom elsewhere in this chapter, while additional information is easily available in regional texts; little further will therefore be said here. Like Germany and France, Britain had the advantage of a substantial domestic market and she enjoyed an expanding colonial market too. In recent years Lancashire, which accounts for by far the greater part of the total output, has been very severely hit by the loss of former markets because of competition especially from Japan and Hong Kong.

THE COTTON TEXTILE INDUSTRY OF THE U.S.A.

During the eighteenth century there was considerable interest in the U.S.A. regarding the possibility of establishing a cotton textile industry which would be a great asset to the newly developing country, but before 1790 little progress was made, largely because England jealously guarded the secrets of her new textile machinery. In 1789, however, Samuel Slater, who had been an apprentice of Arkwright, was persuaded

---

* It was an Englishman, W. H. Perkin, who first succeeded in 1856 in extracting a purple dye from a coal-tar derivative known as aniline, but thanks to apathy in Britain on the one hand and initiative, research and government help through tariff control in Germany on the other, the British aniline-dye industry was vastly overshadowed by the Germans until the First World War.

to emigrate to America by the promise of a bounty for the introduction of Arkwright's techniques into the country. His baggage was searched in England, as was customary, for plans or prints of any description, but Slater carried the necessary information in his head and so eluded the authorities. He built, from memory, the first cotton-spinning machinery in the U.S.A., and in 1790 established the first cotton mill in the country at Pawtucket, Rhode Island, which is located where the Blackstone River plunges over a rock ledge below which the river is navigable. After this there was rapid advance in the new industry.

There are many reasons why the cotton textile industry flourished in New England, and the following are probably the most important:

1. *Physical factors.* One of the early advantages which greatly helped the first manufacturers was the large number of small rapids and waterfalls on many streams. The earliest mills were strung out along streams where small waterfalls occurred, but later on technological developments made it possible to use larger falls such as those at Lowell (first used in 1823), Manchester (1838) and Lawrence (1848), all on the Merrimack River. Paterson (today an industrial suburb of New York) had become a leading cotton centre by the 1830s, the mills developing power from rapids along the Passaic River. After 1850, however, the general move was towards the coast, where coal could be imported cheaply by sea; Boston and the Providence–Fall River–New Bedford areas became important cotton producers for this reason. By 1883 over 40% of all power used in the cotton mills was steam power and by 1900 the figure had risen to 66%. Another important feature was the excellence of the New England soft water for various processes such as bleaching and dyeing. (For further details *see* pp. 130–131 above and Estall (reference at end of chapter).)

The relatively high humidity reduced to a minimum the snapping of threads during spinning and weaving. Under dry conditions the rubbing together of the fibres creates static electricity which causes the threads to become entangled and to snap, but humid air is a good conductor and therefore carries away the electric charge. It is true that the importance of this point has been overstressed in the past, but it is equally true that this was a useful advantage in early days before artificial humidification could be carried out in factories.

We should also make mention of the comparatively harsh environment of New England, with its severe winters and its poor soils. Under such conditions farmers' wives and daughters often undertook spinning and weaving at home to supplement the precarious livelihood wrested from the farm, and when work became available in the mills many were glad to take advantage of this. Their skilled labour proved a great asset to the textile entrepreneurs.

2. *Capital.* Capital was frequently available to textile entrepreneurs from commercial and whaling interests. The comparatively limited amounts of capital needed to establish the small mills of the time were at first secured from merchants and seamen whose activities were curtailed by the Napoleonic Wars, while at a later period the moribund whaling industry supplied additional capital.

3. *Labour.* We have seen that much labour in the early days of the industry was provided by the womenfolk from the farms. As standards of living improved, new workers entered the industry as a result of immigration, at first from Ireland and Britain (especially during the periods of depression which followed the close of the Napoleonic Wars and the famine in Ireland) and after the middle of the century from Canada and Central and Southern Europe. These immigrants, who were frequently accustomed to low standards of living in their homelands, kept wages at a low level, while there was little tendency towards the development of trade unions when the nationalities of the workers were so mixed and when labour turnover in the industry was considerable.

This high rate of labour turnover proved to be a very real difficulty to manufacturers, and it encouraged the introduction of improved machines, thereby increasing the efficiency of the industry. Jones and Darkenwald point out that the ratio of spindles to operatives in Slater's 1790 mill was 8 to 1, but by 1880 it was 62 to 1. In 1919 the ratio for Massachusetts was 107 to 1.

4. *Market.* In the early days of development New England was well placed as it provided the major market region in the U.S.A. itself and it had good contacts with such expanding centres as New York and Philadephia along the eastern seaboard.

5. *Taxes.* The early mills were located in the countryside and for a very long time, therefore, they paid no municipal taxes, while other taxation was low. Some towns even granted tax concessions or exemptions in order to attract the textile industry.

6. *Historical factors.* The development of almost every region has been affected in some measure by historical circumstances peculiar to itself, and this has certainly been the case in New England. The latter part of the eighteenth century and the early part of the nineteenth were troubled in Europe by almost constant warfare and this greatly interfered with trade and slowed up imports of textiles into America from England. This situation naturally encouraged local interest in the establishing and developing of a domestic textile industry.

Another point has reference to the opening up of the American Middle West which went ahead after the War of 1812 and which made possible the large-scale production of foodstuffs. The Mid-West began to influence developments in the north-east, particularly after the construction in 1825 of the Erie Canal (p. 86 above), which made it possible economically for agricultural produce to be sent to New England. This further increased the difficulties of New England farmers and caused many of them to give up farming and to enter the cotton mills; at the same time New England manufacturers could take advantage of the developing market in the interior (p. 16 above and Fig. 1).

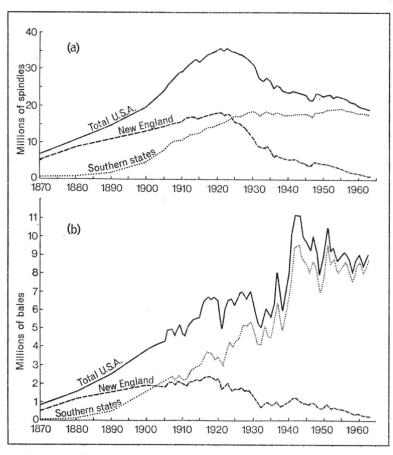

FIG. 49.—The cotton textile industry of the U.S.A.: (*a*) cotton spindles in operation; (*b*) consumption of raw cotton.
Based on Jones and Darkenwald, Boesch, and the "U.S. Statistical Abstract."

The high level of development of the New England cotton textile industry and its subsequent decline following the First World War have already been examined (Chapter VI, pp. 130–132 above). The general course of events in the U.S.A. can be traced in Fig. 49, from which we can see how New England has seriously lost ground to the southern states, both as regards the numbers of spindles in operation and as regards the amount of raw cotton consumed. Two interesting points emerge from a closer study of Fig. 49. The first is that, although New England had more spindles working than had the South until 1924, the South overtook New England in its consumption of raw cotton as early as 1903. This situation obviously reflects the greater efficiency of southern mills, an efficiency which was the result of later establishment of industry in the South with more advanced types of machines. It is a situation comparable to that existing between the United Kingdom and Japan during the inter-war period. The second point is related to the first; although the number of spindles in the U.S.A. has dropped markedly since the early 1920s there has been no comparable drop in the consumption of raw cotton—indeed, consumption has risen dramatically since the beginning of the Second World War. Greater efficiency within the industry must be held to account for this apparently paradoxical situation.

The position today is that the southern states now possess over 90% of all the spindles in operation in the U.S.A. and consume over 95% of the raw cotton used (this figure includes imported cotton). Boesch states that 93% of the cotton grown in the Cotton Belt for domestic use is processed in the South, the other 7% going to New England. The southern industry is located mainly on the Piedmont, which has more than one-half of the mills in operation, but there is a "spilling over" eastwards over the Fall Line on to the inner coastal plain and westwards into the Appalachian "ridge and furrow" province. This is shown on Fig. 50, which illustrates the present distribution of the industry in the South. The main centres of manufacture are named on the map.

As in the case of New England we can detect specific advantages in the South for the development of a cotton textile industry. They are set out below, and the reader might like to consider them in conjunction with what has already been said about the general suitability of the South for industrialisation (Chapter VI, pp. 135–136 above):

1. *Physical factors.* One big advantage lies in the power reserves that are available in the South, which lies 400 miles nearer to coal than does New England. Hydro-electric power also is widely available, and the South was first after 1900 to use electric power in the industry on a large

scale. As in the case of New England, we are dealing with an area poor in agricultural resources, or perhaps it would be truer to say that the original resources had been so ruthlessly "mined," mainly by cotton farmers, that most of the land was ruined for future production until a T.V.A. could restore some of the lost wealth. Gilman, in *Human Relations in the Industrial Southeast*, has graphically painted the picture: he describes the "hill people who came to the mills from their worn-out rented farms.

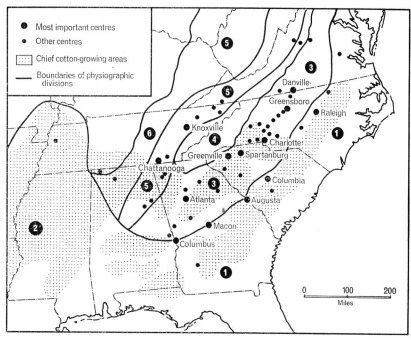

FIG. 50.—The cotton textile industry of the southern states. The most important textile centres are named. The physiographic divisions shown are: (1) the coastal plain; (2) the Mississippi lowlands; (3) the Piedmont; (4) the Blue Ridge; (5) the Appalachian ridge and furrow; (6) the interior plateaus.

They came from the spectre of five-cent cotton and from fields hungry for expensive fertiliser. They came from tables set with pellagra-breeding fare. They came from a countryside wretched and terrifying in its poverty, in the desolation of its gutted red clay hills. They came poor to the point of desperation, ragged and gaunt, uneducated, uncouth, and awkward . . . a farm people using the mills as an avenue of escape from the land that had failed them . . ." This excerpt explains in the clearest terms why the mills in the southern states have never been short of labour.

Another advantage which may be listed as a physical one is the warm climate with its mild winters, for this means that living costs are appreciably lower than in New England. The same is true of operating costs in the mills, which are able substantially to economise on fuel. This point has previously been mentioned (p. 97 above).

2. *Social factors.* The effect of the poor environment, as we have seen, was to drive the workers towards the factories which were looked upon as avenues of escape from rural poverty; they were never regarded as something of an unfortunate necessity as they were in the North, where working in the mill was regarded as a social stigma to be avoided if at all possible. There have been no union-inspired restrictions on production; many workers live in houses provided by the cotton firms for which they pay comparatively low rents. While the South was not so fortunately placed as the North with regard to supplies of capital, this difficulty was largely overcome as many workers combined to produce the necessary capital out of their own very modest means. The result was that in many cases the southern mill, unlike its northern counterpart, was in a sense a social enterprise. In other cases the necessary capital was provided by firms in the North who saw the advantages of a southern location and who set up in the South branches of their northern establishments.

3. *Wages.* Not only were wages lower in the South but working hours were longer, and there were no restrictions on hours of women's work as in the North (p. 104 above). Jones and Darkenwald point out that in 1934 the average hourly wage in the South ranged between 30·4 and 60·7 cents, while corresponding figures in the north were 33·5 and 68·6 cents. Average hourly rates in 1960 were $1·36 in the South and $1·57 in New England. The advantages of a southern location are thus heavily underlined.

4. *Taxes.* Taxes are generally lower in the southern states than they are in New England (but *see* reference to Massachusetts, p. 104 above). In many cases towns in the South were so anxious to encourage the new industry that they provided land and exempted the companies partially or wholly from taxation for specified periods. The situation was, in fact, reminiscent of that in New England in the early days of industrial enterprise.

Despite the present strong position of the U.S.A. in the manufacture of cotton textiles (as the world leader in production the country accounts for almost 25% of total world production) there is some anxiety for the future, largely because of the threat of foreign competition. The story of the United Kingdom shows how economic supremacy in one age may be swept away in the next, and Fig. 49 shows that there has been a noticeable falling away in the consumption of raw cotton since the peak years of the Second World War and the immediate post-war period. On the whole, despite fairly marked yearly irregularities, a slight downward trend seems to have been operating since 1950, though it is not yet sufficiently pronounced to give cause for great alarm. One bone of contention in the present situation is that American raw cotton is sold in foreign markets by the federal government at a price appreciably lower than that obtaining in the U.S.A. itself, and much of this cotton may be made up into cotton goods for subsequent import into the country. In a sense, therefore, the American taxpayer is subsidising foreign manufacturers, and there may well be a move to redress this grievance, possibly through the imposition of higher tariffs on incoming goods. Measures are also being taken to promote the interests of the industry through research into new forms of fibres and combinations of fibres, through the production of goods of higher and more consistent quality, and through the use of more efficient machinery and methods of production.

SUGGESTIONS FOR FURTHER READING

Boesch, H., *A Geography of World Economy*, New York, 1964.
Dury, G., *The British Isles*, London, 1961.
Estall, R. C., "Changing Industrial Patterns of New England," *Geography*, April 1961.
Jarrett, H. R., *Africa*, second edition, London, 1966.
Jones, C. F., and Darkenwald, G. G., *Economic Geography*, third edition, New York, 1965.
Manchester and its Region, British Association Handbook, 1962.
Robson, R., *The Cotton Industry in Britain*, London, 1957.
Smith, J. R., and Phillips, M. O., *North America*, New York, 1942.
Stamp, L. D., and Beaver, S. H., *The British Isles*, fifth edition, London, 1963.

Chapter X

The Manufacture of Motor Vehicles

THERE can be few stories in the whole field of industry as dramatic as that of the rise of the motor-vehicle manufacturing industry. In 1900 the industry was of minor importance only, but in little more than half a century it has developed into one of dominating stature in the major producing countries. It is also a fact of common observation that the motor vehicle has had a profound influence on the way of life of millions of people all over the world, changing personal habits of life and social patterns of behaviour, and we are perhaps only just beginning to realise that this changed pattern of social behaviour is one about which we know little and about which we need to know more. O'Connor has demonstrated, for example, that in East Africa railway construction has not in many cases stimulated economic development to anything like the degree which had been anticipated; this is especially true with respect to lines fairly recently constructed such as the Tanzanian Southern Province Railway (now closed), the Mpanda line and the extension of the Uganda railway to Kasese (Fig. 47). Now this may come as something of a shock to those many authorities who for years have been proclaiming the doctrine that railway construction is needed to "trigger off" general economic development in emergent territories, and that when railways are built they will stimulate advance in the areas through which they pass. One significant omission from this thesis which appears to emerge from O'Connor's analysis is that the contribution of the railways to economic development in the cases cited has been singularly disappointing mainly because the areas concerned are already served by road transport. It seems that a new transport revolution has quietly taken place, and important lessons have clearly yet to be learned regarding the relative contributions to economic development which can be offered by road and rail.

Figure 51 shows that the manufacture of motor vehicles* is now carried on in a small number of countries only; in fact, 72% of all vehicles

* The term "motor vehicles" is used in this chapter to include cars for private use, trucks, lorries and coaches; it does not include motor-cycles or tractors.

manufactured originate from three countries—the U.S.A. (48%), Western Germany (14%) and the United Kingdom (10%). A very high degree of concentration is obviously a marked feature of the industry, and we see here a strong contrast with the cotton textile industry which we examined in the last chapter.

Figure 52 attempts to show something of the economic and social impact made by the motor vehicle; it shows the average number of persons in selected countries per single licensed vehicle. There are clear contrasts between the first four territories (the U.S.A., Canada, France and the United Kingdom) and the other two (the U.S.S.R. and China). It seems likely that we have here another form of index which is indicative

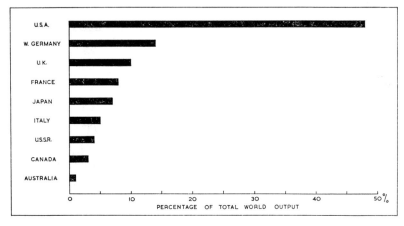

FIG. 51.—World manufacture of motor vehicles, 1963.

of the level of economic development of any given territory (the average *per capita* income for the territory is often taken as an index, while on pages 93 and 47 above it was suggested that the *per capita* consumption of water and energy respectively might provide good indicators). For reasons of space we shall concentrate in this chapter on a study of the motor-vehicle industry in two countries only—the U.S.A. and the United Kingdom.

HISTORICAL DEVELOPMENT OF THE INDUSTRY

The story of the motor vehicle may be said to date in some measure from 1860, the year in which Lenoir invented the first internal-combustion engine, though this engine was not at first used for vehicle propulsion. The fuel used in the early engines was coal gas or producer gas (hence the term "gas engine"), and the new form of power brought about great changes in manufacturing methods in many industries. Until this time the small-scale manufacturer had been at a great disadvantage in at least one respect, for the installation of a steam power plant was frequently too expensive for him, while the running costs of such a plant were almost certainly prohibitive. Since no other form of mechanical power was at that time available he had to do without mechanical power unless he could hire a plant. The new gas engine, however, was very well suited for use in small factories and workshops, and many manufacturers installed the new form of power during the last two decades of the nineteenth century.

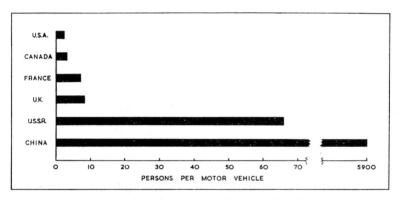

FIG. 52.—Ratio of people to motor vehicles in selected countries.

At a slightly later date came the next step—the propulsion of a road vehicle by the new engine. It is true that road vehicles had earlier been produced which were driven by steam engines, but the *Red Flag Acts* of 1831 and the development of the railways had discouraged experiment. It was in 1863 that Lenoir invented a road vehicle which was driven by a gas engine, the coal gas being ignited by an electric spark, but the problem of fuel supply was a very difficult one; the carrying of sufficient coal gas for any but the shortest journeys was not easily managed. The problem was finally solved by the adoption of a new form of fuel altogether, gasoline or petrol, a fuel which releases great energy per unit of weight. Markus used petrol as fuel for a vehicle he produced in Austria in 1865, while a great advance was made by Otto in 1876 by the compressing of the fuel immediately before ignition; this very considerably increased the explosive force in the cylinder and performance was correspondingly improved. The production of motor vehicles was well advanced by the 1890s as a result of the subsequent inventions of such pioneers as Benz, Daimler and Panhard; the English Daimler Motor Company was founded in 1896.

Subsequent advances in the manufacture of motor vehicles were associated not only with improvements to the vehicle as such but also with developments in related industries. For example, the idea of the pneumatic tyre went back to 1845 though at the time of the initial discovery there was no obvious application for it, but when Dunlop took out his patent in 1888 cycling was fast becoming an important and popular means of transport for business and pleasure. Another patent taken out shortly afterwards by Clincher made it possible to fit pneumatic tyres to motor vehicles. Significant improvements in the metal-working trades towards the end of last century and early in this made possible further advances in the motor-vehicle industry. An important point was that since the first internal combustion engines (like the first steam engines) were designed for stationary use as power plants in factories or workshops little thought was given to their overall weight. More important factors were reliability and the ability to give long service. Engines were accordingly of robust construction, but this feature, which was a virtue for the stationary engine, proved a serious handicap to the motor vehicle engine which had to be carried along by the very vehicle which it powered. Lightness in weight became a very desirable feature under these conditions, and this requirement stimulated research into the possibility of producing forms of metal which are light in weight in proportion to their strength. The production of alloy steels was therefore initiated with far-reaching effects in other industries.

Other industries which have been stimulated by the development of the motor-car include the electrical industry and the manufacture of petrol and lubricants. The manufacturer of the modern car needs a wide variety of raw materials and component parts (*e.g.* tyres, upholstery, safety belts, silencers, windscreen washers, sparking plugs and other electrical equipment) and we shall see later that this fact has had marked effects upon the structure and location of the industry. Meanwhile, we may note that an assembled motor-car may need parts from as many as 5000 different workshops and factories, many of which are normally small establishments manufacturing small-sized parts. We may also note as an indication of the key position held by the industry that in the U.S.A. motor-car manufacturers provide a market for 80% of the rubber, 63% of the leather, 54% of the malleable iron, 20% of the steel and 12% of the copper produced in the country. Miller points out that automobile firms in the U.S.A. purchase such unexpected items as ground corn cobs (for use in the manufacture of a polishing compound), pipe cleaners (for cleaning small holes), razor blades (for cutting such items as insulation components and cartons), babies' feeding-bottle nipples (for covering small parts before larger components to which they are attached pass through paint sprays) and walnut shells (for use in certain polishing operations).

It is very largely the wide selection of necessary raw materials and component parts, and the consequent linkage with other industries which accounts for the high degree of concentration of motor-vehicle manufacture in the world as a whole. Such an industry can be supported only by a highly industrialised country. We therefore begin to see why it is that the industry is so localised (Fig. 51) and we might at this point turn to a more careful examination of the geographical and economic features which have helped to mould the pattern of the industry to its present form. We shall take the examples of Europe (especially the United Kingdom) and the U.S.A. as case studies.

DEVELOPMENT OF THE INDUSTRY IN EUROPE AND THE U.S.A.

After the early experimental stage in the production of the motor vehicle came the necessity to commence production for the market, and from the earliest stages of this we see divergencies in production patterns between the two great world producers—the U.S.A. and Western Europe. In Europe the market was comparatively slow to develop, for the possession of a car was not considered important by people generally. A motor vehicle was usually bought for a specific purpose (*e.g.* by tradespeople or by professional people such as doctors who needed reliable personal transport) or as a luxury. This may have been in part a result of the fact that public transport had reached a high level of efficiency in Western Europe and in part a reflection of the simple fact that people were not accustomed in those days to anything like the amount of travel which today we take for granted. The normal way of life did not demand very much travel, while the habit of travel for pleasure had scarcely developed. Under these circumstances the market was bound to remain limited in size, and the pattern of production developed into the manufacture of a limited range of luxury-type cars. Cars were generally small in size, a probable response in a country like the United Kingdom to narrow, winding roads and lanes in country and city alike.

In the U.S.A., on the other hand, the automobile became the generally accepted vehicle for private transport at a very much earlier stage. Reasons usually given for this include the inadequacy of public transport over much of the country; the need, more pressing than in Western Europe, to travel over long distances; the comparative cheapness of petrol and lubricants in a country which possesses large supplies of native mineral oil; a more concentrated advertising campaign; the early development of car ownership as an accepted social feature; the comparatively high *per capita* income of U.S. citizens; and the readiness of producers to experiment with the new form of transport in a country which possesses considerable talent and initiative, as well as large supplies of raw materials.

The need to develop an industry which could cater for a widely expanding market was felt, therefore, at an early stage in the U.S.A., and it is no accident that it was in that country (at Detroit) that Henry Ford first introduced his ideas of mass production based on the moving assembly line. This was in 1908.* In the early days of the industry it was by no means generally clear along which line the industry was likely to develop, and vehicles propelled by steam and electricity competed with petrol-driven cars. This uncertainty led to one rather unexpected result. It is well known that New England was the first part of the U.S.A. to become industrialised, yet despite this strong advantage very little manufacture of motor vehicles is carried on there today. The reason was given on page 133 above.

The rise of the industry in the U.S.A. in the years following 1908 was remarkable. From a total output of about 65,000 vehicles in that year production rose to almost 2 million in 1917, to nearly $2\frac{1}{4}$ million in 1920, and to an inter-war peak of just over $5\frac{1}{4}$ million in 1929 (Rae). Perhaps the most important result of this activity was to promote the development of large-scale manufacturing units. In the early days of limited production the manufacture of motor-cars was undertaken almost as a sideline by coach- and carriage-building firms which experimented with the fitting of internal-combustion engines on to coach bodies to provide a self-propelling vehicle. Under these conditions it is not surprising that many small firms developed an interest in the industry. Today, however, the manufacture of a motor vehicle is a specialised and expensive business and very large capital investment is required, especially when production

* The principle of the assembly line was understood well before 1908, and it was first used by Eli Whitney, who started to manufacture standard components in 1798 following an order for guns from the American army. Later on the technique was employed in connection with the production of railroad box cars during the 1880s; the "assembly line" was in this case simply a length of railroad track along which the box cars were slowly drawn as the various operations were completed.

is organised according to the assembly line type of production. This limits severely the number of newcomers into the industry and it helps to explain why the number of manufacturers in the U.S.A. has declined sharply though the overall output of vehicles has notably increased. The number of manufacturing firms declined from 300 in 1914 to 108 in 1923 and to 44 in 1927, and of these 44 three (General Motors, Ford and Chrysler) manufactured about three-quarters of the total vehicle output. The number of firms was further reduced by the Depression of the early 1930s, smaller producers suffering more than the larger ones which were able to weather the economic storm more successfully, and by 1939 the "Big Three" were producing 90% of the total output of vehicles (Rae). The present structure of the industry was virtually established.

The general situation in the motor-car industry of the United Kingdom is comparable with that of the U.S.A., as regards both its historical development and its present structure. Until the First World War almost 200 different makes of car were available on the British market but by 1922 this number had dwindled to 96, and by 1939 to 33. Immediately before the Second World War three firms were responsible for two-thirds of the total production of vehicles, and in 1966 output was overwhelmingly in the hands of the "Big Five" (B.M.C., Ford, Vauxhall, Standard and Rootes) who between them controlled about 95% of the total output of motor vehicles, The first two firms named were responsible for about two-thirds of total production. Since that time there have been further mergers (B.M.C.–Jaguar in 1966 and B.M.C.–Leyland in 1968), and it is now probably more in accordance with the facts of production to talk of a "Big Two": British Leyland and Ford. Clearly, the trend evident in the U.S.A. towards domination of the market by a very few firms (oligopoly) is also being followed in Britain.

FUNCTIONAL STRUCTURE OF THE INDUSTRY

It is as true in the motor-car manufacturing industry as in other large-scale industries that very considerable economies of production can be achieved. The manufacture of motor-car bodies provides a case in point, for such bodies can be produced according to a single basic pattern far more cheaply than completely differently styled bodies can be turned out. It has been pointed out that dies for producing five variations on a single design for the former British Motor Corporation cost the manufacturers £$1\frac{1}{2}$ million, but if five separate companies each had produced a single type of car the five separate dies necessary would have cost £5 million.

It is also true that when a single large firm is producing different types of vehicles in great numbers very substantial economies can be achieved by standardising component parts used in the various engines, in other moving parts of the vehicle, and in different accessories; the result is that in fact a wide range of interchangeable parts is used in cars which are different in appearance. Even at the assembly stage of production there is an advantage in scale of output, for a plant designed to assemble, say, 5000 vehicles per week may not cost a great deal more than one designed to assemble half that number.

While it is generally true that motor-car manufacturers normally themselves undertake production of intricate and expensive items such as the chassis, body, engine and transmission, it is also true that it is common for smaller components to be produced by subsidiary companies or by independent firms. Thus we find that such items as tyres, sparking plugs and magnetos are normally bought by the motor-car firms from outside. The manufacture of these parts is a specialised and skilled job in itself and there are clear advantages of scale possible when a company can produce the parts in large quantities for a number of car firms rather than that each firm should itself manufacture them on a considerably smaller scale.

Two other factors which affect the economy of motor-vehicle production may be mentioned at this point. Firstly, it is sometimes argued that for real success in the motor-car market a manufacturer needs to produce a wide range of vehicles which can between them cater for all types of demand, and it may be the fact that the large firms can do this that helps to account for their success. Nevertheless, the success of the Volkswagen in Germany and the Rambler in the U.S.A. shows that this is not always a vital factor. In the second place, a factor which may be more important than is often realised is connected with the market for second-hand cars. With the present-day pattern of production whereby manufacturers produce a new type of car each year, each allegedly a vast improvement on its predecessor, there is tremendous pressure on car owners to "trade in" their cars regularly for newer models; indeed, it is only the "trading-in" system which makes possible the constant output of new models. The economics of trade-in prices could prove an interesting field of study in itself; such prices may be in part a reflection of public taste but it may be that they are set by the manufacturing firms working through their distributors to their own advantage. Whatever the facts of the case, however, it is frequently the case that trade-in valuations are higher on the models of the large firms and this acts as a powerful urge to purchasers of new cars to keep to these models.

The transporting of an assembled vehicle over considerable distances is an expensive affair for reasons which we have touched upon earlier (p. 90 above), and considerable savings can be made if the vehicle can be moved before assembly. Hurley has pointed out that a single railway wagon which can take four completed cars can take unassembled parts closely packed for twelve cars, while the obvious advantage secured in this way is strengthened by the normal practice in the U.S.A. of charging the purchaser for the cost of transport of the assembled car. It is not difficult under these circumstances to understand why new assembly plants have been established in the U.S.A. in areas such as Georgia, Texas and New Jersey which are far removed from the manufacturing centres; Detroit and southern Michigan between them now assemble only about one-third of the total output of vehicles in the country.

An interesting point with regard to the automobile industry in the U.S.A. is the limited export market which the industry commands; an export total of about 400,000 cars in 1937 had dropped to barely 180,000 in 1960. There may be three main reasons for this:

1. The American car is normally a large and elaborate affair with many unnecessary "frills" and an engine far too powerful for ordinary use. Such a car is expensive to buy and to run and is at an immediate disadvantage in an export market. It is significant in this connection that competition even in America itself from compact, comparatively low-cost European cars has caused U.S. manufacturers considerable concern and has stimulated domestic output of this type of automobile.

2. After the close of the Second World War the world-wide shortage of dollars acted as a powerful brake on American exports, including exports of cars.

3. Competition from other sources, especially the Volkswagen, has been severe. The Volkswagen has gained a strong reputation for durability, toughness, reliability and low running costs, all features which are strongly attractive—especially in territories in which servicing facilities are limited. The after-sales service offered by Volkswagen has deservedly proved to be a powerful draw and puts to shame the desultory service which all too often is all that the owner of other types of medium-priced car can expect.

The place of car exports in the economy of American automobile production has largely been taken by heavy investment in subsidiary and associated companies in other parts of the world, notably in Canada, Western Europe and Australia. Profits earned overseas in this way form a substantial part of the the total income of firms such as General Motors and Ford.

LOCATIONAL FACTORS

We should at this point turn our attention to the actual location of producing firms, taking the U.S.A. and the United Kingdom as case studies. The point has already been made that one region in the U.S.A. which we might naturally expect to be a major producer, New England, fell by the wayside at an early date because of an unfortunate concentration on lines of production which proved failures* (p. 133 above). In 1900, of the 4000 cars which were manufactured in the U.S.A., more than half were produced in New England, most of these being electric or steam-driven vehicles, but by 1910 production in New England had virtually ceased and the main manufacturing area was the Mid-West. Detroit, in particular, has developed into a leading centre, while other important centres include Cleveland, Flint, Buffalo, Chicago, Milwaukee and Toledo.

The pronounced concentration of the manufacturing side of the industry (as opposed to the assembly of vehicles, p. 239 above) in the Mid-West inevitably suggests that there are strong reasons which have encouraged such localisation, and we are naturally led to enquire into this. The following may be given as the main points at issue:

1. The early existence of a developing market, based in the early stages on the needs of an expanding agricultural region for appliances and equipment of all kinds, was a powerful factor. The demand for motor vehicles was simply one aspect of a wide demand in the Mid-West for industrial products from a population which was rapidly increasing in size and constantly raising its standard of living.

2. From the topographical and locational standpoints the Mid-West has much in its favour, for no part of the region offers real difficulties of terrain to the construction of roads, while the general absence of steep gradients along roads proved to be an important advantage for the early cars which possessed little power (judging by present-day standards). The Erie Canal (p. 86 above) provided an early link with the eastern seaboard while the Great Lakes themselves have played a vital role, permitting

---

* Failures, that is, in the economic sense. Mechanically, the quietly moving, smoothly running, electric and steam vehicles had much in their favour. One New England firm, in fact, continued the manufacture of steam vehicles until quite recently and was renowned for the quality of its product. The author has been told that the owner of the producing firm would not consent to make a vehicle for a prospective client whom he did not like!

easy assembling of raw materials and distribution of manufactured products, whether completed cars or component parts.

3. The line of cities which extends from Milwaukee and Chicago eastwards through Detroit and other Lake Erie towns to Buffalo is located along the transition zone of hardwood and coniferous forests which lies between the northern coniferous forests and the rich agricultural region of the Mid-West. Wood working of all kinds developed in this zone, from furniture making to wagon production in such centres as Detroit, Flint, Lancing and Pontiac. The hardwood forests provided raw materials for the construction of the early horse-drawn carriages, and the first stage in the production of motor-cars consisted simply in the fitting of an engine in these vehicles to replace the horse, so giving rise to the "horseless carriage." At a later date other primary raw materials such as the iron ore of Lake Superior and energy requirements such as coal from Pennsylvania and West Virginia were easily accessible and have provided a sound basis for industry in general. Conditions were therefore as favourable as they could be for the establishment of trades which could supply necessary parts to the motor-vehicle industry; questions of linkage rapidly came to the fore.

4. Many small metal-working firms were, in fact, established to meet the growing demands of an agricultural community, especially for the production of agricultural implements and machines and to serve the growing needs of transport; cities like Cleveland and Detroit were already producing such essential components as springs, brass items, rubber tyres, paints and varnish, as well as pressed steel and malleable iron, while a marine-engine industry had developed near Lake Erie because of the importance of powered boats on the lake. A tradition of inventiveness and a facility in the metal-working trades were therefore both available to the new industry.

5. There was never any difficulty in securing necessary supplies of labour (which can be very largely unskilled) while capital formation went steadily ahead in a progressive agricultural and industrial society.

6. The initiative of people like Henry Ford played an important part in the early days of the industry, and this perhaps more than any other factor was the reason why Detroit drew ahead of possible competitors such as Cleveland. As a result of Ford's experiments with mechanical inventions and innovations of technique, he was able by 1922 to reduce the price of the famous Model T to less than one-third of what it was in 1909 ($295 instead of $950). Ford's imaginative ideas led to the marrying of the large-scale market with large-scale production, and the later story of the development of the industry shows how successful the marriage proved to be.

7. Not only is the Mid-West well situated with respect to the whole American market, but the region itself needs transport on an unusually large scale. This comes about because of its enormous size and the way in which the population is distributed over the entire area. Distances to be travelled are great, far greater than anything known along the eastern seaboard except in a north–south direction, and people early felt the need of some form of transport which could get them over long distances easily and speedily. It is not surprising that this has become the land of the large and speedy car; this is a natural response to environment.

8. Thanks to the early start enjoyed by the Mid-West and to the eclipse of New England, a region which could have been a strong competitor, manufacturing firms such as Fords and General Motors were able at a comparatively early date to develop substantial economies of scale which put them in a very favourable position as against possible competitors elsewhere. It it very difficult for newcomers successfully to challenge the supremacy of well-established large-scale firms in an industry which naturally favours the large production unit.

Conditions in the United Kingdom are in many ways similar to those in the U.S.A. though the scale of operations is much smaller. Allen has shown that the early history of the motor-vehicle industry in the United Kingdom shows the working in a modified form of the same forces which helped to mould the American industry, and he comes to the conclusion that in Britain the industry became concentrated in the area "where American conditions found their closest parallel." This area is that generally known as the West Midlands, with its focal points of Birmingham and Coventry; at a later date the industry spread towards the South-east, particularly to Oxford, Luton and Dagenham. The following are the main considerations to bear in mind:

1. In the West Midlands as in the Mid-West there were traditionally strong links between agriculture, industry and transport. It was the demand of farmers and travellers which helped to account for the early interest shown by Birmingham in the manufacture of such metal goods as nails, iron tyres for fitting on to wooden wheels, springs and agricultural implements of all kinds, while the specialisation of Walsall on saddlery is well known. Similar considerations encouraged the early interest of Coventry in the manufacture of bicycles,* an industry which spread

* The city of Coventry owes a great deal to James Starley, who settled there in 1861. With the financial backing of a rich employer he began the production of a new and improved form of sewing machine which he had devised. In the later 1860s he turned to the manufacture of bicycles which he improved with the addition of such features as the rear-hub step, brakes of superior performance, and a comfort-

to Birmingham and Wolverhampton. Many bicycle firms interested themselves in the manufacture of motor vehicles as an extension of their original activities.

2. The diversified nature of the metal-working industries proved to be a most helpful factor. Allen mentions the small independent producers of brass goods, screws, nuts and bolts, pressed steel, tubes, general iron goods, paints, varnishes and leather goods, while other workshops busied themselves with the manufacture of springs, with plating and with other crafts. These activities were of the greatest possible assistance to the newly developing industry.

3. The region is well placed with reference to its raw material supplies. As an industrial region it produces much of its own requirements as we have just noted, while such a vital commodity as pressed steel is manufactured at Margam, Ebbw Vale (see p. 292 below) and Shotton (Fig. 43) which are all easily accessible from the West Midlands and the South-east. The industry is therefore just as soundly based as is its American counterpart from this point of view.

4. The largest market for the completed vehicles is the agricultural and industrial region which has been described as "coffin-shaped" in form and which lies between Lancashire to the north-west and the London area to the south-east; thus the industry can be said to be market orientated. We have by implication stressed the considerable advantages of this fact by the reference to the unhappy manufacturer from the north of England (p. 110 above).

5. As in the case of the U.S.A., the British industry owes much to the inventive and organisational genius of outstanding personalities; the selection of Cowley, near Oxford, as a manufacturing centre is inseparably linked with the young Morris, later Lord Nuffield, who has done for Oxford what Ford did for Detroit.

6. Since the region under review includes some of the most densely populated parts of the United Kingdom we shall not expect labour supply to be a difficulty, and neither has it been. The skilled labour available in the West Midlands has always been an important locational feature, but an interesting point has been raised by Beeley. This is that the skilled worker does not take kindly to the monotonous type of work required at the assembly bench, and the argument develops that the location of motor-vehicle industries at Oxford, Luton and Dagenham meant that they were able to take advantage of unskilled labour from surrounding

---

able leather saddle of the modern type. In 1870 he produced, together with William Hillman (whose name is still remembered in the motor-car of that name), the first lightweight all-metal bicycle, which was called the "Ariel."

agricultural areas which was at the same time cheaper than skilled labour and who also were willing to undertake the dull semi-skilled work of assembly (the case is in some respects similar to that already mentioned regarding Banbury, p. 80 above). At the same time, the industries of these towns were not so far removed that they were out of easy reach of the supplies of component parts from the Midlands. In some measure it is perhaps not unreasonable to see in the cases of Luton and Oxford a parallel with the decentralising of assembly plants in the U.S.A., an observation which is largely borne out by Miller's contention that Oxford is "fundamentally an assembly centre," producing vehicles from component parts many of which have been manufactured in the West Midlands.

7. Another pertinent point is that the Midlands and south-east have good access to ports, and this is an important feature in an industry which exports a significant proportion of its total output. London and Liverpool are easily the leading ports of the country, and it is significant in this regard that the vehicle industry has in recent years developed in Lancashire, largely as a result of government policy of denying (in some cases) producers' applications to extend their present locations. Location in Lancashire is undoubtedly favoured by proximity to Liverpool and also to raw materials and component parts.

8. The special case of Dagenham should briefly be mentioned. This manufacturing centre was established by Fords at a site on the north bank of the Thames, and in some measure it represents a new departure for Britain as it takes the form of a fully integrated production unit, with its own blast furnace, steel works and vehicle-manufacturing plant. The aim has been to utilise as far as possible the economies of large-scale vertical integration. From the physical standpoint the site was chosen because it was originally an expanse of low-value riverside marsh which also benefited from a tide-water setting and a situation within easy reach of the important market region previously designated.

FINAL CONSIDERATIONS

An earlier mention of government control reminds us that no government can ignore the economic and social implications of an industry so large in its scale of operations and which has so pronounced an effect upon associated industries. In Britain, for example, this industry consumes about 20% of the domestic output of sheet steel while it is the largest single user of machine tools in the country (this point was emphasised in the case of the U.S.A. on p. 234 above). An interesting case showing how the motor-vehicle industry has stimulated the growth of a linked industry is that of the Pressed Steel-Fisher Company of Oxford, a company which produces car bodies for the Morris works near by. From this activity the company branched out into the production of refrigerators since the processes required to stamp out refrigerator cabinets is essentially similar to those needed in the stamping out of car bodies. The manufacture of refrigerators is now a major industry in its own right.

The main emphasis of government interest today in England is being laid on attempts to locate new works in areas such as South Wales, Lancashire or Scotland which need industrial infusions to revitalise their economies, and in the case of Lancashire some success has been achieved. It has to be recognised, however, that such enforced location can impose considerable handicaps upon the firms concerned.

The big difference between the American and the British motor-vehicle industries (apart from that of size) was the fact that the British industry developed at a later date; something about this has already been mentioned. Enough has probably been said, however, to show why the West Midlands emerged at an early stage as the leading producer of motor vehicles in the United Kingdom and why the region has maintained its leadership. It is significant that later developments have underlined the general suitability of the region though manufacture and assembly have extended over a wider area, but the advantages enjoyed in the West Midlands are also felt in such centres as Oxford, Luton and Dagenham.

SUGGESTIONS FOR FURTHER READING

Allen, G. C., *British Industries and their Organisation*, third edition, London, 1952.
Beeley, M., "Changing Locational Advantages in the British Motor-car Industry," *Journal of Industrial Economics*, 1957–58.
Boas, C. W., "Locational Patterns of American Automobile Assembly Plants," *Economic Geography*, July 1961.
Fryer, D. W., "British Vehicle and Aircraft Industry," *Geography*, 33, 1948.
Hurley, N. P., "The Automotive Industry: A Study in Location," *Land Economics*, February 1959.
Miller, E. W., *A Geography of Manufacturing*, Englewood Cliffs, New Jersey, 1962.
O'Connor, A. M., "New Railway Construction and the Pattern of Economic Development in East Africa," *Trans. Inst. Brit. Geog.*, 36, June 1965.
Rae, J. B., *American Automobile Manufacturers*, New York, 1959.
Scott, P., "The Australian Motor-vehicle Industry," *Geography*, July 1959.

Chapter XI

The Oil-refining Industry

ONE of the most dramatic features of the economic geography of our time has been the enormous increase in the consumption of mineral-oil products. No one can stand near a busy highway in country or in city and fail to be impressed by the numbers of vehicles passing to and fro—and almost every one of them needs some form of petroleum fuel and lubricants. Something has been said in Chapter III of the story of the production of mineral oil and the table on page 74 above gives some idea of the present overall pattern of production of this vital commodity. In this chapter we are concerned with the industry which takes the crude mineral oil and produces from it many commodities for which there is an increasing demand in the modern world; the more important of these commodities appear in the table on page 76 above.

The first point to emphasise is the tremendous scale upon which mineral oil is today produced and refined. Seven major oil companies conduct exploration for new reserves of oil and also engage in the production, transport, refining and marketing of mineral oil and the various products obtained from it; the combined gross receipts of these companies noticeably exceed the entire national incomes of many independent territories—including some which have attained a fairly high level of economic development. Petroleum products form the most important single group of commodities now entering world trade, accounting, according to Fryer, for about 10% of the total value of this trade. The fleet of oil tankers which carries mineral oil and oil products over the seas of the globe accounts for more than one-third of the entire shipping tonnage of the world. (*See* footnote★ opposite).

DEVELOPMENT OF THE INDUSTRY

Any economic activity which shows evidence of such vigorous development must be one of major world interest and one which deserves careful study. It may first of all be helpful to remind ourselves of the main stages by which the production and refining of mineral oil have reached such gigantic proportions. The following are important points to bear in mind.

Historical Development

The existence of mineral oil has been recognised for a very long time; as long ago as 1581 (at about the time when Philip of Spain was assembling his Armada to invade England) an Italian wrote of mountains in Romania which "carry a sort of fuel oil which, if purified, yields on the one hand pitch, and on the other black wax with very good burning qualities." In the eighteenth century small-scale production of crude oil was carried on near Ploești by means of hand-dug wells, and the crude oil was used by the peasants of the area to grease their cartwheels. Similarly, in Burma, oil has been produced from hand-dug wells near Yenangyaung for at least two centuries. This method of production was slow and costly and the oil was used chiefly for medicinal purposes for which "Burmah oil" was in considerable demand. Actual amounts produced, however, were very small.

In 1859 came Colonel Drake's achievement of the boring of an oil well by cable drilling at Titusville, Pennsylvania (p. 72 above).† By

---

* The total world tanker fleet at the end of 1964 and 1966 was made up as follows:

	1964 d.w.t.	1966	1964 %age of World Total	1966
Liberia	15,340,572	21,203,745	18·80	21·84
Norway	12,185,345	14,830,721	14·93	15·27
U.K.	11,749,906	12,348,383	14·40	12·72
U.S.A.	8,714,766	8,450,837	10·68	8·70
Others	33,628,010	40,271,873	41·19	41·47
TOTAL	81,618,599	97,105,559	100·00	100·00

Source: Petroleum Information Bureau.

The United Kingdom dropped from first place in 1962 to third in 1964; in 1962 the rankings were: United Kingdom (15·87% of total), Liberia (15·38%), Norway (14·38%) and the U.S.A. (12·78%). The total world d.w.t. was 70,352,697. It is worth noting that most of the large Liberian registration comprises, in fact, U.S. interests which use this "flag of convenience."

† It is interesting to notice the dramatic effect which this apparently modest discovery is alleged to have had upon the course of the political geography of North America. When the southern states defeated the northern at Bull Run, the first engagement of the Civil War, the southerners hoped for quick final success since they controlled the cotton trade, then America's most lucrative activity. The argument was that the North would be forced to capitulate simply because of lack of money. They reckoned, however, without the growing economic strength of the North, and one feature of this increasing power was the infant petroleum industry; between July 1862 and June 1864, the federal government secured no less than $7 million in duties levied on the production of crude oil and finished products, in addition to large sums received through the channels of ordinary income tax. A prominent British banker of the time, Sir Morton Peto, commented upon the value of this new

1900 the rotary-drilling method was coming into use, and this made possible more efficient drilling and also drilling to greater depths. It then became possible to produce mineral oil on a very large scale with the help of the new equipment, and the stage was set for increased output when demand warranted it.

Sea Transport

Before mineral oil could be used on a global scale some means had to be devised for transporting it over considerable distances, and this was in due course achieved by the development of the tanker on water and the pipeline on land. It is a far cry from the *Elizabeth Watts* and the *Gluckauf* (p. 72 above) to the tankers and giant supertankers of today, but the story is one of a steady increase in the size of the special vessels designed to carry mineral oil. The *Gluckauf*, the vessel which might be called the prototype of the modern tanker, could carry about 3000 tons of oil, but by the beginning of the second World War tankers of up to 12,000 d.w.t. were in use. After the war, the size of tankers increased sharply; in 1961 Japan built one of 131,000 d.w.t., while in mid 1966 a 150,000-tonner was launched and one of 250,000 tons quickly followed (p. 73 above). Orders soon followed for even larger vessels, and one of 312,000 d.w.t. was launched in Japan in 1968. Studies are now in progress regarding the feasibility of building tankers of 500,000 tons; it seems certain that the maximum size for this type of vessel is by no means finally determined. The following table gives some idea of the great change which has come over the tanker fleets of the world since the close of the Second World War.

Sizes of tankers expressed as a %age of total world tanker fleet

	Under 17,000 d.w.t.	17,000– 25,000 d.w.t.	25,000– 50,000 d.w.t.	50,000– 100,000 d.w.t.	Over 100,000 d.w.t.
1951	80	13	7	0	0
1956	52	26	22	0	0
1961	24	26	43	4·7	0·3
1965	16	19	38	25	2

Source: Petroleum Information Bureau

product (which also provided a valuable export and became an earner of foreign currency) to the North, and wrote that "it is difficult to find a parallel to such a blessing bestowed upon a nation in the hour of her direct necessity," while the Titusville *Morning Herald* went so far as to assert that "petroleum alone enabled this country to carry on successfully and terminate our Civil War." While allowance may be made for local pride, it may fairly be conceded that petroleum did indeed exercise a marked effect on the course of the Civil War though the importance of this effect is often overlooked.

In 1966 no less than one-third of all orders booked were for tankers of over 100,000 tons.

It is worth while saying a little more about this remarkable change in tanker policy, and it seems reasonable to argue that there are two main reasons for the sudden upsurge of interest in giant supertankers. These are (*a*) political, and (*b*) economic.

The political reason stems from the closure after the short 1967 Arab-Israeli War of the Suez Canal, through which passed enormous amounts of mineral oil between the Middle East and Western Europe; before the closure four out of every five vessels using the Canal were oil tankers. Middle East oil is of very great world importance, for probably two-thirds of the world's total reserves are located in that region, which in 1967 exported 460 million tons of crude oil (this represented 43% of all world exports for the year). The region supplied almost a half of the total consumption of Europe (530 million tons in all) and almost all of this would in normal times have passed through Suez (some would have been piped directly to eastern Mediterranean ports). Before the closure the Canal was Egypt's second largest earner of foreign currency (cotton was the first), and the reason for this is quickly apparent when we remember that the sea distance between the Persian Gulf and London via the Cape of Good Hope is 11,300 miles, but via Suez it is only 6500 miles.

The importance of the Suez route therefore needs little emphasising, and the implications of the closure are clear enough. The question facing the oil companies after June 1967 was the basically simple one of how to move crude oil to Europe from the Middle East at the lowest possible cost without using Suez. Various expedients were perforce tried at first (chartering tankers, which sent the cost of chartering rocketing upwards, and the imposition of a surtax in Britain to account for the additional costs involved), but it was not long before the companies began to realise that considerable economies could be gained by using giant supertankers. These economies arise through both constructional and operating costs. The following table shows how unit constructional costs diminish generally as the sizes of tankers increase:

Year of construction	Tanker capacity (d.w.t.)	Approx. constructional price per tanker ton
1956	30,000	£52
1962	80,000	£41
1966	120,000	£41
1969	300,000	£28

Sources: *The Economist*, 2 March 1968; *The Times Business Supplement*, 31 July 1968.

Economy in constructional cost develops because it is not necessary to duplicate or triplicate* many of the ship's most expensive items such as electronic and navigation equipment, controls, crews' quarters and engines (though larger engines will be needed than in a smaller vessel). Economy in operating costs arises from increasing mechanisation and automation and from the larger amounts of cargo carried per voyage. An early type of 60,000-tonner would have needed a crew of fifty, but new leviathans of 100,000 d.w.t. and above need a complement of only about thirty-five men, while the larger vessel uses little more fuel to carry about three times the amount of cargo at almost the same speed (probably 16 knots instead of 17). It therefore becomes possible to move crude oil between the Middle East and Western Europe more cheaply via the Cape in supertankers than via Suez in conventional vessels; a 250,000-tonner making the longer trip both ways will move its cargo of oil at a cost of £1 5s. 0d. per ton, while a 75,000-tonner using Suez will run up a cost of £1 1s. 0d. per ton.

Under these circumstances it is not surprising that almost the whole emphasis today is on supertankers, but this emphasis must have serious implications for the future of the Suez Canal. A number of factors have to be considered in this respect. While it seems likely, for example, that in a year or two's time about half the ships of the world's tanker fleet will be too large to carry oil through the Canal, many of them (those of 210,000 tons or less) will be able to make the return journey via Suez in ballast. And, while it is possible that the Egyptian authorities might decide to deepen the waterway, this must be an expensive undertaking, and the power to raise compensatory tolls is severely limited because of the existence of the alternative route around the Cape. Another possibility is that of complementing the Canal with a 207-mile-long pipeline (the feasibility of constructing one between Suez and the Mediterranean is now being studied); while the political instability of the region could weigh against this, plans are in hand for a 42-inch pipeline with storage tanks and pumping stations, and it is hoped that the scheme will be in operation by 1970. It is anticipated that when working at full capacity, the pipeline will be able to handle about half the 200 million tons of crude oil, which on the average passed annually through the Suez Canal before the war of 1967. At the present time, all that can be said is that the future of the Suez Canal must be regarded as extremely uncertain. Further consideration along these lines would take us outside our terms of reference, but the situation is a pertinent reminder of the intimate links which so often exist

* To haul 52 million tons of crude oil in a year needs, for example, three 67,000-ton tankers or one of 190,000 tons.

between industry and trade on the one hand, and social and political geography on the other.

Land Transport

It soon became clear after the Titusville discovery that the transport of mineral oil overland in barrels was far too cumbersome and costly a business, and that some alternative method of distribution was urgently needed. Such a method was developed in the pipeline, a method first tried out in Pennsylvania about 1885 when wooden pipes were used; these, however, were soon superseded by pipes of iron and steel. Today, "trunk" pipelines consist of steel pipes of between 12 and 30 inches in diameter, and the oil is forced along them by pumping stations which are spaced at intervals of between 12 and 50 miles (at even longer distances under favourable circumstances) depending upon topography, climate and the type of oil being pumped. Pipes of smaller diameter are used for tributary and distribution lines. It is true that if it is a matter of despatching less than one million tons of oil a year it may be cheaper to make use of tanker transport, over inland waterways if necessary, but for amounts in excess of one million tons, the pipeline is at present more efficient and comparative advantages increase markedly as increased amounts are pumped.* Thus, for instance, while there is little difference in transport costs as between water and pipeline when one million tons a year are despatched, when the amount rises to three million tons the unit cost by pipeline is only half that by inland water.†

The reason for this increasing rate of economy arises largely because the capacity of a pipe may be increased nearly sixfold as the diameter is doubled; this stems partly from the fact that doubling the diameter more than doubles the area of cross-section, and partly from the fact that frictional drag is proportionately far less in a pipe of large diameter than in one of small bore.‡ It has been found that plastic pipes are more efficient than metal ones since they induce less friction and since they do not corrode and so are virtually indestructible from natural causes. Pipelines are normally very efficient in operation, for losses in transit are negligible, while bad weather creates no operational problems; neither must a pipeline spend time loading and unloading in port and incurring heavy port dues. At the same time, however, pipelines do carry certain

* But see p. 259 below.
† A more detailed examination of the comparative costs of transporting oil by water and by pipeline is included in G. Manners, *The Pipeline Revolution*. See reference at end of chapter.
‡ The actual increase in throughput is dependent also upon other factors such as the nature of the oil concerned, prevailing temperatures and the pressure of the pumping.

disadvantages. The installation of a pipeline, for instance, may well be rendered uneconomic if the pipe is not kept working to capacity; if, to take a possible example, the pipe works at only half its capacity the cost per unit of transported oil may increase by as much as 80%. Another disadvantage of the pipeline is that it is not flexible in operation. It can operate only along a predetermined route and there is no possibility of changing the route as can be done when tankers are employed. Pipelines are also notoriously subject to attack by dissidents in areas of political instability, and can be put out of action with comparative ease.

Present Developments

There are signs today that we may be on the fringe of a revolution in oil transport, a revolution heralded by the development of the supertanker referred to above. In 1966 came the announcement that Gulf Oil were to establish on Whiddy Island, in Bantry Bay, a giant oil terminal and storage depot which will be fed by 300,000 d.w.t. tankers to be constructed in Japan; the first of these tankers (312,000 d.w.t.) was launched early in 1968 and is now in use. The oil is distributed from the terminal by "conventional" tankers of between 50,000 and 80,000 d.w.t. to Gulf Oil refineries, such as those at Milford Haven, Europort, Huelva, in Denmark, and possibly even to the U.S.A. The terminal began operating late in 1968, and it has storage capacity for one million tons of crude oil.

It is interesting to note the reasons why Bantry Bay was chosen for the site of this venture, and the following are the chief points. The number of harbours in North-West Europe capable of accommodating 300,000 d.w.t. tankers is very limited. These will be the largest vessels in the world (though not quite as broad in the beam as the largest American aircraft carriers) with a draught of 75 feet; they will need a distance of between 5 and 10 miles (according to circumstances) in which to stop and a turning circle of one mile on full rudder. Possible locations for the terminal included certain Norwegian fjords (which are subject to ice hazards) and some Scottish sea lochs (which are rather remote), while Milford Haven and south-coast anchorages in the English Channel were ruled out, partly because the tankers would have to lie off-shore during discharge and partly because it was thought better to keep the new tankers away from the busy shipping routes of the English Channel and the Irish Sea. Bantry Bay has the advantage of facing the Atlantic so that the tankers can avoid in-shore shipping lanes, while it offers an 18-mile reach of water 90 feet deep within the shelter of the bay. Furthermore, the weather will permit continuous operation throughout the year while the available land on Whiddy Island is sufficient for storage purposes.

The terminal itself has cost £10 million and the tankers between £6 million and £9 million each, while the first oil-despatching port has been established at additional cost in Kuwait (11,000 miles via the Cape). It is also hoped when circumstances permit, to establish a despatching port in Nigeria (5000 miles distant).

PRODUCTION AND REFINING

Thanks to the various developments referred to above, it has become possible greatly to increase the output of mineral oil when market possi-

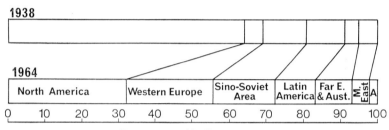

Fig. 53.—The changing location of the oil-refining industry 1938–64. (A = Africa.)

bilities allow. In fact, the output of crude oil has increased during the present century in a most spectacular manner. This is illustrated in Figs. 12 and 53 and in the following table:

World Production of Crude Mineral Oil

1920	99 million metric tons
1938	281 ,, ,, ,,
1950	538 ,, ,, ,,
1960	1100 ,, ,, ,,
1964	1457 ,, ,, ,,
1966	1679 ,, ,, ,,

The reasons for such a spectacular increase are not difficult to find. The market for mineral oil products a century ago was a very limited one; in some areas such as Oil Creek, Pennsylvania, oil which seeped to the surface of the ground was simply scooped off and sold in small amounts for medicinal purposes, but in 1855 it was predicted that some 90% of crude oil could be distilled into useful products. This forecast helped to bring about the interest in oil which resulted in Drake's discovery—which was meagre enough judging by present-day standards. Oil began to come to the surface after drilling had reached a depth of $69\frac{1}{2}$ feet, and the subsequent yield of 8 gallons a day was stored in any available barrels and tubs. Within a year 74 wells were producing oil in the Oil Creek region and in five years 6000 barrels a day were being produced.

Crude refineries were quickly established to deal with the increased output of mineral oil and refinery products were soon being offered to purchasers. It was a fortunate accident that Drake's well yielded an oil from which a high proportion of kerosene could be obtained, since kerosene, used for lighting purposes, was easily the product in greatest demand. Some refineries, however, processed the wastes left behind after the extraction of the kerosene with profitable results. Naphtha, for example, came to be used as a solvent cleanser, paraffin was used in the manufacture of candles and matches, while petroleum jelly found a use in the preparation of lotions, ointments and pomades. But the large-scale market awaited the development of the internal-combustion engine and the consequent high demand for petrol which came about early in this century (p. 234 above), while at a later date came the market for furnace oils and aviation spirits.

World refining of crude oil kept pace with world market demands and with the production of crude oil, and Figs. 53, 54 and 55 illustrate some features of the changing pattern of oil refining since the years immediately preceding the Second World War. Two major features are immediately apparent: (*a*) the remarkable overall increase in world refining activity, and (*b*) the changing locational pattern of the industry. Perhaps the two most notable locational changes are the decreasing comparative (not absolute) importance of the U.S.A. in the refining industry and the increasing importance of Western Europe. The share of the U.S.A. fell from 64% in 1938 to 29% in 1966 (though the actual amount of oil refined more than doubled in amount) while that of Western Europe increased from 4 to 24% over the same period. The Far East and Australasia showed a sharp comparative increase from 4 to 10% (notably in Japan and Australia), while the relative importance of the other major areas remained fairly constant. This changing pattern is emphasised by a consideration of the following table, which has reference to overall imports of mineral oil and oil products into the United Kingdom:

U.K. Imports of Petroleum (thousand long tons)

	1938	1966
Crude oil	2,219 (19%)	70,360 (76%)
Refined products	9,523 (81%)	21,641 (24%)
TOTAL	11,742 (100%)	92,001 (100%)

Source: Petroleum Information Service.

THE OIL-REFINING INDUSTRY

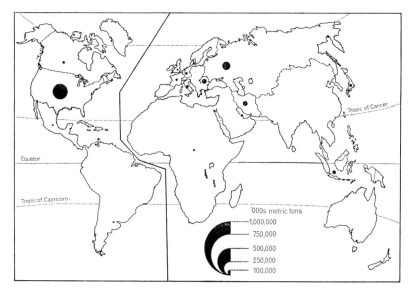

FIG. 54.—World oil refining, 1938.

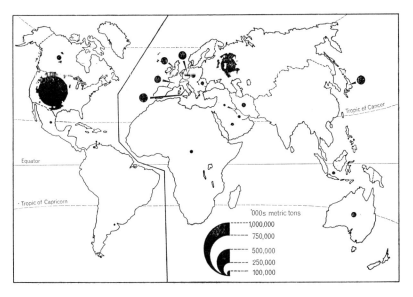

FIG. 55.—World oil refining, 1964.

It will be clear from this table that the United Kingdom has very substantially increased her refining capacity (Figs. 54 and 55), as she now imports over three-quarters of her requirements of petroleum in the form of crude oil, though before the Second World War the corresponding figure was less than one-fifth. In 1947 British refinery capacity stood at about $2\frac{1}{2}$ million tons, though in that year total oil consumption was over 12 million tons. Twenty years later, early in 1967, annual refinery capacity had leapt up to over 78 million tons, while at the end of the year the figure was about 85 million. It is expected that in 1970 the figure will further have risen to almost 115 million tons.

It is also of interest to note that there has been a comprehensive change in the overall pattern of crude oil imports into the United Kingdom as is shown by the following table and in Fig. 56(b):

Sources of U.K. Imports of Mineral Oil

	1938	1966
Middle East	24%	48%
Africa	—	21%
Latin America	49%	13%
North America	18%	very small
Other	9%	18%

Source: Petroleum Information Bureau.

The Changing Pattern of Oil Refining

There will be little need to stress the fact that very important reasons must lie behind such an impressive change in the global pattern of the oil-refining industry as we have commented upon above, and we might proceed to examine this point further. It seems clear that the market is exercising an increasing attraction for the industry, a point well brought out in the following table:

Location of the World's Refining Industries

	At source	Intermediate	At market
1939	70%	—	30%
1959	35%	9%	56%
1962	30%	7%	63%

The changing pattern is too clear and too marked to be accidental and we may now proceed to put forward some reasons to account for the change.

1. In the early days after Titusville almost the only mineral oil products which were in commercial demand were kerosene and lubricants, and the crude oil was simply distilled in large iron retorts to separate out

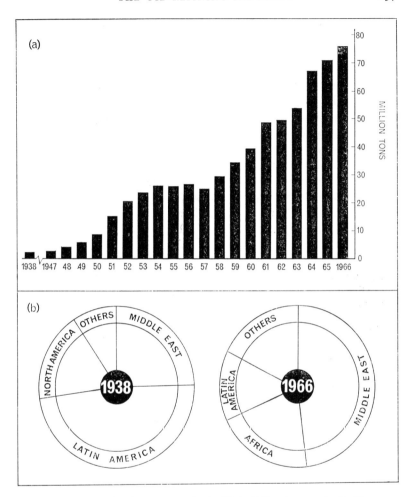

Fig. 56.—(a) Total oil-refinery output in the United Kingdom, 1938–66; (b) sources of United Kingdom imports of petroleum, 1938 and 1966. See table on p. 256.

these two products; later uses were found for certain other products (p. 254). Petrol, however, which was necessarily produced during the distillation, was unmarketable and was burnt as waste or shot out into the nearest ditch. Only about 50% of the crude oil processed in the early refineries in Pennsylvania could be sold, though later improvements in refining techniques and in the extracting of useful products from the wastes raised the figure to 75%. The high proportion of waste, however, meant that it was uneconomic to transport crude oil very far before refining, and it was natural that source refineries should be established.*

2. Near the turn of the century came the first big demand for petrol owing to the development of the motor vehicle, and this meant that a substantial proportion of former "light fraction" waste became an important market product. Much larger amounts of crude oil than had hitherto been produced were needed to satisfy this demand, but unfortunately the market for the heavier fractions could not keep pace. Increased waste was in some cases further augmented by the use of crude oils with a high proportion of heavier fractions. For instance, in the U.S.A. there was increasing use of crude oil derived from oil fields near the middle of the North American continent and from those near the Gulf of Mexico, and these crude oils contain a greater proportion of heavy fractions than do the Pennsylvanian oils. The overall proportion of waste was under these circumstances still considerable, and source refineries were still at an advantage.

3. The proportion of waste was slightly diminished by the development of the "cracking" process which rearranges the molecules of certain of the higher fraction products such as naphtha, so that petrol is produced (for some information regarding the processes involved in oil refining *see* below, pp. 270–271). It was in 1913 that the first "cracked" petrol was sold in the U.S.A. but the process was not widely used until some time afterwards. In the 1930s a new cracking process was developed which made use of a catalyst (hence the term "catalytic cracking" as opposed to the older "thermal cracking"). It was the discovery of the cracking processes which made possible the production of high-octane, anti-knock petrols. Cracking is, however, an expensive process and only a limited amount of the wastes could successfully be treated.

4. Very substantial market changes have come about since the period immediately prior to the First World War, and the demand for the heavier fractions has greatly increased. This came about firstly because of the increasing use made of diesel oils as a source of energy and power, while in more recent years there has been an increasing demand for fuel oils, for use in furnaces in industry and for domestic heating purposes. The overall result of these changes in demand is that today up to 95% of the crude oil can be made to yield marketable products, while of the 5%

* The very limited demand in 1859 for petroleum products helped to produce the first sequence of boom and slump in the industry. Drake's first gusher sparked off tremendous enthusiasm for oil development and barrels of oil sold for over $20; the consequent boom, however, soon flooded the restricted market and the price fell rapidly to five cents.

residue about half can be used by the refinery itself. One of the main reasons, therefore, for the locating of refineries at source—a high proportion of waste—has now disappeared, and it is no accident that in recent years we have seen a marked shift to market location.

5. The series of developments which we have so far examined has brought us to the point at which from a technical standpoint a market-orientated refinery is at no disadvantage as compared with a source-orientated plant. But other factors come into the picture. One is the vastly increased consumption of mineral-oil products today as compared with that of earlier years. When, for instance, Middle East oil products were sold in limited amounts in a market which included Western Europe, South Africa and even areas bordering the Pacific Ocean, it was logical to establish the refinery (Abadan) near the oil field, for the single refinery could supply the combined demands of the market areas; the situation is quite different today, when the consuming areas separately are well able to accept the output of large refineries. Another point is that a very wide range of products is now obtained from the processing of crude oil (p. 76 above)—it is said that a modern refinery may manufacture as many as 2000 distinct products, most of which were never even imagined a quarter of a century ago. Now, while it is one thing to transport crude oil in bulk, it is quite a different matter to transport equivalent amounts of distillates, and on the grounds of transport costs alone a market refinery has much in its favour. We might, however, note that in recent years "finished product pipelines" have been developed (Fig. 57), and these lines now handle a number of refinery products; notable existing examples include the pipelines linking London Airport (Heathrow) with the Kent (Isle of Grain) refinery and with Fawley.

6. Refineries in these days are immensely complicated plants requiring very large amounts of capital for their construction. The Isle of Grain refinery, for example, cost about £80 million to construct. While such amounts of capital are simply not available in many source countries, it is more likely that they will be procurable in highly developed market territories.

7. Social investment often is necessary on a very much higher scale in source countries than in market territories. As an example of this we may take the case of Das Island, part of the Abu Dhabi sheikdom, which lies in the Persian Gulf off the Trucial coast of Saudi Arabia. Das Island is being developed as an assembly centre for the output of mineral oil from the off-shore oil field of Umm Shaif, and a terminal for oil tankers has been constructed there. Before these developments there were virtually no natural resources on this tiny desert island, and all materials,

equipment and skilled labour had therefore to be brought in from Europe and the U.S.A., while unskilled labour was recruited from India, Pakistan and Abu Dhabi itself. Apart from the tremendous costs of assembling labour and materials, £300,000 was spent in the first eighteen months of the project simply on providing quarters for senior staff while a further £250,000 was needed to provide a wireless and telephone communications centre. Further substantial sums were also needed to supply recreational, health and shopping facilities—facilities which would have been generally available in a more highly developed community. While there is no question of establishing an oil refinery on Das Island, the principle remains—that the establishing of highly sophisticated undertakings in developing areas may well necessitate social investment by the companies concerned on a very large scale.

8. Market refineries are more flexible in the sense that they can accept crude oil from competing regions, while the source refinery is virtually tied to using oil from a single source. In the Suez crisis of 1956, for example, when supplies of oil from the Middle East to Western Europe were interrupted, British refineries were able to switch to Venezuelan oil and thereby avert a complete catastrophe. More recently, refineries in France and Britain have been able to take advantage of the new oil reserves now being exploited in Algeria and Libya, with a speed which would not have been possible had these market countries not established refineries of their own. North African oil products have therefore been rapidly made available in Western Europe.

9. Although the modern oil refinery does not need large numbers of workers to operate it, the workers who are concerned with its operation are normally skilled men, and such men are far more likely to be available in developed territories than in an under-developed source region.

10. Crude oil is naturally cheaper to purchase than is an equivalent amount of derived products, and the bulk purchase of crude oil therefore economises on foreign exchange. Melamid has shown that the construction of refineries in Western Europe in the years following the Second World War was greatly stimulated by the pressing need of a continent engaged in rehabilitating itself after the ravages of war to cope with serious deficits in its foreign payments accounts. More recently, the importance of this point even to a "developed" territory like the United Kingdom scarcely needs emphasising after the financial crises which have beset this country. An importing territory which refines its crude oil can even benefit from exporting refinery products; the United Kingdom, for example, today earns more foreign exchange from her export of refined oil products than from her export of coal (£106 million in 1966).

11. Since, as we have earlier observed, an oil refinery is a very expensive piece of equipment to install, oil companies naturally prefer to establish refineries in developed territories where the danger of civil strife and political instability is at a minimum. The expropriation of the Abadan refinery in 1951 by the Persian government of the time, stimulated considerable expansion of the oil-refining industry in Western Europe, and there is a natural reluctance on the part of oil companies to invest the very large amounts of capital required for a refinery in any area which is politically troubled or which may become so. It is not therefore surprising that political and strategic considerations have encouraged the development of market orientated refineries in the post-war years.

12. We have observed in Chapter IV (pp. 88–91) that there is a notable trend for modern industry to seek a market location, and this is true of the oil-refining industry. Mineral oil and the more important of its derivatives are relatively bulky commodities, and difficulties therefore arise with regard to distribution; a market location helps to minimise these difficulties. The strength of the attractive power of the market in this respect, should be in no doubt to anyone who even cursorily notes the establishment of refineries in the Rhine and Ruhr districts of Western Europe, refineries which are fed by pipelines extending from the North Sea and the Mediterranean coasts (*see* p. 267 below). It is instructive to notice the difference in this respect between developments in the United Kingdom and those in Europe. In the former case the ready availability of sea-borne tanker transport has encouraged the development of coastal refineries in a comparatively small country, no part of which is far from the coast, but in Europe tankers cannot supply inland market areas far removed from tide water. There has therefore been a much greater incentive to establish inland refineries in market areas supplied by pipelines running from the coast.

FURTHER CONSIDERATIONS

It should be clear at this stage, that there are indeed strong reasons to explain the changing pattern of the world's oil-refining industry as shown on Figs. 53, 54 and 55, but before we proceed to a closer examination of the various factors which affect the siting of refineries, there are three related points which we should emphasise. The first is that oil-producing territories themselves are not necessarily happy about the latest trends in the refining industry, and some of them are making efforts to secure the benefits of a refining industry for themselves. The increasing competition between the ever-increasing number of producers, however, is a factor which must weaken the bargaining position of any one of them, though some producers have met with some success in their efforts as is shown by the case of Venezuela, who, by the passing of the *Hydrocarbons Act* of 1943, was able to bring about a substantial expansion of her oil-refining industry.

The second point is that more countries than ever before are establishing oil refineries, and some of these countries are not "developed" ones. The demand for mineral-oil products is today virtually world-wide; even newly developing territories need petrol for motor cars and probably for buses, lorries and tractors too, while diesel-powered locomotives and heavy road vehicles are being increasingly used in developing territories. There is, therefore, a natural urge for any country in which the demand for mineral-oil products is substantial, to set up its own refinery with the hope that it will be able to sell elsewhere refinery products in excess of its own needs. We may add to this the point that the possession of an oil refinery is fast becoming something of a status symbol. It is not, therefore, surprising that many territories such as Ghana, Tanzania and Rhodesia have established refineries, and it is perhaps true that the refining of mineral oil is becoming one of the most widely dispersed industries in the world.

The third point is this. We have so far tended to speak of mineral oil as though it were an homogeneous commodity, as though mineral from one region is similar to that obtained from another. This is not the case. Some oils contain proportionately more of the lighter fractions and some more of the heavier, and the actual constitution of the crude oil is an important factor as regards its exploitation and refining. An increased demand, for example, for fuel oils is likely to lead to an increased production of these oils, but since fuel oils and petrol are in joint supply, there will be a corresponding increase in the production of petrol. It does not follow, however, that the market for petrol will conveniently expand to absorb the increased supply; it is by no means easy, therefore, to balance the output of the various fractional products with the market demands. As a result of fairly recent increases in demand for fuel oils in the United Kingdom, the overall pattern of demand accords fairly well with the natural yield of Middle East crude oils—about 20–24% petrol, 18–20% middle distillates (including diesel fuel oils), and 43–45% residual fuel oils; the balance is made up of smaller bulk products and refinery consumption and losses.* It does not follow that crude oils from another source region would fit this pattern at all. Venezuelan oils, for

---

* The proportional importance of petrol in relation to other petroleum products in the United Kingdom is declining (pp. 271–272 below).

example, contain a greater proportion of the heavier fractions and this proved to be a troublesome feature during the Suez crisis, while North African oils contain more of the lighter fractions, a point which counts against them in the markets of Western Europe (*see* next paragraph). Neither does it follow that an oil refinery (a very delicate piece of precision plant) which is equipped to handle crude oil from one source, can equally efficiently deal with crude oil of a different constitution from another region.

It might be appropriate to observe that a fairly distinct pattern in the consumption of mineral-oil products can be detected in states at different stages of development, though exceptions to the general rule inevitably exist. Developing territories for the most part show a comparative preference for paraffin and a low consumption of petrol, while highly developed countries consume proportionately more petrol. The pattern of consumption in North-West Europe is interesting since it shows a relatively high consumption of fuel oil; this is a reflection of the increasing importance of fuel oil in industry, and is also in part a measure of the increasing use of this fuel for domestic purposes—a reflection of the unpleasant character of the winter climate.

SITING OF OIL REFINERIES

We might now turn to a consideration of the various factors which affect the siting of oil refineries, and the following points may be emphasised:

1. Most importing countries receive their supplies of crude oil by sea, and since a break of bulk necessarily occurs at the coast, it is often advantageous if a refinery is sited at the point of discharge. The modern tanker needs a sheltered, deep-water anchorage with a minimum of interference from such impediments as strong currents or other shipping for safe and speedy unloading, and the coastline adjacent to such an anchorage can sometimes provide a useful site for the erection of an oil refinery. Examples which come to mind include Southampton Water, the Thames Estuary and Milford Haven, though the coastlands of Milford Haven are too steeply sloping to be ideal. A minimum of eight fathoms of water at low tide is desirable in order that the movement of tankers is not interrupted, but as tankers increase in size this will not be sufficient. The only oil ports in the United Kingdom capable of handling 100,000-ton tankers a year or two ago were Milford Haven and Finnart, but today Finnart takes 200,000 tonners, Milford Haven 250,000 ton vessels, the Port of

London those of 200,000 tons, while 100,000 tonners can discharge in the Humber near Immingham.

Another factor is that berthing wharves of half a mile or more in length may well be needed to accommodate more than one supertanker at the same time, while these wharves must be immensely strong. A vessel of 100,000 tons or more moving even "dead slow" at a speed of one knot will register a tremendous impact if it touches even slightly a berthing jetty.

These and other features are being allowed for in the new installations at Europoort, now the largest oil refining and petrochemical complex in Europe, and also at the Rotterdam terminal of the Rotterdam–Rhine crude oil trunk pipeline. Other points to notice concern the fixed-point mooring device which is in use at Mersa el Brega in Libya for loading. Crude oil from shore storage tanks is pumped via a submarine pipeline into the tanks of a vessel moored against the structure some distance offshore; the structure embodies a loading arm which is free to rotate according to the direction of prevailing winds and currents. Some device such as this may in future be used for off-loading. Off-loading at sea is, in fact, being considered by the Shell Company, who are investigating the possibilities of pumping crude oil from supertankers into smaller vessels in Lyme Bay, off the Dorset coast.

The importance of some of the above factors can be underlined by means of a British example. The Milford Haven anchorage, to which reference has been made, suffers from notable disadvantages: it is not near any significant market, existing communications are not good, the coast is cliffed along much of its length, and the actual refinery site in its natural form included sharp gradients of as much as 1 in 6. Expensive terracing had to be undertaken before installation of the plant could begin. Despite these shortcomings, however, the Esso refinery was established there in 1960, the Regent plant in 1964, while the Gulf Oil refinery came into operation late in 1968. This fact emphasises the strong advantages to be derived from an *anchorage* which offers favourable physical characteristics.

2. An oil refinery needs very large quantities of water for processing and possibly for cooling. A plant with an annual capacity of 5 million tons needs at least 4 million gallons of fresh water *per day* for processing, and, if water is used for cooling, additional substantial amounts (which may be fresh or salt) will be required.* The Baton Rouge (Louisiana)

* It is interesting to note that about twice as much cooling water is needed if the initial temperature of the water is 85° F (29·4° C) rather than 60° F (29·4° C). The importance of climate in this respect will easily be recognised. In areas of high humidity the efficiency of cooling towers is notably reduced because of a lowered rate of evaporation.

refining plant pumps 260 million gallons of water daily from the Mississippi. For the refining of one gallon of crude oil at this plant 23 gallons of water are needed; of this, half a gallon is used in boilers, fire-protection services and for direct use by the employees, while the other $22\frac{1}{2}$ gallons are used for cooling of the various products as they leave the distillation plants. Process and boiler water must be of good quality, and this is sometimes a limiting factor because in some areas the water is too alkaline or it contains too much suspended matter for efficient working. Many new refineries are now using air-cooling systems.* The disposal of wastes which carry chemicals and refuse oils is also an important matter, since extensive pollution can result from unregulated discharge into conveniently placed streams and rivers.

3. A modern refinery needs an extensive site of reasonably flat or gently sloping land, since unit refining costs diminish with increasing scale of operation up to an annual capacity of about 10 million tons. A very small plant needs over 100 acres, but for a large plant a site of more than 1000 acres may be necessary in order to accommodate the heavy and extensive plant and to allow for storage and safety requirements. A gently sloping site is advantageous since a gravitational flow of oil from one production unit to the next can then be arranged; the crude-oil tanks can be set up on the highest ground and the loading facilities for the finished products on the lowest.

4. It may be noted that a refinery does not make heavy demands on labour, and labour requirements do not normally constitute a locational factor. With increasing use of automation we may expect that even less labour will be needed. A labour force of less than 500 is usual, even in a large refinery and most of these workers are skilled men of the type who normally have no great objection to moving away from their home areas. Even in a fairly remote location the accommodation of such a labour force does not present particularly big problems.†

Suitable sites for oil refineries are not easy to come by in view of this formidable list of requirements. Figure 57 shows the position in the United Kingdom in 1968. Points to notice include the limited number of deep-water sites and accessibility to markets.

It is today becoming more usual for refineries to be sited nearer to markets than coastal sites may allow, crude oil being piped from the

* See table, p. 96 above for the effect this can have upon water consumption at refineries.

† Note the difference between establishing a sophisticated establishment like a refinery in an undeveloped territory (Abu Dhabi, p. 259 above) and in a fairly remote part of a "developed" country. In the latter case substantial social costs will be borne by the state and will not fall on the oil company.

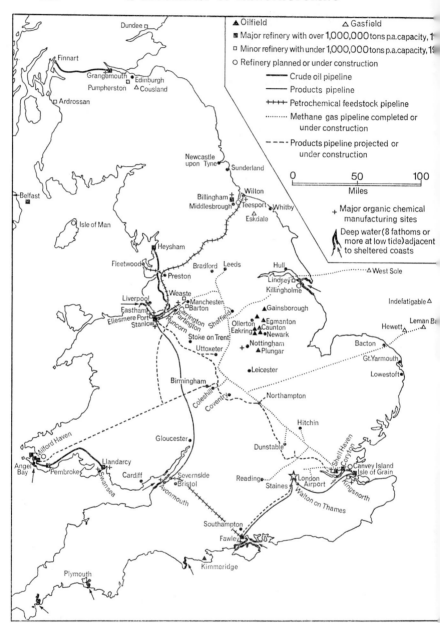

FIG. 57.—Some features of the oil-refining and chemical industries of the United Kingdom, 1968.
(Deep water for England and Wales only.) *From James, Scott and Willats.*

points of entry on the coast. This is a situation much in evidence in Western Europe, where refineries in the Ruhr area are fed by pipelines running from Wilhelmshaven and Rotterdam, while near the Rhine new plants at Karlsruhe and Strasbourg are now fed by a 500-mile-long pipeline from Lavera, near Marseilles. This line makes it possible for Saharan crude oil to flow into Germany for refining. A pipeline from Genoa supplies refineries in Switzerland and Southern Germany.

The refineries located at Grangemouth and Llandarcy shown on Fig. 57 may appear from the map to fall into this "market-orientated" category, but these cases are rather special ones. Both were developed when oil tankers were significantly smaller than they are now,* and the ports serving them could not be deepened to meet modern requirements. New receiving installations were therefore constructed at Milford Haven and at Finnart, on Loch Long, and the crude oil is now piped to the refinery in each case.

ASSOCIATED INDUSTRIES

An important feature of the modern oil refinery is the manner in which it can trigger off the development of associated industries which use refinery products as their raw materials; questions of industrial linkage are clearly involved. For instance, a wide range of organic chemicals is now manufactured from refinery products, some of which were originally produced from coal or molasses and some of which are completely new discoveries. The petrochemical industry was first developed in the United Kingdom in the early 1950s, and until 1964 it expanded at the astonishing rate of 24% per annum; the growth rate after 1964 was about 8% per annum. In 1950 less than 9% of the organic chemicals produced in the United Kingdom were derived from mineral oil, but by 1962 the figure had risen to 61% and by 1965 to 70%. As a result of this there is a marked tendency for minor industrial complexes concerned primarily with petrochemical enterprise to be located near refineries; this point is brought out in Fig. 57.

As an example of the siting of a petrochemical industry we might take the case of Avonmouth, where a sulphuric acid plant was established during the First World War. Other chemical industries followed, including the production of metals (especially zinc) with the aid of chemical processes. In 1960 construction began of a large petrochemical plant which is now producing ethylene oxide (used in the manufacture of

---

* The Llandarcy refinery has been in operation since 1922; it was the first major refinery in the United Kingdom. Grangemouth opened in 1951.

polythene) and ethylene glycol (used in the manufacture of artificial fibres), while ethylene, the petrochemical feed stock and raw material is pumped to the plant along a 78-mile long pipeline from Fawley (Figs. 57 and 58).

Other forms of industry, too, are developing in the Avonmouth area. As a development of earlier enterprise, for example, early in 1968 came

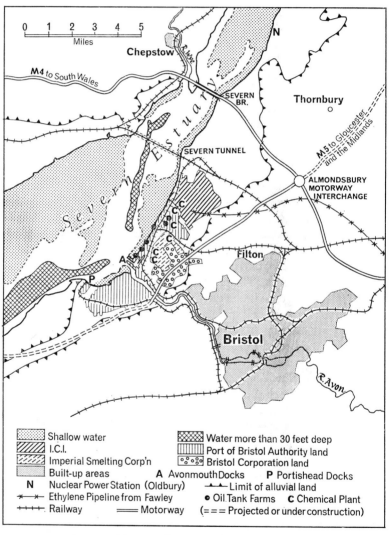

FIG. 58.—The Bristol region: industrial development.

the opening of Imperial Smelting Corporation's £15 million zinc plant, which will manufacture zinc on an unprecedented scale. Lead is produced in the furnace simultaneously with the zinc, and the new process makes it possible for Britain to retain an interest in zinc production. Had it not been for this new process British companies might well have been forced out of the industry as they are precluded by relatively high energy costs from using the comparatively cheap electrolytic process which is generally employed in other countries.

Apart from the availability of feedstock through the Fawley pipeline there are four main advantages enjoyed by the Avonmouth site (Fig. 58) for the petrochemical group of industries:

1. The proximity of Avonmouth Docks, made possible by the juxtaposition of deep water off-shore and a sheltered inlet (the Avon estuary). In addition, non-resistant alluvial deposits (*see* point 2 below) made possible the excavation of the docks with comparative ease.

2. The presence of flat, ill-drained alluvial land along the southern shore of the Severn estuary. Opportunity costs for this unattractive land were low, and the existence of this low-cost land made possible the purchase of considerable areas for industrial development.

3. The broad Severn estuary can supply large amounts of water which are needed particularly for cooling.

4. The water in the main estuary has a very considerable tidal movement (the high tides and the "bore" of the Severn are well known), and this movement is valuable as it makes possible the disposal of effluent. Effluent from one petrochemical plant is discharged via a mile-long outlet pipe.

In addition to these factors we might add:

5. Two new motorways, the M4 linking London and South Wales and the M5 from the Midlands, converge near the area, and when completed they will offer first-class road connections with other parts of the country.

It is not surprising, in view of these advantages, that industry, including the petrochemical industry, is expanding at a very rapid rate in the Avonmouth area, and further developments can almost certainly be expected.

As a final note, we may remind ourselves that petrochemical products are extensively used in the manufacture of plastics, synthetic rubber, artificial fibres, solvents, fertilisers, pesticides and detergents, to name a few of many (*see* also p. 254 above). The great importance of the petrochemical industry to Britain is illustrated by the fact that petroleum chemicals are now earning for this country about £150 million (net) annually in foreign exchange.

REFINING TECHNIQUES

The main operational features of an oil refinery are brought out in Fig. 59; we have incidentally referred to some of the processes involved, but we might now examine them more carefully.

In the first place, the crude oil enters the fractional distillation plant. After heating in the furnace until the whole of the oil is vaporised the vapours pass into the bottom chamber of the tower at a temperature of about 700° F (370° C) prior to condensation; as they rise to the higher chambers the temperature falls and the various fractions condense at the different levels, heavier ones near the bottom of the tower and lighter ones higher up. The lightest fractions which condense near the top of the tower include petrol, while gases are drawn off at the top of the plant.

The distillates which are obtained from this process generally need further processing before they can be marketed. Part of the reason for this is that the demand for petrol is normally high (even in 1951 petrol was the most important distillation product in the United Kingdom in terms of tonnage, accounting for about one-third of the total), and this has led to the development of various supplementary processes designed to increase the output of petrol.

One of the earliest of these processes was that known as "thermal cracking," a process which subjects certain higher fractions to great heat and pressure and so converts them into the lighter fractional products such as petrol. The process is very expensive but a great advance was possible after 1930, when it was discovered that the use of a catalyst made for greater efficiency in the cracking process; thus "catalytic cracking" was developed. A further step forward came with the introduction of catalytic reformers which convert naphtha and paraffin into high-quality petrol (Fig. 59). By means of these processes the proportion of petrol secured from the crude oil can substantially be increased.

Another process more recently developed is that of polymerisation which converts very light distillates and gases (including gases thrown off by the cracking plants) into high-octane petrol; it also produces by-products from which detergents, plastics, synthetic fibres and synthetic rubber are manufactured. As a result of refining improvements, the yield of petrol from crude oil in the U.S.A. (where the demand for petrol is comparatively heavy) increased from 26·1% in 1920 to 44·8% in 1963.

The refining industry has been marked by spectacular development during the course of this century. Broadly, it may be said that the big advance of the 1920s was that of continuous distillation which permitted a much greater throughput than the earlier process in which "batches" of crude oil were dealt with one after the other. The interruptions to working which necessarily occurred under this system markedly limited the scale of operations, but these interruptions were obviated with the introduction of continuous distillation. Following this came thermal cracking and reforming, and in the 1930s and 1940s catalytic cracking

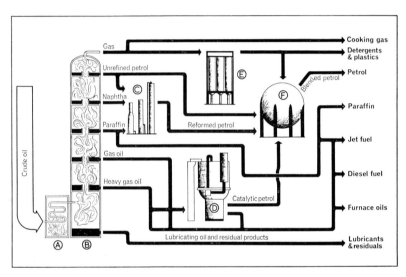

FIG. 59.—An oil refinery: the main processes and products: (A) furnace; (B) distillation plant; (C) catalytic reformer; (D) catalytic cracker; (E) polymerisation plant; (F) petrol-blending tank.

came into general use. The 1950s saw the adoption of "platforming," a process made possible when the catalyst maintains its activity without regeneration over a long period, so that the cracking process is more continuous. In the 1960s there has been an increasing use of hydrogen in the more sophisticated hydro-cracking and hydrofining techniques, the purpose of which can be varied to produce, for example, mainly petrol in the U.S.A. and furnace oils in Europe.

PRESENT AND FUTURE TRENDS

In recent years, the proportional importance of petrol in the United Kingdom has declined; this is illustrated by the following figures:

K

	Petrol consumption as a %age of total consumption of mineral-oil products in the U.K.
Before 1939	over 50
1951	32
1961	21
1964	16
1966	13

In absolute terms, however, the volume of petrol consumption increased by about two-thirds in the ten years 1951–61. Figure 60 shows the

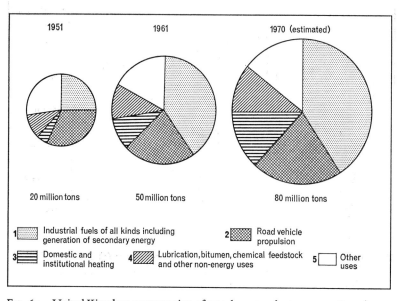

FIG. 60.—United Kingdom consumption of petroleum products, 1951, 1961 and 1970 (estimated). *Based on information supplied by the Petroleum Information Bureau.*

changes in the pattern of consumption of the end-products derived from mineral oil in the United Kingdom over recent years, and also an estimate made by the Petroleum Information Bureau of the likely situation in 1970.

Perhaps the most interesting feature brought out by Fig. 60 is the rapid advance in the consumption of fuel oil in the United Kingdom in the period shown (categories 1 and 3). Fuel oils are obtained from the heavier fractions, which also yield diesel oils. In 1964, while about 16% of the consumption of mineral-oil products consisted of petrol, no less than 57% fell into the categories of gas or diesel oil and fuel oil; corresponding figures for 1967 were 15% (petrol) and 58% (oils). The lubricating oils which are derived from the heaviest fractions form the bases of the many types of lubricants in use today,* while a wide range of products including heavy fuel oil, bitumen and wax come from the same source. Other products of increasing importance now manufactured by refineries include liquefied petroleum gases (propane and butane) and feedstocks which form the raw materials of the petrochemical industry. Altogether, more than 5000 different products are now obtained from mineral oil, and 2000 of these are manufactured by the refineries themselves.

The oil-refining industry is a comparatively new one, and as it is changing rapidly it is difficult to present an up-to-date picture of it. Figure 57 shows the position in the United Kingdom in the early part of 1968, but there is little doubt that future developments will quickly modify the pattern shown there.† Recent exploration beneath the floor of the North Sea has met with some limited initial success in the discovery of mineral oil and spectacular success in the discovery of very large reserves of natural gas; utilisation of these reserves could result in the saving of a great deal of expenditure of foreign exchange on imported crude oils—£426 million in 1966 of which probably £100 million can be saved (*The Times*, 13 August 1968)—while the whole of British industry could immeasurably benefit from such a low-cost energy source. Contracts have been signed between producing companies and the Gas Council, and North Sea gas should be reaching all area gas boards by 1970 (Fig. 57). These developments promise to be of tremendous significance to Britain's fuel economy and may well result in a changing pattern of industrial expansion; already, observers are predicting the development of a major industrial region in eastern England near the shores of the North Sea.

SUGGESTIONS FOR FURTHER READING

Fryer, D. W., *World Economic Development*, London, 1965, Chapter 11.

Petroleum Information Bureau, *Newsletter*, London. An indispensable source of information for those interested in matters related to mineral-oil development, though unhappily the *Newsletter* is no longer published. Especially useful in connection with this chapter are the issues dated March 1963, August 1963, January 1964, July 1964 and November 1964.

* There are several hundred different types of lubricant on the market today, most of them being produced for a particular type of use. In the United Kingdom the annual *per capita* consumption of lubricants works out at almost five gallons.

† Future developments are projected or under construction at Teesport, Canvey Island, Killingholme (a second), Glasgow and on the Isle of Man.

Elkins, T. H., "Oil in Germany," *Geography*, 1960.
James, J. R., Scott, S. F. and Willats, E. C., "Land Use and the Changing Power Industry in England and Wales," *Geographical Journal*, September 1961.
Luckas, M. R., "Recent Developments in the United Kingdom Oil Industry," *Geography*, April 1965.
Manners, G., "The Pipeline Revolution," *Geography*, April 1962.
Melamid, A., "Geographical Distribution of Petroleum-refining Capacities," *Economic Geography*, 1955.

PART THREE
THE PRESENT SCENE

Chapter XII

Some Problems of Industrial Development

It is the purpose of this chapter briefly to examine some of the pressing problems which are associated with industrial development. For convenience, we may classify the problems which we shall consider under four headings:

1. Problems of a technical nature which can (at any rate in theory) be solved by technical means.
2. Problems of development relating to industrialised regions.
3. Problems of development associated with individual industries.
4. Problems associated with the industrialisation of developing territories.

Under the first heading we shall pay some attention to problems concerned with the production and disposal of unpleasant and noxious wastes, oil pollution and the preservation of scenic amenities. Sections two and three will examine some of the difficulties encountered by industry in its endeavours to adapt itself to changing economic and social circumstances, while in section four we shall discuss some of the problems met with in developing territories which are seeking to establish their own industries. This section is illustrated by a brief examination of the situation in Mexico, which offers a case study of a developing territory.

PROBLEMS OF A TECHNICAL NATURE

It is the essential nature of industry to take raw materials and to fashion from them useful goods. It is, however, a normal thing that waste products are formed during the manufacturing process, and while it is sometimes possible to make use of these waste products (which then become by-products) it is quite often impossible to do this. The problem which then faces the industry concerned is that of disposing of the waste products efficiently yet as inexpensively as possible. Sometimes this disposal is effected with the aid of chimneys which discharge smoke and other impurities into the atmosphere; sometimes waste can be excreted in the form of effluent which may be discharged into some conveniently placed stream or river; while sometimes the waste is in the form of bulky solid matter which can be disposed of only with considerable difficulty and expense.

Pollution

We have had examples of each of these three categories of waste earlier in this book. For instance, in Chapter III we saw something of the generation of thermal electricity, and we realised that the disposal of noxious gases and of bulky ash is a very real matter of concern (p. 68 above), while an example showing the urgency of the problem of disposing of liquid effluent was given on page 96. This whole matter is a widespread problem in industrial regions, and one which admits of no easy solution. In some cases a cement works or a steel works will scatter heavy dust as well as fumes over a wide area; a synthetic-fibre factory can render nearby areas downwind almost intolerable with the stench of the fumes which it emits, though these fumes are not noxious; while an atomic power station produces a waste which is mildly radio-active and the greatest possible care must therefore be exercised in the disposing of it. And these examples could be multiplied many times over. Garnett has shown something of the problem of atmospheric pollution in the Sheffield region, but the problem is met with wherever manufacturing is carried on on anything but the smallest scale. The dangers of air pollution have been the subject of close study in New York, a city which possibly suffers from the highest degree of atmospheric poisoning in the world. The city itself pours 600,000 tons of sulphur (mainly in the form of sulphur dioxide) into the air every year, while neighbouring cities probably direct as much again over the suffering metropolis. The amount of carbon monoxide from vehicle exhausts sometimes reaches dangerous levels at busy road junctions, and at times the smog and haze generated by the city extend over an area 400 miles in diameter. Possible ways of dealing with this pressing problem are now being discussed (including the passing of anti-pollution laws and the processing of much of the waste into useful products); its urgency is emphasised, perhaps rather dramatically, by the editor of the *Saturday Review*, who has argued that within a decade most of the large cities of the United States could become uninhabitable. The point is that they could slowly be smothered beneath the load of their own excrement.

This whole problem has come to public notice in the United Kingdom in recent years largely because of the startling manner in which many

British streams and rivers have been turned into virtual sewers by uncontrolled discharge, and while it is only fair to add that this malodorous situation is not entirely due to direct industrial activity (much domestic sewage finds its way into streams and rivers, *see* for example, the table on p. 284 below) it is also true that it has arisen as a result of the working and living problems of an industrialised society. Even rural communities can add a quota which is by no means negligible. This may come about through water used to wash out cowsheds and piggeries and from the washing of insecticides, sheep dip and weed killers into watercourses—to mention a few examples. There is today an increasing tendency to flush out animal wastes instead of spreading them as fertiliser over fields, and this practice results in the creation of vast amounts of odorous slurry which is frequently led into conveniently placed rivers. The National Farmers' Union has estimated that if existing river anti-pollution laws were enforced British farmers would need to spend £250 million in measures to control disposal of their farm wastes. Yet these same farmers continue to purchase large amounts of subsidised fertilisers from chemical factories!

When we have said this, however, we must still admit that industry is largely to blame for the unsavoury condition of many streams that some years ago were sweetly flowing and clean. The "odious Tame" which flows into the Trent from the Birmingham Plateau is a case in point; two-thirds of the water which passes down this river on an average day has been used either domestically or industrially, and such industrial effluents as cyanides, sulphides and phosphoric acid are discharged into the stream. It is said that there is enough cyanide in the river at any one time to kill off half the population of West Bromwich, and on one tributary a water sampler burned the skin on his hand simply by dipping it in the water. It is not easy to imagine that in years past, Thames water was considered to be of such high quality that barrels of it were shipped to Lisbon for brewing purposes! The present situation is that the tidal Thames receives something like 440 million gallons of sewage every day, much of which flows up and down with the tides for three months before it finally struggles through to the sea.* This turgid liquid is constantly stirred by ships' propellers so that it more readily releases the hydrogen sulphide which produces the evil smell that characterises the river, especially on hot days. Paint is destroyed so efficiently by this mixture that Thames police launches have to be repainted every three months. It is true, and again we recognise the fact, that this effluent is partly domestic, but it is equally the product of a society which is an industrial one, and the whole situation is one thrown up by an industrial society. J. W. Day states that the

* G. Lucy, "Filth in Our Rivers," *Reader's Digest*, April 1963.

situation on rivers in the North of England and the Midlands, polluted for years by industry, is almost "past redemption," and that poisoned rivers include the whole or parts of the Mersey, Irwell, Don, Trent, Tyne, Severn, Avon and Ribble.

Other examples of the same sort of thing are easy enough to find, and little purpose would be served by proceeding to serve up a gruesome compilation of case studies; the point at issue is clear enough. One fact which is not always realised, however, is that stream pollution when once achieved is not always easily eradicable; the River Ystwyth, for instance, is still contaminated with zinc from old mines and dumps though mining operations ceased thirty years ago, while it has been estimated that pollution in the lower reaches of the River Hudson by the city of New York has reached such staggering proportions that even if all forms of pollution stopped immediately, the river would not be able to wash itself clean during the present century!

The immediate causes of stream pollution are not difficult to understand. Under normal circumstances a healthy river is purified by the activities of algae and bacteria, some of which are anaerobic and some aerobic. The former cannot live in the presence of oxygen while the latter cannot live without it. The anaerobes, therefore, make their home on the bed of the river, where they consume solids which settle from the water above, while the aerobes feed on suspended solids. As these solids (which may include waste products from human sources) pass through the aerobes they are changed from complex organic substances into inorganic material which cannot putrefy and which provides food for the algae, tiny plant organisms which help to supply the river with oxygen through the process of photosynthesis. Oxygen is also added to the river as the water tumbles over rocks and rapids and around obstacles in its path. As the aerobes consume the waste solids the supply may dwindle, and in this case the population of aerobes will decline proportionately, but if the waste matter is continually supplemented the aerobes can deal with a surprisingly large amount.

If, however, the load becomes too heavy a significant change occurs. As the aerobes attack the masses of waste their numbers grow, to such an extent that they use up increasing amounts of the available oxygen and the river is not able to replace the loss. This results in a kind of mass slaughter of aerobes since they cannot live without oxygen, and their place is taken by large numbers of anaerobes which move up from the river bed into the now oxygen-less waters. These, too, feed on the waste matter, but with different results from the feeding of the aerobes, for they combine hydrogen extracted from the water with sulphur derived from the wastes and eject the resulting compound in the form of sulphuretted hydrogen, the evil-smelling gas which recalls rotten eggs. Under these conditions algae die and even fish become fewer and may die off altogether. When this stage is reached the river is dead and is a sewer rather than a river. Some industries such as the synthetic-fibre industry can accelerate these processes because they eject wastes which, though not in themselves noxious, absorb oxygen from water so that the anaerobes can take over even more rapidly.

There is in theory no great technical difficulty about the problem which this situation poses; more and more authorities in Britain are treating their sewage and some are even turning it to profit by selling the treated material as fertiliser. The main difficulty is one of the capital expense incurred. The Tyne region offers a case in point. The situation there is so bad that a medical officer has alleged that a person falling into the Tyne is in more danger of contracting broncho-pneumonia or intestinal infection than he is of drowning; to deal with this situation would have cost about £6 million in 1935—today the cost is put somewhere around the £20 million mark.* It is understandable that local authorities are reluctant to attempt to wring even more money from already disgruntled ratepayers; it may be that a drastic overhaul of our rating system is a prerequisite to a full-scale tackling of this problem.

The contribution made to river pollution by individual industries can be quite startling. The case of waste from a synthetic-fibre factory has been mentioned, and we might note that waste from a single sugar-beet-processing factory in Suffolk rapidly killed almost all the fish along a seven-mile expanse of the River Lark. Gregory has shown something of the complexity of the situation in Lancashire. But this is a situation which will have to be dealt with, and drastically dealt with, if the population of the United Kingdom is to continue to live under reasonably healthy conditions in the future.

A matter allied to that which we have just been discussing is that of the oil pollution of our seas. This is a particularly difficult question since it is beyond the powers of any single country to deal with it; the first international conference called to discuss it was held as long ago as 1926, and others have been held since that time. A conference which met in London in 1954 was followed by an International Convention, which urged governments to introduce domestic legislation designed to control the activities of vessels with respect to the discharge of oil of any kind; the legislation would have effect over vessels of the country concerned at all times, and over vessels of all nationalities operating within the territorial

* G. Lucy, op. cit.

waters controlled by the legislating government. Great Britain was the first country to take action by the passing of the *Oil in Navigable Waters Act* of 1955, and since that time fifteen more countries have acceded to the request of the Convention; unfortunately, the Pan-Lib-Hon countries (Panama, Liberia, Honduras), which together account for about one-fifth of registered tanker tonnage,* have so far taken no action, and this tardiness goes far to lessen the effectiveness of the legislation passed by other governments.

It is interesting in this connection to note that the major oil companies have themselves made a significant contribution since 1963 towards the resolving of the sea pollution question. In that year Shell devised the "load on top" system, which was later adopted by other European and American companies. Briefly, the system developed from the fact that a prime cause of sea pollution is the discharge into open waters of washings from tankers which have had to clean their cargo tanks at sea. Such discharge cannot often be carried out at ports of call. Under the "load on top" system tanks are washed out during a voyage, but the waste is all collected into a single tank and there left to settle. Contained oil slowly moves to the surface and the final result is that a layer of oil floats on clean sea-water. As much of this water as possible is pumped back into the sea and the oil is retained (probably together with some water), while the next consignment of crude oil is simply loaded on top of the residue. The whole cargo is discharged at the refinery after the following voyage. If the tanker carries finished products the residues are discharged at the loading refinery before the next cargo is taken on board.

It is becoming increasingly clear that the pollution of the surface of our planet is approaching a critical stage, and the effects of this pollution have become unexpectedly widespread. Rainfall over much of Europe, for example, is now increasingly acid because of the emission of sulphurous and other gases into the atmosphere, and this appears to be adversely affecting tree and other vegetative growth in Scandinavian forests—and probably elsewhere. Traces of pesticides are found in almost all living creatures—including Antarctic penguins—and this can render eggs sterile and have other unfortunate effects. We are only just beginning to understand the seriousness of pollution on a global scale, and there is no doubt that this is now a problem of the first magnitude.

Preservation of Scenic Amenities

The question of the preservation of scenic amenities in industrialised territories is obviously a far-reaching one, and one to which we can here

* *See* table in footnote, p. 247 above.

SOME PROBLEMS OF INDUSTRIAL DEVELOPMENT   283

only refer in passing. It ranges from the maintenance of open spaces (*see*, for example, p. 108 above) and footpaths in the countryside (including coastal walks, many of which have been lost through the activities of landowners and builders) to the preservation or institution of open spaces for recreation in cities. Neither should we overlook the parallel question posed by the necessity to provide sufficient water storage to cater for the needs of industrial populations. This can lead to high feelings among lovers of the countryside as the experience of Manchester, which met with bitter opposition in its efforts to increase available supplies of Lake District water by developing Ullswater, strikingly testifies. It is a simple matter to condemn such opposition as emotional rather than rational, but the fact remains that the establishment of large reservoirs in rural uplands can greatly affect not only the appearance but also the economy of the areas concerned. Good farmland, all too rare in an upland environment, may be drowned—even whole villages may disappear beneath the waters in extreme cases—and this can completely upset a rural economy which may be based upon effective use of valley floors as well as hillsides at higher levels. The loss of one may render the use of the other uneconomic. Possibilities such as the replacing of traditional forms of land use in affected areas by afforestation schemes may have much to commend them but they are bound to meet with sturdy opposition from those who can see only that traditional ways of life are being destroyed.

A Case Study

S. Gilewska has studied the effects of industrialisation upon environment in the Upper Silesian industrial district, broadly the area shown on Fig. 61, and the story is a depressing one. Changes in terrain have come about as a result of accumulation of wastes, surface removal of minerals and subsidence. Accumulation takes place whenever "dumping" of any kind goes on, and there has been much dumping in this region; it includes dumping of useless rock and gangue from zinc and iron-ore mines, and the heaping up of slag dumps. All types of dumps are harmful because they can produce chemical changes in nearby soils and in groundwater, because they impede natural movement of surface water, and because they may damage roads and drainage systems since their weight sometimes causes dislocation in surface rocks. Surface removal of material occurs whenever there is opencast mining; useful land has thus been destroyed and scenic beauty has suffered.* Subsidence has taken place

* One is reminded of the situation in parts of the East Midlands of England where good farmland has been stripped for opencast mining of iron ore. In this case, however, the soil was stored after the initial stripping and spread over the lowered surface

in this area as in other mining regions over sub-surface mine workings and this has led to the formation of swamps in the areas of subsidence and a lowering of the water table on adjacent higher ground with consequent killing of trees and of other forms of vegetation.

The subject of water pollution occupies an important place in Gilewska's report, and the gloomy overall situation is represented on Fig. 61. It will immediately be apparent that only headwaters and small streams in and near the industrial region are free from pollution, while the middle and lower courses of larger rivers are polluted "to a very high degree." Along the Vistula itself the pollution is felt as far downstream as Cracow. Waste matter comprises ammonia, nitrates, sulphates, chlorides, iron compounds, phenols, oils (both heavy and light) and tar, and the origins of the pollution are given as follows:

City sewage	15%
Mine waters (including wastes from washings)	40%
Waste water from steel plants	25%
Waste water from other industrial plants	20%

Changes which are perhaps less noticeable than the above include climatic and vegetational modifications and soil deterioration. Slight climatic changes come about on the one hand as a result of the large-scale felling of forests and the considerable constriction of grassland and swamp, and on the other hand as a result of the development of built-up areas, of heat originating in spoil heaps and furnaces (both industrial and domestic), and because of the addition of impurities to the air. Large amounts of volatile compounds of lead, arsenic, zinc and other metals are released into the air, as well as other substances in gaseous form and large amounts of dust, and in many areas there are sufficient impurities to render the atmosphere toxic. In many such cases the disappearance of vegetation has produced small-scale deserts, while residents in such areas suffer unduly from complaints induced by breathing in the impure atmosphere. In addition, the "fall-out" of impurities from the atmosphere leads to increased soil salinity and deterioration. There is no difficulty, on the basis of Gilewska's report, in pointing to problems produced by industrial environments of the kind which we are considering in this chapter.

---

again after the mining was complete, and farming could then be resumed. Cf. also Wooldridge and Beaver (1950).

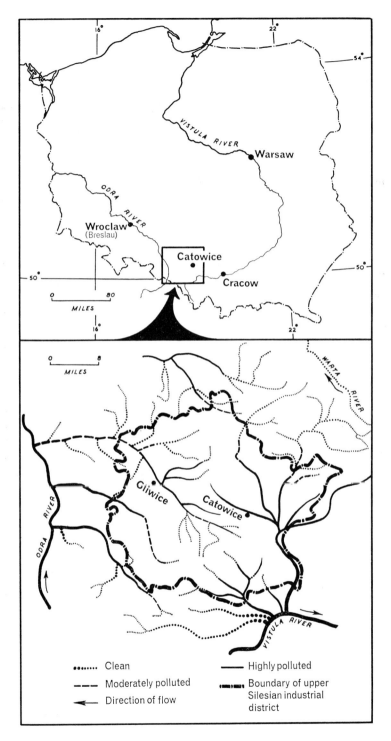

FIG. 61.—Stream pollution in the Upper Silesian industrial district. *Based on S. Gilewska.*

LIVING IN AN INDUSTRIAL SOCIETY

Although the subject of town planning as such is one which needs a separate volume for any reasonable treatment, a brief reference to one aspect of the topic might well be made here as an indication of the type of problem which industrial societies must face. We have seen earlier (pp. 107–109 above) that town planning and industrial location are not infrequently controlled in some degree by legislation, but the point with which we are here concerned is a different one. It arises in large measure from the well-known "pulsating" nature of industrial and city life. At certain times of the day (normally in the fairly early morning) cities and industries "breathe in" and draw towards themselves large numbers of workers, while at other times (normally in the early evening) they "breathe out" and expel workers of all kinds. This pulsation gives rise to two major peaks of tremendous traffic congestion along important roads, while even at other times in the day trunk roads and link roads can carry a very great burden of moving vehicles.

Movement of this kind immediately raises the question as to where all these workers and their families can live. It must be within access of their work, but the days have gone when it was enough simply to erect dwellings along the various roads radiating outwards from urban centres, for such roads at best are noisy and at worst are hazards lethal in character. Crossing them can be dangerous even for the physically active, while in some cases they can become virtual sewers for the disposal of exhaust fumes from cars, buses and lorries.

Various ways of tackling this problem have been devised; an early development in Britain was the housing estate which consists essentially of a network of residential roads removed from main highways. These roads are bordered with houses, while shops, schools and other amenities may be established to serve the estate.

More imaginative, however, is the grouping of houses and necessary amenities in "superblocks" or "neighbourhoods"; a possible layout of a superblock (a neighbourhood is essentially similar, but is on a larger scale) is shown in Fig. 62. Notable features include the positioning of houses facing footpaths, with access roads to garages in the rear, and also the arrangement of amenities such as shops, school and bus-stops in such a way that those approaching them do not have to cross any road at all. Underpasses are constructed to take footpaths beneath roads where necessary. No through traffic passes through the superblock, which is self-contained to quite a high degree. It is only for larger-scale expeditions to the city centre or to places of employment that families or workers

SOME PROBLEMS OF INDUSTRIAL DEVELOPMENT 287

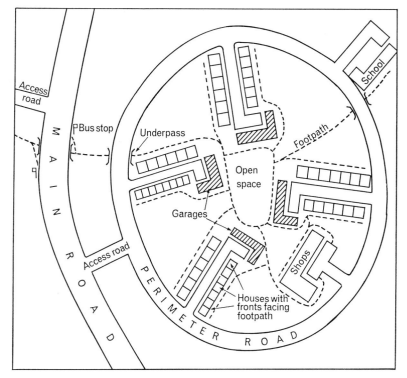

FIG. 62.—Schematic plan of superblock housing layout.

need to use the main road. It is also noteworthy that no residences are constructed along the main road itself (*see* p. 88 above for a further note on the function of such main roads).

Further discussion along these lines would take us beyond the scope of our present enquiry, but enough has perhaps been said to show that industrial communities throw up very real problems of living for their citizens. A great deal of thinking and experimenting is going on in attempts to find ways of dealing with these difficulties.

PROBLEMS OF DEVELOPMENT RELATED TO INDUSTRIALISED REGIONS

An industrial region is a highly complex form of human association, and it would indeed be surprising if the growth and development of such a region were not attended by many problems. Such problems may be expected on quite general grounds; for example, it would be a remarkable coincidence if the amount of work offered at any given time exactly matched the available labour force, and if changes in labour supply over a period matched changes in the demand for labour—in quantity as well as in quality (as, for instance, between male and female labour and as between skilled, semi-skilled and unskilled workers); if the volume of manufactured products coincided with total market demand; if the volume of replacement investment and new investment enabled the region exactly to adjust itself to economic fluctuations; and if manufacturers, even for a short time, managed to foresee market changes with any high degree of precision and were therefore able to orientate their investment accordingly.

In some cases there has been a very high degree of maladjustment between elements such as those mentioned in the previous paragraph, and the results in such cases have been unhappy indeed. An outstanding example, and one which we may usefully take as a case study, is that of South Wales,* and while it is not possible here to give anything like a full account of the developing economic geography of this region some of the main features may be mentioned. Fuller systematic accounts are available in regional texts.

A Case Study

The earliest industry to establish itself in South Wales was the manufacture of iron, an industry which first was located along the north-eastern rim of the coalfield where such towns as Hirwaun, Aberdare, Merthyr Tydfil, Dowlais, Ebbw Vale and Nantylgo in later years developed either in the strike vale associated with the outcropping lower coal measures or in valleys leading southwards from it. There were clear reasons for this location. Shallow and accessible iron-ore beds lay very near the surface, and this was a vital point when mining techniques were crude in the extreme; the area was forested and supplies of charcoal were therefore assured, while, at a later stage, coal was available for coking. The coal could be worked with comparative ease in drifts along the strike vale and in adits in the valleys to the south.

This was the area in which production of iron took firm root, but the industry was to suffer blows in the second half of the nineteenth century following the introduction of the new steel technology. The local ores proved to be unsuitable for use in the early Bessemer converters and open hearths because of the presence of impurities, especially sulphur, while the

---

* The term "South Wales" is a somewhat equivocal one, for much of the region normally included in the area thus designated lies, in fact, in Monmouth. Whether Monmouth is part of Wales is open to doubt—it certainly is not in all respects.

increasing scale of operations could not be matched by the local supplies of ore which were showing signs of exhaustion. Great efforts were made to secure supplies of ore from elsewhere, particularly from Northamptonshire; Beaver (1951) has pointed out that towards the end of the third quarter of last century the South Wales coal measures were being rapidly depleted, production falling from $1\frac{1}{4}$ million tons in 1872 to half a million tons in 1875, and that the consequent demand for ore from elsewhere was partly met by an influx of ore from Northamptonshire. This influx was greatly helped by the completion in 1879 of a railway from Blisworth, in Northamptonshire, through Towcester and Stratford-upon-Avon, to Broom Junction near Evesham, so providing a much more direct route than had been previously available. Unfortunately for the established iron-producing parts of South Wales, these measures proved to be simply delaying operations. Steel manufacture was increasingly attracted to the coast because of increased reliance upon foreign ores (p. 183 above) and severe depression struck the established iron centres. Stamp and Beaver, writing in 1933,* drew attention to the fact that "the huge [iron and steel] works, as at Dowlais, Ebbw Vale, and Blaenavon, have been closed down, probably for good." This prophecy has turned out to be incorrect, but only because of factors which could not have been foreseen by the authors in 1933; we shall return to this point later.

During the second half of the nineteenth century, however, when the iron industry was feeling the first sharp effects of the change to steel, another activity in South Wales, the mining and export of coal, was booming. A small export trade in coal had been developed in pre-industrial days, but as the Industrial Revolution got under way in Britain and elsewhere the demand for Welsh coal increased sharply. Not only did the demand for coking coal expand, but large amounts were required for bunkering on the ocean routes of the world which were increasingly being maintained by coal-burning vessels (Chapter I, p. 7 above). It is well known that Welsh steam coal is almost without equal for steam-raising purposes, and as a result of these developments, Cardiff, Newport, Barry and (later) Swansea became large-scale coal exporters. Unfortunately this happy state of affairs came to a calamitous end during the first quarter of the present century. The reasons for this well illustrates the point at issue—that it is sometimes almost impossible accurately to forecast events which will affect the economic pattern of an industrial region.

The period of large-scale coal exports from South Wales came to an end during the First World War. One reason for this undoubtedly was to be found in the increasing use of oil for powering ocean-going vessels,

* In *The British Isles*, first edition, 1933.

for this change sadly affected the market for Welsh steam coal, while on land Welsh coal met mounting competition during the inter-war period from thermal electricity, fuel oil and hydro-electricity. Even more recently natural gas and nuclear power have provided additional competition. Moreover, the European coal trade was seriously upset during the period following the First World War. French mines which had been wrecked during the war were re-equipped with modern machinery; Germany learned how to utilise her extensive lignite deposits and one effect of this was to release for export bituminous coals which entered into competition with Welsh coals; in some countries such as Sweden, Norway and Italy (and even in some non-European countries like Argentina) domestic resources of hydro-electric power were developed and the need for imports of coal therefore dropped considerably; post-war reparations imposed on Germany called for deliveries of German coal to creditor countries, so further inhibiting British (including Welsh) exports; political events such as the French occupation of the Ruhr, 1923–25, made for wild fluctuations in the demand for British coal as German output was subject to marked irregularities; and, finally, the periods of alternating boom and slump, that fearsome Scylla and Charybdis of the economic scene in the inter-war years, produced great uncertainties in the market for coal. It is not surprising, in view of this depressing list of obstacles, that the Welsh coal-mining industry as well as the export trade in coal suffered from extreme depression during most of the inter-war period.

The story so far has been bad enough with two of the major interests of South Wales savagely hit by circumstances mainly outside the control of the region itself, but even this does not complete the saga of distress. The stories, for example, of the tin and copper industries of South Wales might usefully be considered because they well illustrate the influence of political factors (which are apt to be very unpredictable) upon industrial development. The bases of the refining and smelting of the ores of tin and copper are to be found in the available coal near at hand and in the metallic ores of south-west England, where tin and copper have been produced since very early times, while other non-ferrous-metal industries include the production of galvanised sheets and the refining of nickel (which comes from Canada). The latter activity is centred at Clydach, but the non-ferrous-metal industries generally established themselves for the most part in the western half of the South Wales industrial region, particularly in the valleys of the Loughor, Tawe and Neath at such centres as Llanelly, Gorseinon and Neath.

Now the interesting point is this. While Swansea had developed by the 1880s into the world centre of copper smelting and refining, and while the processing of tin ore had also become a leading activity, today the branches of the copper industry which deal with the initial processing have completely disappeared (no copper ores have been brought to Swansea for smelting since 1924) though the refining of imported impure copper continues. On the other hand, the tin-processing industry is still a leading one. The reason for this apparently odd situation appears to be largely political. During the early years of the present century the copper industry experienced increasing competition from the U.S.A., a country which has become progressively more dominant in the mining and processing of copper ores; even the important Chilean copper mines have been developed with the help of American capital and skill. In these circumstances it was probably inevitable that South Wales (which possesses no native copper) should fall behind. Once again, a circumstance over which the region had no control proved to be a dominant one in the economic life of the area.* The story in the case of tin, however, was different because the U.S.A. has no domestic reserves of tin ore, and of the three leading world producers Malaya (now part of Malaysia), Nigeria and Bolivia, two were under British influence during the important developmental period. The South Wales industry therefore enjoyed a form of protection which enabled it to survive and develop. Another important related factor is that the refined tin is extensively used in South Wales in the tin-plate industry.

Apart from the tin-processing and using industries, however, the story remains on the whole one of general gloom relieved only by a few bright spots, and it is not surprising that during the inter-war years South Wales was savagely hit by economic depression. The public conscience was deeply stirred and it was realised that conditions were so bad that help from outside, probably government-sponsored help, was essential to revitalise the region. This feeling was given expression in the *Special Areas Act* of 1934, and after a slow start the amount of aid given directly and indirectly from government sources has become very considerable.† The following examples of comparatively recent developments help to show how South Wales has been enabled to climb slowly from the industrial depression of the inter-war years:

1. The National Coal Board has made big efforts in recent years to run

* Another important point is that it has become customary to smelt copper ore, which is normally a very low-grade ore, in the producing areas, so making possible large economies on transport (*see* Chapter II, p. 23 above).

† For further information regarding the question of government help see Estall and Buchanan, pp. 129–40, and map, p. 113.

the coal-mining industry on more rational and economic lines than has been the case in the past. Many small collieries have been closed down, and modernisation of techniques and equipment has been introduced in mines wherever possible. New collieries have been opened up, a notable example being that at Maerdy at the head of Rhondda Fach valley, where coal reserves estimated to be sufficient to last for the next hundred years are believed to be accessible. Other mines have been established at Nantgarw, to the south of Pontypridd, and at Cwm, a few miles west of Nantgarw, also on the southern part of the coalfield.

2. There have been many developments in the field of heavy industry, and the case of the Abbey Margam steel works was mentioned on page 100 above (*see* also Fig. 63). This works comprises blast furnaces, coke ovens, a coal- and ore-handling plant and a continuous-strip mill, the whole integrated works extending for about four and a half miles along the low coast of the Bristol Channel. The location of this works illustrates the pull experienced by much modern industry to tidal water, a pull reinforced in this case because of the import of iron ore from such overseas sources as Algeria, West Africa, Spain and Sweden.

A strong attempt has been made, however, to revive the heavy industry of the north-eastern part of the coalfield, in the area where the iron industry first took root. The main reason for this was social and not economic; there was a deliberate attempt to take work to the people rather than compelling the people to move to find work elsewhere. This was the new factor which falsified the prophecy of Stamp and Beaver referred to earlier. The first continuous-strip mill outside the U.S.A. was established at Ebbw Vale in 1937, despite an anomalous situation far from tidal water and on a narrow valley floor which makes for difficulties of layout and operation. Even so, there is a good supply of coal at hand and there are good road and rail communications with the English Midlands, one of the main markets for Ebbw Vale steel; nearly 40% of the bodies for motor-cars manufactured in England are made from this steel. In 1965 British Railways announced a scheme greatly to improve the rail link between South Wales and England by the innovation of special liner trains to move rapidly and efficiently steel from South Wales (and from other steel centres) to the Midlands.

3. Tin-plate works have been established at Trostre and Velindre in the western part of the region. In many ways a location nearer the Abbey Margam plant would have been more economic since that plant supplies the steel sheets used in the industry, but the development of these new large works involved the closure of many small works in the Llanelly-Gorseinon area. This closure would have given rise to considerable un-

SOME PROBLEMS OF INDUSTRIAL DEVELOPMENT 293

employment but for the establishing of the new mills in the same area. We see social rather than economic factors at work in this case.

4. One pressing need in South Wales has been to develop a wider range of industries than in the past so that undue dependence upon the mining and metal-working interests can be avoided. One way in which

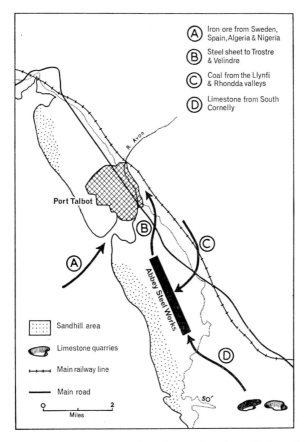

FIG. 63.—The Abbey Margam steel works. *Based partly on Hughes and James.*

this problem has been tackled has been in the setting up of government-owned industrial estates such as those at Treforest, Hirwaun, Bridgend and Fforest-fach (near Swansea). The first of these to be established was that at Treforest, located seven miles from Cardiff at a point where the Taff Valley opens out to offer more level land than is usually available in the valleys. After a slow start the estate picked up largely because of the interest in the years before the Second World War shown in the project by industrialist refugees from Hitler's Germany. A period of war production followed, and after the end of the war the manufacture developed of a wide range of light goods such as paper, clothing, confectionery, leather goods, tyres, soft drinks, glass and light metal goods such as zip-fasteners. Comparable developments have taken place on the other industrial estates.

5. The large factory of British Nylon Spinners at Pontypool represents another example of the attempt to diversify industry in the region. Until fairly recently the Pontypool plant manufactured all the nylon yarn made in the United Kingdom. It is near coal supplies yet is not on a "dirty" coalfield such as those of northern England, while good communications exist with the English Midlands and with north-east England (the raw material comes from Billingham on Tees—*see* Fig. 57).

As a result of these various developments the industrial pattern of South Wales has greatly changed since the First World War. Before that time about 66% of the total workers of the region were coal-miners but today the figure is only 20%. A much healthier situation therefore exists in which there is not undue emphasis on any single activity. Rawstron and Coates (1966) refer to "the quiet revolution that has occurred in South Wales, converting a very narrowly specialised economy . . . into one that is diverse beyond the hopes of the greatest optimist of 30 years ago." But the main point at issue in this analysis is simply that, while we can in retrospect trace the various causes which have helped to produce the present pattern of industry, it would have been impossible for anyone to forecast future lines of development with any notable accuracy at the beginning of the First World War or even later. We have been able to see from this case study that developing economic patterns in industrial regions are highly complex, and this in turn means that problems associated with that development are complex too. The reader might with profit turn back to Chapter VI, where he will find other examples of industrial regions which have had to face very difficult developmental problems.

PROBLEMS OF DEVELOPMENT ASSOCIATED WITH INDIVIDUAL INDUSTRIES

It is part of the essential nature of geography to study problems of development on a regional scale as we have done in the preceding section, but we should always bear in mind that in the industrial field problems arise on a regional scale because problems exist on a cellular scale—that is, in individual enterprises. These are the problems which must be faced by individual managers and firms, and attempts must be made to solve them if private enterprises are to be kept going.

By way of drawing further attention to this very important point we take as a case study the example of the coal-mining industry of the U.S.A., and since this is in itself an enormous topic we must confine ourselves to a particular set of issues—problems which have faced the industry since the close of the second World War and the ways in which these problems have been met. During the Second World War the demand for fuel was so great that almost any grade of coal, however poor in quality, found a ready market. Coal with an ash content of as high as 30% was being sold, and because demand continually outstripped supply, the total output, pit-head prices and profits all rose to unprecedented heights; in 1947, for example, the pithead price of coal was more than double the pre-war figure. This situation continued for a few years after the end of the war but then there came a dramatic change; production by 1962 had dropped by 27% despite an increase in productivity in the national economy as a whole of 50%, and further retrenchment seemed inevitable. The future for the coal-mining industry seemed far from bright. It will be helpful if we set out at this stage the main reasons for this depressing state of affairs, and the following main points may be noticed:

1. An activity like coal mining has very great difficulty in adjusting itself in the short run to changes in demand. By its very nature coal production is a long-term activity, and when once embarked upon is comparatively inelastic in operation. The most obvious form which this inelasticity takes is, of course, an inability to produce anything other than the single product—coal. There is no question of switching over to some alternative form of production.* Another important consideration is that it often takes as long as four years before a new pit can be brought into production, and quite early on in the establishing of a new mine the management has to decide upon the techniques which it will employ to mine the coal and bring it to the surface. Now, when once the appropriate decisions have been made and underground activities are shaped accordingly, it may not be possible to make any radical alterations to the pattern of

---

* This is a statement which may need modification in the not very distant future as ways are being discovered of increasing the number of uses to which coal can be put. Already we have seen at Sasolburg, 15 miles from Vereeniging in the Republic of South Africa, the establishment of the world's largest oil-from-coal plant. This is based upon extremely low-cost coal and has given rise to South Africa's largest chemical industry. It reflects the absence, as far as is at present known, of mineral oil reserves in South Africa.

production agreed upon, while even minor modifications to the plan may prove exceedingly costly. For example, new mines were opened up in Britain soon after the Second World War but they were not really needed at all by the time they were coming into production because of the rapid and unforeseen increase in the output of mineral oil in the Middle East and because of the increasing use of oil instead of coal. It is very difficult under these conditions to know whether it is best to go on producing coal uneconomically for a time, so further overloading the market but hoping that the situation will improve, or whether it is wise to close down all activities and so suffer exceedingly heavy capital losses. There is no easy answer to this problem.

2. The first real shock felt by the coal-mining industry in the U.S.A. after the Second World War came as a result of the adoption by the railways of diesel oil as their prime form of fuel rather than coal. The railways had previously been the mines' best single customer but between 1945 and 1960 their consumption of coal dropped from 125 million tons a year to a mere 2 million.

3. This same general period also saw a marked falling off in the use of coal for home heating. The home-heating market had previously been the industry's second largest customer but demand fell from 119 million tons in 1945 to 28 million tons in 1962.

4. The coal-mining industry suffered heavily because the distribution of its product was tied to one form of transport—rail. In view of the falling markets frantic efforts were made by the mines to cut down costs of production, but these efforts were constantly made of no avail because of rising transport costs. Between 1948 and 1962, for example, the price of coal at the pithead *fell* by 45 cents a ton, but over the same period transport costs *increased* by 57 cents. The railways always had carried most of the coal between mines and consumers and they saw no reason why there should ever be any radical change; they took the coal traffic for granted and saw no reason to offer special concessions to attract coal haulage.*

5. Meanwhile, the threat of possible new competitors continued to keep mine managers awake at nights. In particular, there was a real danger that another substantial customer, the thermal-electricity industry, might make the change to fuel oil. The demand for coal from this industry had been steadily increasing in contrast to the demand from most other customers (Fig. 64), and it was vital for the coal industry that this large market be retained. The demand in other branches of industry generally remained fairly constant but there was the ever-present threat from other energy sources to be reckoned with; we have seen, for example, how the

* The canal carriers adopted a similar attitude in their heyday; *see* p. 6 above.

SOME PROBLEMS OF INDUSTRIAL DEVELOPMENT 297

iron and steel industry is now using mineral oil injection in blast-furnace operation (p. 172 above) and this is by no means an isolated instance.

This, then, was the general situation which threatened the coal-mining industry in the U.S.A. a few years ago, and a very gloomy situation it was. Yet today there is a new feeling of confidence in the industry, and the situation has been transformed, largely by the efforts of the enterprise itself. The following are the main reasons for the dramatic change:

1. The fundamental point is that the industry has subjected itself to a complete overhaul, new techniques and a high degree of mechanisation making possible far more efficient production. Former techniques of

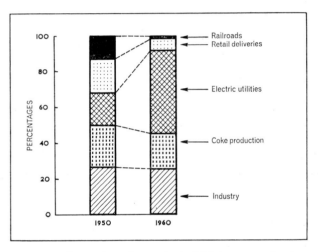

FIG. 64.—Changes in coal usage in the U.S.A., 1950 and 1960.

drilling and blasting have gone, while the miner armed with pick and shovel has almost entirely disappeared except in a few marginal pits. In their places are such appliances as the large-scale electrically driven continuous-mining machines which eat into the coal with the help of circular cutters; the coal is mechanically broken into fragments and carried automatically to the rear of the machine. Here it is loaded (again mechanically) into a shuttle car which carries it to an endless belt conveyor, and this conveyor carries the coal to the mine cars which haul it out of the mine to the processing plant. The formerly difficult problem of getting the coal out of the mine is thus solved. This is the general pattern of operations at a drift mine in Pennsylvania in which 440 mine cars and 17,000 feet of conveyor belt carry coal mechanically over the 19 miles between the coal face and the processing plants, and this kind of working is a normal one today. About one-third of coal mined underground in the U.S.A. today is produced by continuous-mining machines, about 90% is loaded mechanically, while about one-third is transported out of the mines by belt conveyors and two-thirds by mine cars.* As a corollary to the increased mechanisation the number of underground workers has sharply fallen.†

Giant power shovels have greatly increased the possibilities of opencast mining (known as strip mining in the U.S.A.). A six-foot seam at the Sinclair mine in Kentucky, for example, is now being mined with the help of a giant shovel, the boom of which reaches as high as a twenty-storey building, which rips away 173 tons of overburden each time it scoops into the ground, and which digs its way forward at a rate of about 600 feet a day. Another excavator at Cuba, Illinois, can move no less than 3500 tons of overburden an hour! Machines of this type strip off the overburden and expose the coal-seam; smaller shovels then remove the coal and load it into trucks. One signal advantage of stripping is that it can extract over 90% of the available coal as against a possible maximum of 70% in underground working. As a result of mechanisation, therefore, it is now economic to mine seams which it would not have been possible to exploit a few years ago, and almost one-third of all coal mined in the U.S.A. is produced from opencast operations.

In fairly hilly terrain a stripping machine can sometimes follow the line of an outcrop of coal along a hillside, possibly along a contour, but it may not be practicable to dig into the hill to follow the seam because of the rapid increase in the amount of overburden. In these circumstances the auger, first tried out in 1945, may be used. Under this method a battery of giant mechanically operated Archimedean augers with diameters of up to 60 inches may be screwed as far as 200 feet into a seam from the open face,

* It is fortunate that the mode of occurrence of coal in regular seams over much of the eastern U.S.A. (Fig. 14) permits the maximum use of mechanised forms of coal mining. In other areas where seams are folded and fractured there is less future for this method of production.

† Comparable developments have taken place in Britain, where the proportion of coal produced by mechanical means rose from 3% in 1947 to 78·2% in 1965. It is anticipated that the figure will be 100% by 1970. The introduction of electronics is having marked repercussions. Thanks to electronic equipment it is now possible to produce coal by remote control, no workers being needed at the coal-face while the remotely controlled machines are actually in operation. At Bevercotes Colliery (Nottinghamshire) all operations from coal-face production to coal preparation (cleaning, washing, grading) are remotely controlled; the manager of the mine has a complete picture in his central control room of all activities in the colliery. Coal-mining has traditionally been a labour-intensive industry; it is fast becoming capital-intensive.

the coal being forced back along the rotating screw. Rather special conditions are needed for augers to be used successfully, but almost 5% of the coal now produced in the U.S.A. is auger mined, much of it in West Virginia. Auger mining can extract up to 75% of the available coal.

The foregoing are simply selected examples to illustrate the sort of technological revolution which has come over the coal-mining industry in the U.S.A. in recent years and which has helped to revitalise the industry. As a result of changes like these the number of miners employed in the country has sharply fallen from 436,000 in 1948 to 120,000 in 1965. At the same time, however, the output per worker has soared from 6 tons daily in 1915 to 25 tons in 1963 in opencast mines, and from 4 tons daily in 1915 to almost 11 tons in 1963 in underground workings; in auger mining the comparable figure works out today at about $30\frac{1}{2}$ tons per day per worker. Average productivity per miner in all types of mine is now more than 11 tons daily, and it has more than doubled between 1940 and 1963.

2. It is less usual today to sell "raw" coal to consumers; whereas in 1940 less than a quarter of all coal produced in the U.S.A. was cleaned, the corresponding figure today is more than two-thirds. Mining interests now have to exert themselves to give the customer the kind of product he requires, for there is still the very real danger that if a customer is not satisfied he will turn to some alternative form of fuel and energy. There is also the further point that the greater mechanisation of the mining industry means that more waste is produced since the large shovel is less selective than the miner with pick and shovel, and this waste must be eliminated before coal is sold. In some mines as much as 35% of the raw product consists of waste.

Cleaning of the coal is normally effected by means of gravitational or centrifugal methods which depend on the fact that coal is lighter than other rocks. In one gravitational system, for example, the raw coal is thrust into a heavy magnetic fluid on which the coal floats while the impurities sink and can be drawn off. Froth flotation is used to clean the fine grades of coal. Large quantities of water are needed for cleaning purposes, and after cleaning the coal must be dried, graded and (where necessary) blended. The need for grading arises because coal of different sizes is required for different forms of use, and screening techniques are used for separating out the various grades. Increasing amounts of coal are now being crushed to very fine grades because of the greater use now being made of automatic stokers; in 1940 only 8% of all coal produced was crushed whereas in 1963 the figure had risen to 44%. Blending is sometimes necessary to produce a coal more suited to consumers' needs;

for example, a steam coal which gives off great amounts of smoke during combustion may be made more acceptable by blending with a low-volatile coal.

Further processing may take place to extract useful products from the coal, and this form of processing is on the increase. Gasification processes, for example, which involve the passing of steam and hot air through heated coal, produce gases which can be used for direct burning in factories and in homes, while hydrogenation at high temperatures and under high pressures can yield synthetic petrol and useful petrochemicals from coal. This is a clear indication that coal interests today are not content with simply maintaining existing markets but are eagerly seeking new uses for their product. It is hardly too much to talk of a technological revolution in the coal industry since the close of the Second World War.

3. Since 1950 the coal-mining industry has succeeded in bringing about a complete transformation in the methods of distribution of coal to consumers; the bad old days which we commented upon earlier, when the railways took advantage of their monopolistic position with regard to the transport of coal, have gone for ever, thanks very largely to the efforts of the coal-mining industry itself.

The first stage in this campaign waged by the coal companies took the form of an increasing use of coal barges, especially on the Monongahela, Ohio and Mississippi Rivers, the Great Lakes and the Erie Canal. Between 1950 and 1960 the tonnage of coal carried by water increased by 70% while that carried by rail fell by 27%; in 1940 only $6\frac{1}{2}$% of all coal transported in the U.S.A. was water-borne, but in 1963 the figure was $13\frac{1}{2}$%. The Monongahela River now carries nearly 30 million tons of coal a year and the Ohio about 25 million, most of which is destined for the by-product coke ovens of the Pittsburg area, while the Great Lakes carry up to 60 million tons a year. Each group of barges towed by water carries the same amount of coal as four 150-truck coal trains—a by no means negligible loss to the railways.

A second innovation came about as a result of an invention developed by the electric companies, as they too were keenly seeking new methods of cutting down on the price of fuel. The invention was known as Extra High Voltage—EHV—and essentially this takes the form of a technique for carrying so much power along EHV transmission cables that one of the new transmission lines can carry as much power as six orthodox ones. This made it possible to consider the setting up of generating stations near the coalmines, so cutting down transport costs to a bare minimum, for it became much more economic to despatch the electric power with the help of the EHV line rather than to transport an equivalent amount of coal. The first such power station was established near the Sinclair mine which we referred to earlier, and the giant shovel which we mentioned was erected to produce coal for the T.V.A. Paradise generating plant. This was in 1957, and this development alone meant a further loss to the railways of a complete trainload of coal every day—a very substantial loss in freight revenues. And this could have been only the beginning of such enterprises!

The most recent development, however, is perhaps the most startling of all—a pipeline to carry coal. It took a long period of experiment and research before this proved to be a practicable expedient, but success was finally achieved by crushing and powdering the coal and mixing it with equal quantities of water to form a "slurry" comprising 50% coal and 50% water. It was found to be possible to transmit this slurry by pipeline. In 1959 the experiment was tried of pumping such a slurry into barges on Lake Erie and despatching it via the Erie Canal and the Hudson River to South Amboy, New Jersey. From there the slurry was hurried through a pipeline directly to the furnaces of a generating plant and was burned as if it were oil.

The experiment proved to be a great success, and the lesson of it was not lost upon the railways, especially when further developments of the coal pipeline technique materialised. The most notable example was the ten-inch pipe which has been erected in Ohio between Georgetown and Cleveland to carry slurry to a generating plant in Cleveland; this 110-mile-long pipeline each year carries about 1·2 million tons of coal and saves about one million dollars in transport costs as compared with more conventional methods of transport. More recently it has been shown that a "stabilised slurry" comprising 65% coal and only 35% water can be transmitted successfully, and there are possibilities that longer pipelines will be constructed to carry coal over even longer distances. It is not therefore surprising that in 1963 the Baltimore and Ohio Railroad announced a dramatic change of policy with regard to the transport of coal in a desperate attempt to retain this valuable traffic.

The new policy was effected with the help of the "unit train," a distinctively marked coal train which runs to as firm a timetable as a passenger express. Trucks are fully loaded at the mine and there are no halts between the mine and the consumer. The customer is fined if he keeps a truck longer than twenty-four hours and the empty return journey is made to just as firm a timetable as the outward journey. Coal transport has thus become more efficient than ever before, while transfer costs have been effectively cut.

This study of a mining industry will serve its purpose if it succeeds in emphasising that individual industries have their own problems of development, sometimes problems which are not of their own making but which must be solved if continued success is to be achieved. Other examples of the same sort of thing can be met elsewhere in this book (*see*, for example, Chapter VIII, The Iron and Steel Industry). It may also serve to illustrate the contention which is often made that a new industrial revolution which we might term the "Technological Revolution" is now in being, and the results of this new revolution may turn out to be just as far-reaching in our day as the results of the old Industrial Revolution were in an earlier time.

PROBLEMS ASSOCIATED WITH THE INDUSTRIALISATION OF DEVELOPING TERRITORIES

The question of the raising of the living standards of the millions of people who inhabit the "third world"* is perhaps the most pressing problem facing the world today, and possibilities of establishing industry in the territories concerned must seriously be considered as a possible means towards a solution. Developing countries themselves are quick to see that the more prosperous parts of the world are on the whole those which are highly industrialised and it is only natural that these poorer territories should expect increased prosperity if only they can develop their own industries. Unfortunately, however, the matter is by no means a straightforward one as we can see from a consideration of three important issues involved—capital formation, labour and the pattern of industrial development involved. We shall proceed to examine each of these points in order.

Capital Formation

It is merely expressing the obvious to state that the development of industry depends (among other things) upon capital formation, but at the same time it is perfectly true. Now the dilemma of the poorly developed territory is that it can provide only very limited capital from its own resources because the *per capita* income of its people is so low; since this is the case there is a corresponding limitation upon capital investment and this in turn means that productivity and therefore incomes remain low. This "vicious circle of poverty" was recognised by Nurkse (1955) and it is expressed in diagrammatic form in Fig. 65(*a*). The very poverty of a poorly developed territory is a fair guarantee of continued poverty.

* The "third world" comprises the territories which are aligned politically neither with the Western nor with the Eastern (Communist) Powers; these "non-aligned" territories include most of the developing countries of the world.

SOME PROBLEMS OF INDUSTRIAL DEVELOPMENT

This is one main reason why for so long poor countries have remained poor; they are in the grip of an economic situation which is very difficult indeed to change. It is true that with the help of increased diligence and more efficiency some increase in productivity may be expected, but this will be true only to a limited degree; the vicious circle will still operate,

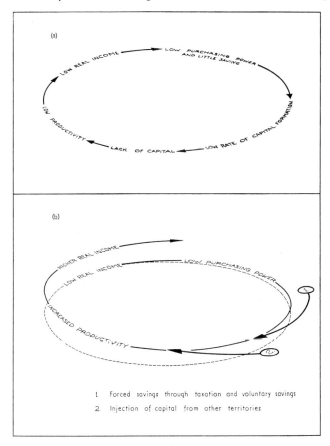

FIG. 65.—(a) The vicious circle of poverty; (b) the spiral of developing production.

albeit at a slightly higher level. An external impetus is needed radically to transform the situation, and this stimulus can only come about as a result of the application of *increasingly skilled* labour made possible by *increased capital investment.*★

★ One is reminded of the old distinction between doing hard work and working hard. It is said of Henry Ford I that he detested the first but loved the second. But

L

Mountjoy states that most developing territories save only 5 or 6% of their net national income; this represents a very limited amount of money in view of the low level of incomes involved. It is said that the average *per capita* annual income over much of the "third world" is less than 150 American dollars (£62 10s.). Now since new capital must come out of savings the inevitably low rate of saving under these conditions very seriously inhibits *increased* investment, not only because the rate of saving itself is low but for two other reasons:

1. Much of the new capital will, in fact, be required simply to replace existing worn-out capital. Machinery wears out, buildings fall into disrepair, railways and roads need periodic overhaul—all these things mean that to remain effective capital must constantly be renewed.

2. In nearly every territory of the modern world population is increasing, in some cases in a startling manner. This means that an increased rate of investment is necessary simply to maintain any given *per capita* volume of investment.

The fact is that an investment ratio of about 5% of the net national income will only just about suffice to keep the *per capita* investment ratio constant, and in such a case little if any increase in this ratio can be expected. The vicious circle of poverty will remain unbroken, and the situation expressed in Fig. 65(*a*) will be maintained. If any significant increase in this ratio is to be achieved a notably higher level of investment is essential. Rostow has pointed out that a developing economy may be viewed as progressing from a "traditional" type of structure through a "take-off" stage to a period of maturity; a later development into a stage of high mass consumption may follow. The chief difficulty is to enter the take-off stage, but this entry can normally be attained when the rate of productive capital investment rises from 5% of the net national income to over 10%.

Now this is not an easy target for a developing territory to reach, and the situation in the contemporary world is rendered much more difficult because everyone is in a great hurry. We are apt to overlook the fact that the industrialisation of Britain was a slow and continuing process which took place over a period of at least a century, though it does not follow that developing territories today must anticipate such a lengthy transition period. What it does mean, however, is that the transition from a traditional to an industrial society is not an easy one; the vast amount of human

---

there is no alternative to doing hard work if there is little capital investment to make possible the use of machinery and labour-saving devices generally.

suffering which accompanied the transition in Britain in factory and in mine can hardly be estimated, and while we can safely say that thanks to the lessons we have learned the amount of suffering in developing societies today need not be so great we dare not say that these societies can anticipate a painless period of development.

The question still remains—how is the supply of capital necessary to ensure progression into the take-off stage to be secured? It must at once be conceded that there is no easy answer to this question, but this is not the same thing as saying that there is no answer at all. It can, for instance, be argued that the saving, even by a poor country, of 10% or slightly more of its net national income is by no means an impossible task if there is the will to do it. This does not mean, of course, that a 10% rate of saving must be expected from everyone—that would manifestly be absurd. What we must remember is that in developing territories it is normal to find extreme inequalities of wealth; differences of income experienced between the rich and the poor are far in excess of what is normal in the developed countries, where inequalities are to a large extent modified by such devices as taxation. In the third world it is not unusual for excessive affluence to exist almost side by side with utter poverty.

The unfortunate feature is that great wealth in developing territories is often directed into unproductive channels; not infrequently it is locked up in the precious jewels and gold which form the hallmark of the wealthy Eastern prince, while luxury goods which are commonly purchased from overseas by the wealthy citizens of some territories of the third world do nothing to help along the economy of the country as a whole. The crying need is for this wealth to be channelled into productive enterprises, but there are commonly two major obstacles in the way of this; firstly, there is no tradition of financial enterprise, and secondly, there are few, if any, financial institutions to guide and offer assistance to the potential investor. In developed countries there are many enterprises which will accept money, invest it with the help of specialised knowledge, and pay the investor interest on his investment, while insurance companies too assemble large sums of money which become available for investment. The lack of financial guidance in developing territories is a very real handicap to productive investment. One method, however, by which such investment can be achieved is by means of taxation and the subsequent initiation of government-sponsored projects. This is part of the point of arrow 1 on Fig. 63(b).

In many ways this is an unhappy argument, though a person's reaction to it will largely depend upon his political philosophy. Many will consider it unfortunate for the influence of any government to be extended, as is inevitable if the government enters directly into economic activities, and there will be those who will loudly aver that any governmental participation in economic affairs is a sure guarantee of future inefficiency. But it is difficult to know what alternative is available, and in fact the issue is by no means as clear-cut as we may so far have implied. The reason for this is that in developing territories the possession of wealth confers power, power on a scale probably unknown (in a comparative sense) in democratically governed industrial societies, and this power may well be used to block the initiation of such measures (*e.g.* high taxation) as have here been suggested. Again, much of the wealth in poorly developed countries stems from the possession of land and it normally takes nothing less than a civil revolution such as has taken place in Egypt, the U.S.S.R., Mexico and elsewhere to secure the land reform which must be an initial step towards a broader economic development.* Even so, the situation is not an impossible one and Rostow mentions territories such as Argentina, Turkey, India, Colombia and Venezuela which have moved into the take-off stage since the 1930s'. The comparable example of Spain was referred to in Chapter VI (pp. 137–140 above).

A second possibility remains—that of securing supplies of capital from more developed territories, in the form of either loans or outright gifts; this is the point of arrow 2 in Fig. 65(*b*). Very substantial grants have in fact been made by governments such as those of the U.S.A. and the United Kingdom to developing territories, and these territories have also succeeded in attracting loan capital on a far from negligible scale.

Unfortunately, foreign capital is generally viewed with considerable disfavour in developing territories. This can easily be understood. The fact that a person (or a country) may recognise an urgent need to receive help in the form of grants or loans to tide over a period of difficulty does not in itself easily reconcile the recipient to the situation. Indeed, a person or a community with pride is more likely to resent the whole set of circumstances which makes the receiving of aid necessary, for no mature person likes to feel himself indebted to another. This is in large measure the

* One is reminded of the enclosure movement in England which was a forerunner of the Industrial Revolution. In many countries the situation is that the large-scale landowner waxes fat while the cultivators dependent upon him remain poor; they have no incentive to develop the land because of their subservient social and economic status. Note, however, that wholesale sequestration of land and the parcelling of it out among cultivators is of little use in helping along economic development if it simply results in the formation of a "petty" peasantry, *i.e.* a peasantry whose individual members have only very small plots of land to work. Such peasants must be impoverished and will be in no position to enter into the commercial life which a developing territory must try to build up.

SOME PROBLEMS OF INDUSTRIAL DEVELOPMENT   307

background to the contemporary agitation against "neo-colonialism." It is also true that a territory which is financially indebted to another may well fear that the economic bonds may lead to political ones at a later stage.

There is, however, a second reason for the general feeling of hostility towards foreign capital. In years past most private investment in poorly developed countries took the form of investment in extractive industries and to a lesser degree in agricultural activities such as the establishment of plantations. At this time markets for consumer goods in those territories were extremely limited and possibilities for investment of profits in other sectors of the economy were almost non-existent; almost the only thing an investor could do with his profits, therefore, was to take them out of the country, possibly for investment elsewhere where more financial opportunities presented themselves. The benefit to the borrower country under these circumstances was therefore minimal, and it was natural that resentment should develop against the foreigner who apparently simply lent his money to enrich himself while rendering minimal aid to the borrower. The situation today is changing rapidly, and in some developing territories has changed completely as opportunities for domestic investment now exist which provide openings for the foreigner to reinvest profits. It is also true that investing companies today spend large sums of money in the territories in which they operate. Wages must be paid, local produce bought, and in many cases provision is made for social services such as hospital treatment, education and transport facilities; many of these are services which in a more developed country would be the responsibility of the state (p. 260 above). Jarrett (1956) has examined the situation at Lunsar, Sierra Leone, where an iron-ore mining company in 1952 spent almost £2½ million within the country—a total which represented a sum of about £1 2s. 7d per head over the whole population.

But feelings of resentment die hard, and as a result we find a paradoxical situation developing as many territories which desperately need foreign capital create circumstances which help to frighten it away. Fear of civil disorder or of some form of expropriation will always discourage overseas investors, but so will restrictions legally imposed on the repatriation of profits or of capital. The same is true of any form of penal or discriminatory taxation; of legislation which requires that foreign investors shall employ a certain proportion of local labour; of directed investment; and of the fear of import and export trading licences which can seriously impair efficient working. All of these restrictions have been imposed by developing territories (though not all at once!) and one is sometimes tempted to marvel at the amount of foreign investment which actually does take place; with ever-increasing demands for capital in the "developed" territories themselves, it could be that the "developing" countries may find it increasingly difficult in the future to borrow capital from overseas.

If this situation does develop it will be a most unhappy thing for the developing nations as Lewis has pointedly reminded us (p. 102 above). The fact is that with increasing applications of capital in the poorer parts of the world the way becomes open for a movement away from the vicious circle of poverty towards the spiral of developing production (Fig. 65(b)); increased production means higher real incomes which in turn can lead to a higher rate of saving and investment as the spiral develops.

Before we leave this point there are three further matters which deserve attention. The first is that Fig. 65(b) is very misleading in one respect since it suggests a smooth and even progression of the economic spiral throughout the whole economy. This will not be the case. Investment will take place selectively, in certain industries and in particular places, and in the first place a comparatively limited number of people will benefit. The point is, however, that this benefit will in time spread over a wider sector of the community according to the principle of the "multiplier" theory, which has reference to the fact that, when once generated, money wealth produces well-being over increasing circles. It might work in this way. A factory or a group of factories is established in a certain community and the necessary workers are employed. Extra money therefore flows into the community because these workers are earning money which previously they were not able to do. This money is spent and shopkeepers feel the benefit in their turn fairly quickly. But the benefit does not stop at that point because shopkeepers begin to place larger orders with wholesalers and the wholesalers in turn step up their orders to manufacturers. Other retailers may start to build up larger stocks of foodstuffs in anticipation of increased demand, and this stimulates farmers to greater activity. Notice that neither wholesalers, manufacturers nor farmers may operate in or near the community which gave rise to the initial economic impetus. This community, meanwhile, begins to find that its own resources are increasing as its income from local rates steps up and it finds itself able to undertake additional public work; for instance, new roads may be laid and old ones repaired, while schools may be built for the extra numbers of children who are moving into the area with their parents. These parents need homes in which to live and so the building and constructional trades are further stimulated, while tertiary services are extended as public transport may be developed and as more doctors, lawyers and teachers move into the community. All this means that the wealth generated initially by the new factories is slowly spreading as the multiplier process operates. The scope of the benefits received by the community as a whole will, of course, depend fundamentally upon the size of the primary industrial basis; a small enterprise can produce only limited and fairly local benefits but large-scale enterprises or a significant number of small ones will result in more far-reaching effects.

The second point to which we should draw attention is this. Earlier reference was made to the markedly uneven distribution of wealth in developing territories, a situation which stands in contrast to the more egalitarian state of affairs which prevails in most developed countries. The natural tendency of the inhabitant of the egalitarian society is to decry marked inequalities of wealth as wrong, even as immoral. There are strong grounds, however, for arguing that a fair degree of maldistribution of wealth is beneficial since it gives rise to saving and investment by the wealthy, and, therefore, in time it produces increased production and generally rising incomes. Clearly, this point must not be pressed too far, but it is difficult to deny that the wealthy industrialists of nineteenth-century England, derisory as they often appeared, living comfortably in their newly erected mansions, sending their children to the "right" schools and marrying off their daughters to impoverished sons of noblemen, rendered a real service to the community as they ploughed back the greater part of their profits into ever-expanding enterprises so that industry was able to develop in a manner previously unknown anywhere in the world. It was very largely thanks to their efforts that Britain has been able to engage successfully in two costly world wars and to develop the welfare state, and the pace of industrial development in Britain could never have been so rapid if these men had been taxed in their time as their successors are today.

The third point is, in fact, a word of caution. Vital as capital investment is to the growth of industry it is only one of the factors involved out of many, and there can be no guarantee that productivity in a developing territory will rise in proportion to any given scale of investment. Other variables involved include labour, raw materials, power, transport and managerial skill, while political unrest can sometimes react upon industrial output with devastating effect. (For an early example of this *see* p. 12 above.) Neither must the question of market demand be overlooked, for in a developing territory restricted demand for the products of industry can notably inhibit production. This point is examined further in Chapter XIII.

Labour

This is a matter of very considerable complexity and it can only briefly be touched on here; we shall consider simply a few relevant points.

The amount of labour available in any community is obviously closely related to total population, and we shall do well to bear in mind that in many developing territories, notably in Africa and Brazil, it is frequently impossible to secure adequate supplies of labour for industrial enterprises. The Rand region of South Africa and the copper belts of Zambia and Katanga, for example, have never found it easy to recruit all the labour which they need. This is a situation which in part has its origin in the troubled history of the regions concerned and partly in the high mortality rates which are so often met with in inter-tropical regions. This point is elaborated by Gourou (1953) but it would take us rather far off our main theme to dwell on it here.

The question at issue, however, is not simply one of the quantity of labour; its quality is of the highest importance. The present writer has pointed out elsewhere (in *Africa*, 1966) that it is by no means an easy matter for a young man fresh from a traditional environment to adjust himself to life in an industrial setting, and it is all too frequently found that moral standards and quality of work both suffer. The custom of working long hours at a repetitive job is in itself strange and unattractive, and a complete reorientation of mental outlook is necessary before a man can adjust himself to such work after the comparatively easy-going life in the villages. We should also bear in mind that workers are often prevented from giving of their best by the scourges of malnutrition and disease, and it is not a simple matter to eradicate these troublesome features of inter-tropical life.

Masser has published a study relating to employment in Southern Rhodesia (now Rhodesia), and he has some very pertinent points to make regarding labour turnover. This is clearly an important issue, for maximum industrial efficiency demands some measure of labour permanency, while a high labour turnover rate must impair efficiency. Admittedly, the case selected for detailed study, a tobacco factory in Salisbury, is not entirely representative because of the seasonal nature of the work. Work in the factory commences soon after the start of the tobacco auctions in March and continues until November, but between November and the following March no work is available except routine maintenance. This break in work continuity has undoubtedly affected adversely labour stability, but on the other hand the enlightened management of the enterprise has made great efforts to retain workers by improving conditions of service.

SOME PROBLEMS OF INDUSTRIAL DEVELOPMENT 311

Perhaps the most important of Masser's findings appear in statistical form showing the "survival rate" of African male staff at the factory; the numbers given are adjusted so as to give rates of labour survival per 1000 workers taken on at the beginning of each season. Two marked features emerge from a study of the statistics, the first being that the "survival" rate is very low. For an adjusted figure of 1000 workers employed at the beginning of the 1951 season only 642 completed even one season's work; 185 completed two seasons, 67 three and only 41 four. The last group of complete figures is that beginning in 1957, when 780 workers completed one season, 267 two, 153 three and 111 four. This particular finding adds point to what was said above regarding the disposition on the part of workers unaccustomed to industry not to apply themselves to it over long periods.

The second feature is far more encouraging, however, and it appears in the figures given in the previous paragraph. It is that there is an increasing tendency for workers to remain for longer periods. Whereas in 1951, for example, only 642 workers per thousand remained for even one season, the comparable figure for 1960 was 833; while 185 per thousand beginning in 1951 remained for two seasons, the corresponding number of those beginning in 1959 was 540. The exact reasons for this change are not known, but relevant factors appear to include an enlightened management, increasing population pressures in the rural areas, loss of cultivation rights following the *Native Land Husbandry Act* of 1951 and a slowing down of economic development in Rhodesia after 1957 which meant that jobs were harder to find; workers already in employment therefore tended to remain in their jobs.

As far as they go, the above conclusions are encouraging, and they underline the need for a new outlook in many developing territories if industrialisation is to continue to develop. We spoke earlier of the frightening amount of suffering which accompanied the rise of the Industrial Revolution in Britain; poverty was rampant and working conditions for men, women and children in the mines and factories were often unbelievably bad (*see* Fig. 7 and p. 33 above, for example). It seems inevitable that periods of transition in human affairs are painful times for many people, and, while we may have learned enough to be able greatly to mitigate the suffering, it is unlikely that we shall ever be able to banish it altogether.

And this, after all, is only the negative aspect of the situation. On the positive side men must be prepared actively to work to bring about the economic transition and to undergo strain and uncertainty in the process. Financial leaders must be prepared to take the risks of manipulating available resources of capital and must learn to apply them where they are most needed, while entrepreneurs must be ready to bear the uncertainties involved in the marrying of technology and capital in productive enterprises. Others must be willing to lend money for lengthy periods, sometimes in risky ventures, and this is quite a different activity from the money lending and the speculating in real estate which are by no means unknown in traditional and little developed societies. Considerable numbers of workers must make themselves available for serious training in the new industries and must put behind them the irresponsible traditional ways of life which are natural enough to the nomad or the unskilled cultivator but which are simply drags on progress in any developing economy. In short, a complete revolution in social and economic thinking is required, and this is something which is not easily brought about.

The Pattern of Development

Mountjoy has examined at some length the various factors which should influence a developing territory which is seeking to select the type of industries she will encourage, and we shall here mention simply one or two main points. The first is that development, to be healthy, must be balanced; a developing industry should, for example, march alongside a developing agriculture. This point was brought out in our study of Spain in Chapter VI. The ideal situation is comparable with that which we briefly examined in Chapter I, where we saw how the development of the prairies of North America helped, and was helped by, the developing industries of North America itself and of Europe. A prosperous agricultural society will be in an advantageous position to purchase many industrial products such as fertilisers, insecticides, petrol, lubricating oils, machinery of all kinds (including tractors, lorries and motor-cars as well as combine harvesters), constructional materials of all kinds from roofing felt to barbed wire and cement, and many other items besides. Such a domestic market must help to produce a favourable environment for industrial development, while the workers in the towns and factories can in their turn purchase goods from the countryside. This is one important reason why great encouragement is today being given to the developing of improved systems of agriculture in the developing territories.

Some forms of industry are probably not very suited to the economies of developing territories. Heavy industry, for example, by reason of the very large amounts of capital required and because of the need for large-scale working in order to secure efficiency of operation (p. 181 above), is not generally well adapted to the needs of these territories, though in some cases there has been development along these lines for reasons of prestige

(p. 108). This point well illustrates an important feature of many forms of industrial development—they have to be established on a large scale or not at all; there is no question of steadily progressive development. We noted in Chapter VIII (p. 191 above) the failure of the fairly recent Chinese effort to produce pig iron in a large number of small blast furnaces. The output of twenty small furnaces is by no means equivalent to that of one large furnace of comparable producing potential as far as volume is concerned, for vital matters of technology are involved.

On the other hand, there are many forms of industry which are well suited to establishment in developing territories. It is perhaps possible to recognise five groups, though the following list is not meant to be exhaustive:

1. Those processing materials of local origin for which there is a strong demand in world markets; these include the extraction of vegetable oils, cotton ginning, plywood manufacture and tobacco processing.
2. Those which utilise materials of local origin in the manufacture of goods for which there is a buoyant domestic demand; these include the production of such commodities as soap, textiles, cement, foodstuffs and beverages.
3. Those which make use of imported component parts in the assembly of comparatively sophisticated end products such as bicycles and motor-cars.
4. Those which use imported raw materials which form the basis of processing industries; examples include flour milling from imported grain and oil refining.
5. An industry such as oil refining can be used as the basis for a chemical industry, and such an industry can be a most valuable asset to a developing country which needs large amounts of fertiliser to improve its agricultural output. It is possible to argue that a country such as India has made a mistake in emphasising so strongly the establishment of heavy industry, and in paying so little attention to the manufacture of the fertilisers which are so desperately needed on its farms. This point is referred to again in Chapter XIII (p. 328 below).

The foregoing are simply intended as examples of the sort of thing which is possible; the important thing to bear in mind is that development is cumulative, and wisely selected initial enterprises will stimulate further development, so hastening progression into and through the take-off stage of industrial development.

MEXICO: A CASE STUDY OF A DEVELOPING TERRITORY

To close this chapter we shall take an example of a developing territory as a case study—that of Mexico. The example of Mexico has been chosen because this is a country which has been traditionally backward, but which has in recent years deliberately embarked upon schemes designed to raise the standard of living of its people. Mexico is a developing territory rather than a developed one. Incomes are still on the whole low; for instance, the average annual *per capita* income was still only about £173 in 1965, but even this modest figure represents a real economic advance over the past thirty years in a nation of more than 42 million inhabitants.* In the ten years 1945–54 the index of industrial production rose by 38%, while the current rate of economic growth in the country is a very impressive 6·3% (7·0% in 1966) which compares very favourably with the 4·6% of the U.S.A.

Relief and Climate: General Survey

It is not the aim of this study to present a regional geography of Mexico, but it will add point to the discussion if we bear in mind a few salient features of the general geography of the territory, which extends over $17\frac{1}{2}°$ of latitude, from 15° N to 32° 30′ N; the Tropic of Cancer cuts through the country almost midway between the northern and southern extremities. This fact of location alone guarantees a wide range of environments, which extend from the hot deserts of the north-west to the lush rain forests of southern Yucatan. The deserts include such areas as the Sonora Desert (which has been compared to the Mohave Desert of the U.S.A.) and the peninsula of Lower California (which has been described graphically as the "fleshless arm of Mexico"). Southern Yucatan, on the other hand, experiences heavy summer rainfall brought by the north-east trade winds, and while the northern part of Yucatan is usually classed as semi-arid, largely because surface water drains rapidly away through the underlying limestone, the southern parts support dense forests. Between the hot deserts and the tropical rain forests lie various intermediate types of environment.

But it is by no means latitude and location alone which determine the quality of environment in Mexico, for relief plays a very important role. Mexico is a mountainous country and over half of it lies at elevations of

* 1966 estimate. About 70% of the people are *mestizos* (a people of mixed Indian and European stock, the Indian influence being normally dominant), 28% are Indians, just over 10% European, and barely 1% Negro. Negroes were brought in during the early colonial period to provide slave labour for the *haciendas*, but by the end of the War of Independence most of them had either become absorbed racially or purchased their freedom and left the country.

SOME PROBLEMS OF INDUSTRIAL DEVELOPMENT   315

more than 3000 feet above sea-level. In general terms, an extensive central tableland (which is by no means flat and which is broadest in the north) is flanked on the west and the east by mountain ranges, on the west by the Sierra Madre Occidental and on the east by the Sierra Madre Oriental. These mountain systems rise fairly abruptly from coastal plains facing the Pacific Ocean and the Gulf of Mexico respectively. The peninsula of Lower California is a continuation of the Coast Ranges of California, while the Gulf of California continues the line of the Central California Valley and the depressed Salton Sea region.

The central plateaux become progressively narrower southwards until they descend sharply towards a narrow corridor, which cuts across the Isthmus of Tehuantepec and which links the Gulf and Pacific coast lowlands. East of the Isthmus lie the Chiapas Highlands which continue into Guatemala, while most of the Yucatan Peninsula consists of a low-lying limestone plateau.

Mexico is noted for its many volcanoes (including the famous Popocatepetl), and the country has suffered considerably in the past from volcanic action and from earthquakes. One startling volcanic manifestation was the sudden appearance in a field of a new volcano, Paricutin, in 1943. Paricutin is about 20 miles from Uruapan and about 250 miles west of Mexico City; it rapidly reached a height of several hundred feet, spread devastation in the surrounding countryside within a radius of 60 miles, and destroyed a small market town. It is now quiescent. We should, however, note that volcanic areas are fertile and productive, and they support the densest populations in the whole country.

Such marked variations in elevation as are met with in Mexico give rise to well-defined types of environment based on altitude, and four chief types are generally recognised—the *tierra caliente*, the *tierra templada*, the *tierra fria* and the *tierra helada*. The *tierra caliente* extends upwards from sea-level to an elevation of about 3000 feet. Mean annual temperatures range between 75° and 85° F (23·9° and 29·4° C). The *tierra templada* lies between 3000 and 6000 feet above sea-level and has mean annual temperatures of between 65° and 75° F (18·3° and 23·9° C); this is the most favourable zone for human settlement and development. Above 6000 feet the *tierra fria* reaches upwards to 9000 feet and mean temperatures between 55° and 65° F (12·8° and 18·3° C) are normal, while above 9000 feet the *tierra helada* (the "frozen land") experiences mean annual temperatures below 55° F (12·8° C). Much of the *tierra helada* consists of bare rock and snow-capped summits, and very few people live at these altitudes.

A weakness of such generalised descriptions as the foregoing is that an unintended impression is conveyed, that the climates concerned are more equable than is actually the case. For example, in the northern deserts temperatures on the lowlands are quite frequently above 100° F (37·8° C) during the summer months, while during the winter unusually cold spells are associated with the *norte*, known locally as the *papagayos*, a southward extension of the "northers" of the U.S.A.; the *norte* is usually unpleasant and it is sometimes dangerous as it may cause considerable damage to crops. Temperature conditions are less extreme in the more tropical and humid southern parts of the country.

Rainfall is everywhere seasonal in Mexico, the rainy season occurring in summer, except in the extreme north-west where summer is a period of drought as in neighbouring parts of California. Much of the north, especially the north-west, is very dry, and there are many years when no rain at all falls. Heavy summer rains are experienced in the southern parts of the country, however, and over 100 inches fall on an average each year on some of the dripping rain-forests in the extreme south. It is true to say that much of Mexico is an inhospitable country, partly for reasons of terrain and partly because of climatic rigours; one-third of the total area has an annual rainfall of less than 15 inches and another third has between 15 and 30 inches a year. From an agricultural point of view these are low figures for a country situated across the tropic. Barely 10% of the total area is at present under cultivation while another 40% is given over to grazing, often grazing on pastures of poor quality. Only about 15% of the total surface area is cultivable.

It may still be true to say that the picture of a typical Mexican conjured up in the outside world is that of a barefoot *mestizo* or Indian who wears a gaily coloured shawl (a *serape*) and a sombrero and who spends much time enjoying a siesta in the somewhat limited shade cast by an odd-looking cactus. This picture is fast becoming obsolete, if it is not already so. Changes are rapidly affecting all walks of life, even agriculture—that most conservative of all occupations.

Agriculture

We may say that the modern phase in agriculture began in 1934 under the government of President Cardenas (1934–40). Prior to this period the old *hacienda* system inherited from the colonial era* still flourished, but

---

* Mexico secured her independence in 1821, after which came a lengthy period of political instability and repression. The *hacienda* system conferred very favourable living conditions upon the comparatively few landowners (the *caballeros*, who formed about 1% of the total population) and less favourable conditions of the landless peasantry (the *peons*), who were forced to live in a perpetual state of servility. The revolution against this tyrannical system was won in 1915 and agrarian reform began in 1916, but it made little headway until the Cardenas régime.

SOME PROBLEMS OF INDUSTRIAL DEVELOPMENT 317

under Cardenas the *haciendas* were broken up and the land was assigned to the people in the form of *ejidos*, properties owned by rural communities. About one-half of the agricultural land of Mexico today is owned by the *ejidos*. Some are run as collective holdings but there is evidence to suggest that these co-operative farms are not popular and that the people prefer to own their own land. This does, in fact, happen on other types of *ejidos* where individual farmers own plots of about 44 acres, about 10 of which are normally classed as land suitable for crops. In cases where a

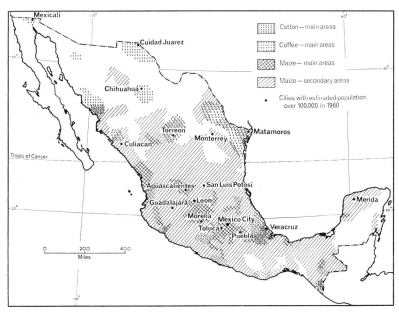

FIG. 66.—Mexico: agriculture and chief towns.

farmer makes good use of his land he is left to farm his plot without government interference.

Signs are not wanting that an agricultural revolution is developing in Mexico; between the years 1939 and 1953, for example, the volume of crops grown generally more than doubled while the output of industrial crops trebled. Great attention is now being paid to land development, and techniques such as contour ploughing, terracing and crop rotation are all in evidence, while tractors and other forms of mechanical aids to efficient farming are seen on the farms in increasing numbers. Cotton and coffee, the two chief export crops, are being more extensively cultivated, cotton being grown mainly under irrigation in the north and coffee in the *tierra templada* zone of Veracruz and Chiapas (Fig. 66). The output of the traditional crop, maize, is increasing largely thanks to greater use of improved hybrid strains of seed; even so, yields of maize per acre are scarcely one-third of those in the U.S.A. and the native crop is not sufficient to meet the domestic demand although maize occupies more than half of the total cultivated land. Beans form a leading food crop in the south, while a very significant recent development is the great increase in the output of wheat. Total production of this crop almost trebled between 1930 (587,000 tons) and 1966 (1,640,000 tons), and some wheat is now available for export, but the notable point is that this development indicates a rising standard of living as more inhabitants are eating wheat rather than maize. The increasing consumption of rice (production 187,000 tons in 1950, 354,000 in 1966), grown on reclaimed land in the humid tropical areas, is also an indication of this. Sugar cane, grown on the hot coastlands of the Gulf of Mexico and the Pacific as well as along many valleys of the south, is another leading crop (production $9\frac{1}{2}$ million tons in 1950, 23 million tons in 1966).

An important development on the agricultural front is the increasing use of irrigation, an obvious need in a territory which includes so much desert and in which rainfall is almost entirely seasonal. It has been, in fact, on the irrigated areas of the north and north-west that most of the increases in agricultural output have been achieved. Eighty-five per cent of Mexico is seasonally dry and needs irrigation if agriculture is to be carried on successfully. A number of large schemes was initiated in 1926, and by 1946, 2 million acres had been brought under irrigation; since 1946 the programme has been greatly speeded up and today more than $7\frac{1}{2}$ million acres are irrigated. More than fifty projects are playing a part in helping to water farms in the desert areas. Since irrigation projects are very costly, the tendency today is to initiate multi-purpose schemes which at the same time provide water for irrigation and also water power for the generation of hydro-electricity. Such a scheme is that which is to develop the basin of the Rio Papaloapan, an area of 17,000 square miles, along the lines of the T.V.A. project; attention is being paid to the provision of drainage and irrigation facilities, to the generation of hydro-electric power and to the construction of roads to open up the area. Five dams in all (the largest being the President Aleman Dam near Temasul, Oaxaca State) are being constructed and will provide enough water to irrigate 400,000 acres as well as to generate electricity. The recently completed Dieguez Dam at Santa Rosa (Jalisco State) is a massive wall of concrete nearly 200 feet high blocking a gorge half a mile broad in the Sierra Madre, so harnessing the waters of a turbulent river, storing 16,000 million cubic feet of water, and supplying power for use in nine states, while the giant turbines associated

SOME PROBLEMS OF INDUSTRIAL DEVELOPMENT 319

with the new dam at El Infiernillo on the River Balsas (Michoacan State) have helped to increase the generating capacity of the territory to twice that of seven years ago. Other examples could be cited of the strides which Mexico has taken in this direction, but enough has been said to show that a real revolution has taken place as regards the provision of irrigation and of hydro-electric projects in recent years. Much has been done but much remains; it is estimated that if the country is to become self-supporting in agricultural products about 17 million acres in all must be brought under irrigation.

Mexicans are becoming increasingly aware of the opportunities which exist for the export of agricultural produce. The main roads leading to the U.S.A. are today carrying large numbers of refrigerator trucks whisking such commodities as tomatoes, melons and strawberries to San Francisco, Denver and other American cities; as many as 3 million lb of strawberries alone may be exported to the U.S.A. in a single year. And, since the political troubles in Cuba, Mexican farmers have increasingly taken over the supply of special light-leaf tobacco to American cigar manufacturers. Further developments are likely as a result of schemes such as that known as the "March to the Sea," which envisages the moving of many inhabitants from the fairly densely populated central plateau where most farms are small, poor and backward, and the resettling of these people along the, as yet, undeveloped but fertile and coastal lowlands.

Mining and Industry

Agriculture in Mexico is clearly being freshened up by the winds of economic and social change, and the same is true of mining and industry. During the colonial period and for some time afterwards the economy of Mexico depended very largely upon exports of silver, and the silver export is still an important one accounting for about one-fifth of the world's total production. Petroleum, lead, copper and zinc are all now more important exports in terms of value than silver, however, while in recent years Mexico has become the world's second largest producer of sulphur (p. 33 above). Reserves of coal and iron ore exist (Fig. 67) and form the basis of domestic industry.

Robinson has given an up-to-date picture of the petroleum-producing industry of Mexico. The main producing areas lie along the Gulf Coast (Fig. 67), and it was in the Tampico area that mineral oil was first discovered in Mexico in 1901, while the Tuxpan field was first tapped in 1908. These coastal deposits proved to be so productive that by 1921 Mexican production accounted for about a quarter of the total world output, and during the heyday period of the early and middle 1920s Mexico

M

ranked second only to the U.S.A. as a world producer. After this period output declined; the industry was nationalised in 1938, and since that time all oil-producing activities have been under the control of Petróleos Mexicanos, a government monopoly commonly known as Pemex. With the help of technical and financial aid from other countries, especially from the U.S.A., Pemex was successful in raising output after 1945, and Mexico now holds tenth place among world producers, while the fairly recently established petrochemical industry is vigorously expanding. It is an interesting measure, however, of the rate of economic growth

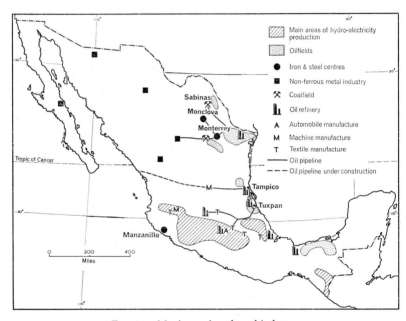

FIG. 67.—Mexico: minerals and industry.

within the country, that despite the installation of fourteen oil refineries Mexico still needs to import refined products such as petrol and kerosene, and she helps to pay for these by the export of crude oil. This is obviously an uneconomic situation, and one which calls for remedy as economic conditions allow.

The sulphur-producing area is located in the Minatitlan region in the state of Veracruz on the northern slopes of the Isthmus of Tehuantepec, and we have already said something about sulphur production in Mexico in Chapter II.

By this time the reader may well have in mind a picture of a country which has greatly broadened the scope of its agriculture in recent years, and which is no longer dependent upon the export of a single mineral (silver). Concurrently with these spectacular developments runs another one, the establishment of modern manufacturing industry, and Fig. 67 shows some of the main features of industrial development in Mexico today. Most widespread is the processing of mineral ores, but the oldest industry is the manufacture of cotton textiles; it is, in fact, thought that the pre-Aztec Indians developed a comparatively sophisticated textile industry, and Cortez found a flourishing cotton industry which produced well-woven and brightly coloured cotton cloths. The modern phase of the industry, however, began in 1830 when textile machinery was imported from the U.S.A. and was installed in a mill near Puebla; within ten years there was quite substantial development of the industry in the Puebla area and fifty-nine spinning and weaving plants were in operation. The industry also established itself during the same period in the Orizaba Valley, power for the mills being obtained from the Rio Blanco. Cotton manufacturing continued to make progress during the remainder of last century, and most mills were electrified before 1900. Unfortunately, many mills are suffering today from this early start since so many of the machines are of types now obsolete. At present, the main centres of the cotton industry are Mexico City, Guadalajara (the second largest city of Mexico with a population approaching a million), Puebla and Orizaba, while the output of woollen and worsted goods is notably increasing. A recent development has been the manufacture of man-made fibres, 4000 metric tons being produced in 1948, 23,000 tons in 1960 and 39,000 tons in 1966.

There has been for some time limited development of industries concerned with the production of consumer goods such as clothing, footwear, furniture, soap, foodstuffs, drinks and general household wares, and during the present century the manufacture of consumer goods in general has greatly increased; at the same time there has been considerable expansion of the industrial base of the country and a very significant feature is that, whereas before 1945 industry in Mexico was concerned almost entirely with the production of consumer goods, since that date the manufacture of capital goods has become steadily more important; between 1954 and 1960 the output of capital goods actually increased by 66% as against the 10% increase in the manufacture of consumer goods.

In 1903 Mexico became the first Latin American country to install a modern integrated iron and steel plant; this was at Monterrey, and until 1945 this was the only plant producing pig iron in the country. In 1945, however, a second integrated plant came into operation, this time at

Monclova, and since that time heavy industry has been more broadly established though about three-quarters of the output of iron and steel comes from Monterrey and most of the remainder from Monclova. The industry is based upon coking coal from the Sabinas coalfield, iron ore from Durango, local supplies of limestone and manganese from northwestern Chihuahua. Plans are well advanced for greatly increasing the output of iron and steel; in fact, production of pig iron rose from 227,000 tons in 1950 to 669,000 tons in 1960 and to 1,137,000 tons in 1966, while that of steel ingots increased from 390,000 tons (1950) to 1,491,000 tons (1960) and to 2,763,000 tons in 1966.

Two further points regarding the iron and steel industry are worthy of note, the first being that this is one of the very few areas in Latin America where the raw materials for the industry can be assembled at low cost and which is within economic reach of a substantial market; the Sabinas coal is of good coking quality though limited in amount, while the Durango iron ore is obtained by low-cost opencast mining (p. 54 above). The second point is that the Mexican iron and steel industry relies heavily upon imported scrap, and it produces therefore considerably more steel than pig iron per annum.

The motor-car industry is one which has shown spectacular development in recent years, largely as a result of the establishment of branches of parent firms in the U.S.A. Many assembly plants have been set up and in 1966 no fewer than 85,000 motor vehicles were assembled in Mexico itself. Prohibitive tariffs have now virtually halted imports of completed vehicles. Plants are now being installed which will permit the manufacture within the country of most automobile parts, and the formerly quiet market town of Toluca, 40 miles from Mexico City but today linked with the capital by a four-lane motorway over the mountains, has been described (perhaps rather grandiloquently) as the new Detroit of Latin America. Bicycles have been manufactured since 1954 in sufficient numbers to satisfy domestic demand.

The establishment of oil refineries has previously been mentioned and an associated petrochemical industry is rapidly establishing itself. The manufacture of fertilisers is fast expanding—an important point for the developing agricultural activities mentioned earlier—while production of bleaches (for the paper and textile industries) and dyes (for the textile industry) is also increasing. A recent visitor emphasises the current expansion of industry when he writes that in a reconnaissance of Mexico City he counted more than one hundred new manufacturing enterprises,* while today no fewer than 70 industrial firms from the United Kingdom alone

* E. Tomlinson, "There's New Life in Old Mexico," *Reader's Digest*, April 1965.

are active in the country. Some of these firms own assembly of manufacturing establishments jointly with Mexicans, while some have entered into licensing agreements with Mexican firms. There are probably now in operation well over a thousand industrial enterprises which enjoy the help of foreign capital, and it has been said (*Manchester Guardian Weekly*, 27 April 1967) that, if industrialisation is to continue at its present rate of expansion, investment of foreign capital will be needed to the extent of about £180 million a year. Since 1955 industrial production has risen at the rate of 10% per annum, and in 1966, 21% of the total working population was engaged in manufacturing industry. While Mexico City, with a population of almost 5 million people, is still the great industrial centre of the country, industry is increasingly taking root in other towns; as a leading Mexican has said, "The industrial revolution is spreading from Tijuana to Yucatan" (quoted by Tomlinson). Some measure of the overall growth of industry may be observed in the fact that the output of electricity has increased from 2480 million kWh in 1937 to 4423 million in 1950, to 10,636 million in 1960 and to 19,022 in 1966. Mexico is, in fact, trying to double her output of electrical energy every seven years.

The present position in summary is that there are two main industrial regions in Mexico, the largest lying in the southern part of the plateau and extending from Guadalajara to Cordoba; nearly one-half of the total industrial population of Mexico live in this region, which is probably the leading industrial region in Latin America. Because of the absence of coal or petroleum in the area power is supplied in the form of H.E.P. from the rivers of the adjoining mountains and is utilised in the manufacture of such goods as textiles, motor vehicles, leather goods, wireless receivers, petrol and metal goods, and in food processing. The second region centres on the towns of Monterrey, Monclova and Saltillo, the first two of which account for the manufacture of 80% of the iron and steel of Mexico. Other industries include the manufacture of metal goods, building materials, furniture, cigarettes and liquor. Manufacturing is in addition carried on in all the larger cities in the country and in state capitals.

Effects of the Mexican Industrial Revolution

We might try at this stage to sum up the lessons of this case study by pointing out that the industrial revolution in Mexico fits into the general pattern of development in the country in at least four ways:

1. It creates employment for former workers on the land who find themselves displaced by the new farm machinery and by other technical developments. One result of this is that the proportion of urban dwellers

(42·6% in 1950, 50·6% in 1960 and 55·3% in 1966) is constantly increasing at the expense of the numbers of rural inhabitants (57·4% in 1950, 49·4% in 1960 and 44·7% in 1966). The proportion of agricultural workers fell from 58·3% in 1950 to 54·2% in 1960, while the figure in 1966 was 50·8%.

2. It creates employment for the steadily increasing work force which is a natural concomitant of an increasing population. This is an important feature in a territory which recorded an increase in population of over 34% in the ten years between 1950 and 1960—an annual rate of increase of 3·43%, one of the highest in the world (the average rate of increase in the "third world" is about 2·5%). Between 1960 and 1966 there was a further increase of 21% (3·5% per annum).

3. It provides the mechanical and other forms of assistance which are making it possible for agriculture to modernise itself and to make itself more efficient. Farmers are requiring larger numbers of tractors and agricultural machinery of all kinds while they also need the petrol and the lubricating oils which the new refineries can help to supply. The fertilisers manufactured by the new petrochemical industry are indispensable to the modern farmer. The new irrigation and multi-purpose schemes upon which agriculture so largely depends are voracious consumers of concrete, while they also need the electrical machinery which the electrical industry can supply.

4. It produces goods of many kinds which help to make possible a higher standard of living, goods which range from textiles to television sets and motor-cars. The fact that there is a steadily mounting demand for such goods shows in itself that the economic revolution (of which the industrial revolution forms part) is now well under way. And this, after all, is the prime purpose of industry, to supply goods which will help to make living more pleasant and gracious, and the interested reader who cares to glance back to the first three paragraphs of this book will immediately see that we are now back at the point from which we began our survey—our circle of investigation is complete as far as the limits of a single book of this nature will allow. Neither should we forget the cultural benefits which industrial progress is making possible. When, for example, the revolutionary period in Mexico ended, the vast majority of the population was illiterate; by 1960 the proportion of illiterates had fallen dramatically to 44% while in 1966 it was 29%.

The main point of this case study is to emphasise the manner in which the development of agriculture, mining and industry has gone ahead, each separate development forming part of a broad economic advance. It is this feature which encourages many observers to view the future of Mexico with considerable optimism. Teichert has noted that "Mexico is

SOME PROBLEMS OF INDUSTRIAL DEVELOPMENT 325

one of the few republics that has achieved a considerable degree of industrial development without neglecting its agricultural sector. Its economic development has advanced on all fronts simultaneously and at the same rate." Much remains to be achieved, but it may fairly be argued that the example offered by Mexico to the underdeveloped territories of the world is one which they would do well to follow.

SUGGESTIONS FOR FURTHER READING

Beaver, S. H., "The Development of the Northamptonshire Iron Industry, 1851–1930," in *London Essays in Geography*, London, 1951.
Cline, H. F., *Mexico*, London, 1961.
Dury, G. H., *The British Isles*, London, 1961.
Estall, R. C., and Buchanan, R. O., *Industrial Activity and Economic Geography*, London, 1961.
Focus, June 1959 (Mexico) and November 1959 (Coal in the U.S.A.).
Gilewska, S., "Changes in Geographical Environment brought about by Industrialization and Urbanization"; and Gregory, S., "Some Aspects of Water-resource Development in Relation to Lancashire." Both papers published in *Problems of Applied Geography*, II, Warsaw, 1964.
Gourou, P., *The Tropical World*, London, 1953.
Highsmith, R. M., *Case Studies in World Geography*, Englewood Cliffs, New Jersey, 1961.
Hudson, F. S., *North America*, London, 1962.
Hughes, M. E., and James, A. J., *Wales*, London, 1961.
Jarrett, H. R., "Lunsar: A Study of an Iron-ore Mining Centre in Sierra Leone," *Economic Geography*, April 1956.
Jarrett, H. R., *Africa*, second edition, London, 1966.
Manners, G., "The Pipeline Revolution," *Geography*, April 1962.
Masser, I. F., "Changing Patterns of African Employment in Southern Rhodesia," in *Geographers and the Tropics* (Liverpool Essays), London, 1964.
Mountjoy, A. B., *Industrialization and Under-developed Countries*, London, 1963.
Nurkse, R., *Problems of Capital Formation in Under-developed Countries*, Oxford, 1955.
Rawstron, E. M., and Coates, B. E., "Opportunity and Affluence," *Geography*, January 1966.
Reddaway, W. B., "External Capital and Self-help in Developing Countries," *Progress*, vol. 51, no. 286, 1965–66.
Robinson, H., *Latin America*, London, 1965.
Rostow, W. W., *The Stages of Economic Growth*, Cambridge, 1961.
Stamp, L. D., and Beaver, S. H., *The British Isles*, first edition, London, 1933.
Stone, R. G., *Mexico*, Overseas Economic Surveys, H.M.S.O., London, 1956.
Teichert, P. C. M., *Economic Policy Revolution and Industrialization in Latin America*, University of Mississippi, 1959.
Wooldridge, S. W., and Beaver, S. H., "The Working of Sand and Gravel in Britain," *Geographical Journal*, March 1950.

Chapter XIII

Postscript

THE scope of modern industrial development is immense, and the ground covered in this book is very limited. It is hoped, however, that the book will help some of its readers to begin to understand something of the greatest economic and social phenomenon of this age (or probably of any other)—the development of the technological society.

INDUSTRY AND AGRICULTURE

It is upon an ever-increasing development of manufacturing industry that the future well-being of the peoples of this globe very largely depends. It has even been suggested that it is mineral supplies rather than food resources which are vital for the further development of the human race, and while this extreme point of view can hardly be defended it does contain a substantial kernel of truth. It is industrial development which has enabled the peoples of the "Western" societies to achieve their present high standard of living; even territories of the Western world such as Denmark which have for long relied heavily upon their agriculture, depend very largely for their prosperity upon the markets offered by the industrial countries.* The wide inequalities in wealth which exist between the "developed" and the "developing" territories are brought out by the following table:

	%age of world population	%age of world domestic product
Asia	52.7	12.1
U.S.S.R. & E. Europe	11.5	18.0
W. Europe	11.0	20.6
Africa & Middle East	10.7	4.0
N. America & Oceania	7.3	39.9
Latin America	6.8	5.4

Based on U.N. statistics, 1955.

* The story of how Denmark reorganised her agriculture along large-scale cooperative lines and the reasons for it are well known. Since the Second World War,

The point has more than once been emphasised in this book that if the poorer countries are to begin to catch up with the more wealthy ones, the development of industry and agriculture must go hand in hand. It is a fact not easily to be missed, that regions which are purely agricultural and which have few contacts with the world of industry, are in the main regions of poverty. This is surely the point at issue. If we are to depend on agricultural development *alone*, then the peoples of the world can expect slowly deteriorating standards of living; higher standards can only come about through greater industrialisation, and greater industrialisation is dependent upon increasing supplies of raw materials—including minerals.

At the same time we must recognise that the question of increasing agricultural output is of the greatest possible importance, especially in view of the contemporary "population explosion"; this is a feature of the modern world which has been commented upon many times, but it is hardly possible to overestimate the serious character of the unparalleled increase in population which is taking place at the present time. It is estimated that the population of the world may well grow from its present 3500 million people to nearly 7500 million by the year 2000 (considerably more than one-half of these people will be living in Asia). At the same time we might recall that one country, Japan, has in two decades cut its birth-rate from 34·3 per 1000, one of the highest in the world, to 17·5, the lowest recorded in any major territory today;* something like this is essential the world over if large-scale calamity for the human race is to be averted. United Nations statisticians believe that birth control measures will keep the global population figure in the year 2000 down to 6000 million people, but even this figure represents a very great increase in numbers over those obtaining today.

In the face of this situation, no one can deny that the improvement of agriculture in the developing territories is a matter of the highest priority. Professor W. Schultz, of the University of Chicago, has shown how applications of capital have transformed farming in Western Europe and in parts of Mediterranean Europe, areas not inherently particularly fertile. Italy, Austria and Greece, for example, are countries which have less *arable* land *per capita* than India and they have soils of generally poor quality, but they have recently increased their agricultural production at rates of 3·0%, 3·3% and 5·7% per annum respectively, as opposed to the

---

however, "Danish manufacturing . . . has gained so rapidly that it now contributes more to the national income than does agriculture" (Nystrom and Malof). Today only about 13% of the Danish work force is employed directly in agriculture though more workers are engaged in processing agricultural products.

* *See* p. 304 above.

2·1% increase in India. The point is that industry and agriculture have gone hand in hand in the achieving of these gratifying results in Europe (*see also* the case of Spain, pp. 137–140 above) and farmers are using modern implements and machinery, new forms of chemicals to fight pests and diseases, and artificial fertilisers, all of which are products of industry. The scarce factor, land, is in a sense replaced (shades of Ricardo!) by the variable factor of capital. It has been estimated that India could increase her output of farm products by 50% over the next five years if she adopted improved farming methods, and used three times as much fertiliser per unit area as she now applies to her soils. Since she cannot afford to purchase such large amounts of fertiliser from other countries, however, her only hope is to manufacture her own—which can only be done by the establishing of a chemical industry. With increasing agricultural output on the scale theoretically possible, India might even begin to export limited amounts of foodstuffs and other farm products and thereby earn additional foreign exchange which could be used to finance further industrial development. This example reminds us again that there is no real justification (except for purposes of convenience for the student) for the recognition of any dichotomy between agriculture and industry; they are closely interwoven and developments in one march alongside developments in the other. This was in large measure the theme of Chapter I of this book.

PROBLEMS OF CAPITAL UTILISATION IN DEVELOPING TERRITORIES

There is little need further to emphasise the vital necessity of expanding industrial production *of the right kind* in the "third world" (p. 313 above), and it is hardly too much to say that future world peace and therefore the entire future existence of the human race might well depend upon the successful prosecution of this programme. Yet it is foolish to underestimate the problems which lie ahead, some of which have been set out above (pp. 302–313), though we cannot pretend fully to have set out the situation in the limited space available. It was suggested, for example, that greater applications of capital should result in greater productivity, and as a general principle this is true, but several important caveats should be introduced:

1. It has taken the developed countries of the world a very long time to establish ordered forms of government which are at the same time fundamentally honest and open to a minimum of corruption; even now it

would be a bold man (or a singularly naïve one) who would declare that we have reached anything like the degree of order and incorruptibility theoretically possible. The newer states* have had very little opportunity to develop and exercise these essential public virtues, and it is a great temptation to point to examples of extravagance and mismanagement in the "third world" and to argue that economic help should be withheld until evidence is forthcoming, that help afforded will be wisely used. But what is to happen in the meantime? There is a case (which will not appeal to everyone) for arguing that some corruption and mismanagement are inevitable concomitants of the early stages of social and economic advance, that they have been features of our own national life in the not too distant past, and that we should bear with them in the hope that a greater degree of future prosperity in the "third world" in years to come will encourage them to disappear.

2. A vital point is that the need for additional capital in the developing countries is on such a vast scale that it is quite beyond the power (even given the will) of the developed world to satisfy more than a small fraction of it. And it is unfortunately true that what a developing country may gain on the roundabouts of loans and gifts it may well lose on the swings of price depression as world prices for its primary products fall. A case in point concerns Ghana, whose prosperity very largely still depends on her cocoa crop. The world price for cocoa, however, fell from £352 a ton in 1957 to £120 in 1966, though in 1968 the price rose again to £280 with expectations of a further increase. Such price fluctuations bring uncertainty and perhaps poverty, not only on a personal scale for the cocoa farmers but also for the state as a whole. It is reckoned that between 1950 and 1962 the terms of trade moved against the primary producers by about 25% for reasons which will be discussed later, and as a result the value of primary exports in total world trade fell from 31 to 21%; and this at a time when the richer territories have become richer than ever before! In the face of this situation it is difficult not to agree that it seems a harsh joke that the developed territories are apparently content to offer gifts and loans with one hand and to take back raw materials at depressed prices with the other, while there are not lacking those who point to the pitifully small scale of giving when contrasted with other forms of spending. In 1964, for example, the United Kingdom allocated £190 million for aid to developing countries, yet in 1966 the inhabitants of

---

* The term "nation" which is sometimes applied to states which have recently secured their independence is avoided since these states are not nations in the usually accepted sense of the word. The term "state" is less debatable since it refers simply to an area under a common government.

that same country could afford to place bets to the estimated value of at least 2½ million on the outcome of a single general election; and this was just part of an annual betting bill of possibly £1000 million.* It is a sad commentary on human affairs that while this state of affairs existed a Labour government, apparently unable or unwilling to control earning and spending at home, could ruthlessly cut aid to overseas territories by forcing cuts in overseas investment and in grants to developing territories. "In other words, the needs of the poor [countries] must not be allowed to threaten the growth of the high standard of the rich" (Perham). One may perhaps be forgiven for wondering just what brand of socialism this is.†

One fact which emerges from the above line of thought is that developing territories must adjust themselves to finding ways of accumulating *domestic* supplies of capital, a point made on page 305 above. It is also important that such capital as is available should be used to the best advantage, which brings us to our next point.

3. Capital investment will achieve disappointing results unless the investment is wisely made and adequately used, and it is often not easy in developing countries to ensure the full use of equipment for reasons already noted (p. 309 above). There is therefore a very great need to ensure that any investment is made to the best possible advantage; there must be fewer national airlines and iron and steel plants, and more industries of the kind enumerated above (p. 313). Reddaway has argued that in developing countries it is capital rather than manpower which is the scarce and therefore expensive factor, and that in the efficiency drive the emphasis should be on raising output from any unitary amount of capital rather than on raising output per worker, though the two aims are not necessarily mutually exclusive.

An example of the allure of the large-scale investment project is that of railway construction. Railways can be very glamorous forms of capital

* It is sometimes argued that such a fact is irrelevant to the point at issue because betting money remains within a country. Protagonists of this dreary view overlook two points. Firstly, the moral argument. It can hardly be denied that a country overmuch addicted to gambling is in no position to exercise moral leadership or constructive example to developing territories or to anyone else. Secondly, betting money admittedly remains within a country but even such money could be used for various kinds of overseas aid. Money collected for philanthropic purposes is collected within the donor country and is then forwarded to the recipients either through the foreign exchanges or in kind. It can be used, for example, to purchase capital or consumer goods within the donor country, and these goods can be shipped overseas as part of a general aid programme.

† Since this was written the British Labour government has introduced various measures designed to control in some measure earning and spending in Britain (1966 and 1967), not, however, from any moral motives but because of compulsion arising from the hard facts of Britain's economic position in the world.

investment, yet it was mentioned early in Chapter X that this form of investment (which is very expensive indeed) can sometimes yield very disappointing results.* It will be most interesting to see to what extent the new much publicised railway to Maiduguri, in north-eastern Nigeria, succeeds in opening up the country through which it runs and in stimulating economic growth. Unfortunately, the installation of a railway has to be carried out on a comparatively large scale or not at all, but it is not always possible in a developing country to ensure that a railway system is fully used as there may be insufficient traffic. On the other hand, roads can be constructed in more of a piecemeal fashion as circumstances warrant, and the quality of construction can be varied in anticipation of the amount of traffic to be carried. Wide variations are possible between what is merely a superior form of track and an arterial highway. Road transport, moreover, is more flexible than rail, for the number of cars and lorries in operation on a road can be varied to an almost indefinite degree. The glamour which surrounds the initiating of a new railway is an example of the dangerous fascination of enterprises more notable for prestige than for productive capacity.

4. While attention is understandably often focused upon the large-scale enterprise which needs large capital sums to finance it, it is quite possible that there is a greater future for the small-scale enterprise, both industrial and agricultural, than is often supposed. This is a point stressed by Debenham after work in Central Africa. Debenham argues, for example, that the small-scale dam which can be constructed by local labour using local materials and operated by local people (all after initial help and guidance) may often be a more worthwhile project when multiplied many times over than a single large multi-purpose scheme such as the New High Dam at Aswan, while there would be tremendous savings in costs. Already, before the Aswan enterprise is even completed, voices are being raised to warn us that the whole scheme is inefficient and outdated. In the industrial field pertinent examples of the sort of thing possible have previously been mentioned in connection with the textile industry and on page 313.

5. A related point may be briefly emphasised again; it was referred to on page 311 above. Unless there is sufficient will and determination on the part of the people generally, schemes for economic, including industrial, progress will inevitably prove disappointing in their results. The question immediately arises—where is this will to be found? This is a question to

* The quickest and surest returns from an investment in railway construction occur when the track is built for a specific purpose, for example for the transport of mineral ore or concentrate from a producing area to a port.

which no firm answer can be given but perhaps it is pertinent to point out that nineteenth-century Britain developed it from a religious foundation,* and it is possible to argue that as this foundation has weakened, so the determination of the nation has weakened. It seems likely that in the modern world the Chinese people have the will to succeed, and it can be argued that this, too, stems from a religious foundation though the religion in this case is Communism.

FURTHER PROBLEMS OF INDUSTRIAL DEVELOPMENT IN DEVELOPING TERRITORIES

There is no doubt that one of the most pressing needs of the developing territories is to get away from their present dependence upon too few products, most of which are primary products. The present writer has given elsewhere examples of the extent to which certain African territories rely upon too narrow a resource base (Jarrett, 1966), and there is no question but that this situation greatly weakens the bargaining power of developing countries. As a result, these countries are often hard put to secure enough foreign currency to purchase overseas the necessary equipment for industrial development and expansion. A further difficulty arises because under modern conditions developing countries not infrequently find that the terms of trade move against them, and they can under these circumstances expect to receive lower unit prices for their exports. Crawford has examined some of the main reasons for this economic situation, and the following are the main points he makes:

1. In developed territories industries tend to use smaller amounts of raw materials to produce any given output of manufactured goods because of technological improvements. Demand for raw materials correspondingly slackens.

2. As incomes in the developed world rise, demand for primary products (including foodstuffs) does not rise in proportion because of increasing expenditure in the tertiary sector.

3. Export of foodstuffs from the "third world" is also restricted by agricultural protection in developed countries which wish to safeguard their own farming interests. Both European countries and the U.S.A. practise this form of restriction.

4. Some raw materials (wool and rubber, for example) have to meet

---

* Trevelyan remarks that the Victorian era was characterised by "interest in religious questions and was deeply influenced by seriousness of thought and self-discipline of character, an outcome of the Puritan ethos."

increasing competition from synthetic substitutes manufactured in the industrialised countries.

5. Increasing production of primary products and an expanding volume of exports of these goods have between them lowered unit prices.

It is not surprising that developing countries which seek to develop their own industries find the going very hard. A further handicap commonly arises from the fact that any attempt on the part of a developing country to speed up its rate of industrialisation leads to a severe balance-of-payments problem because of the rapid acquisition of capital equipment from developed territories, and the corresponding need to pay for this equipment within a comparatively short space of time. It is, of course, in such circumstances that grants or loans from external sources are particularly valuable.

It is sometimes argued that as the "third world" develops its own industries, the developed territories will suffer because many manufactured goods which they now export will be produced in the newly industrialised countries. Enough has been written in this book to show the fallacy behind this argument. It is true that as the emerging states develop their industries, the *proportional share* of the import market in which manufacturers from the developed territories can expect to sell their goods will diminish, but it must be remembered that this smaller proportional share will be of a much larger total industrial cake. Under these circumstances the actual amount of manufactured goods produced in and exported from the developed territories may well increase rather than decrease.

But this is by no means the whole story. It is a fact of observation that most of the overseas trade of the developed territories is carried on with other developed territories; it is the U.S.A., for example, which is the largest single export market of the United Kingdom. There is a simple reason for this. The size of the market in the developed territories with their comparatively high standards of living is far greater than it can be in countries which are as yet poorly developed and in which personal incomes are low; the numbers of television sets, washing-machines, computers and tractors (to take only a few out of many possible examples) which can be sold in the developing territories are very limited. It is a simple fact that with increasing economic development the poorer countries will not only produce more and sell more, but also *demand* more; their exports *and their imports* will therefore increase in volume and in value.

What must happen, of course, under these circumstances, is that the overall global structure of industry will change, just as the structure of industry in, say, New England has changed in recent years.* Developing countries will take over an increasing share of the production of the simpler types of manufactured goods† while the more highly developed states will be forced to concentrate to a greater extent on more sophisticated products. It is idle to pretend, however, that such a change can come about without considerable strain (in the developed countries as well as in those which are advancing), and there may well be distress and hardship involved during the period of adjustment.‡ The final result, however, could well be a higher general level of economic production in the world as a whole and a far higher standard of living for many millions of the world's inhabitants.

During the past years we have witnessed what is nothing less than a political revolution in the "third world" as many emerging territories have achieved independence; this independence has been won on the whole (with one or two unhappy exceptions) remarkably peacefully. What is desperately needed now is an economic revolution, equally peaceful, to match the political one, for upon such a revolution depends the future well-being of many millions of our fellow inhabitants of this planet.

THE PROBLEM OF MATURITY

The problems facing developing territories have been accorded considerable publicity in recent years, and their general nature is now fairly well recognised; what is perhaps less well realised is the fact that the developed countries have now come to a critical point of development and that the course of their future progress is by no means clear. The uncertainty inherent in this situation is referred to in the heading above as the "problem of maturity," and it is rather different in character from other problems at present facing developed countries which were examined in Chapter XII.

The problem is complex and we can do little more in the space at our disposal than accord it brief mention, but we might approach it from the premise that today great wealth is produced in the developed territories

* See p. 131 above.
† See, for example, the suggestion made above, p. 216.
‡ Sayers has shown how the pattern of manufacture in Britain changed during the very difficult period between 1913 and 1938. In 1913 the most important industries were concerned with the manufacture of cotton goods and steel, shipbuilding, and coal mining (p. 137 above), but by the late 1930s the production of all kinds of electrical appliances, motor-cars, chemicals and radios had developed in a spectacular way. Sayers refers to the "battering" which the economy of Britain received during this period, and it was certainly a time of very considerable hardship for many inhabitants and of accompanying social unrest.

by a minority of the workers (using the term in a broad sense and not in the restricted sense of factory worker). A small number of these workers initiates improved productive techniques; these are the research workers who are responsible for innovations and inventions of all kinds, which make possible more efficient and more economic production. Other workers carry out the work of organisation, of construction and of manufacture. Some of these workers are classed as "skilled," some as "semi-skilled" and some as "unskilled" according to the type of job they do and the training which they have (or have not) received. It is not, of course, argued that this minority of production workers pursues its occupations in isolation from the rest of the community. This is manifestly not the case, and considerable numbers of ancillary workers (especially those engaged in the public services) help to make it possible for production workers to carry on their work. It is, however, true that the number of workers employed in the tertiary sector of the economy in the developed countries is far higher than is needed simply to make possible the efficient prosecution of industrial activity.

The paradox of the whole situation is that, although the greater part of the wealth-producing capacity of the developed territories is in the hands of a minority of the inhabitants, there are comparatively few really wealthy persons in these communities. Why is this? The main reason is that the greater part of the wealth produced is scooped off by the state in penal forms of taxation, and most of the wealth thus secured is used for purposes which are in many cases worthy enough in themselves (and which, indeed, are in some measure essential), but which are not directly productive. Little criticism can be made of this feature in general terms, though bitter controversy often rages over the patterns upon which the public services should be moulded (this is especially true of education) and over the exact limits which should be set to them. In so far as this economic and social system leads to a greater degree of sharing out of the wealth produced than would otherwise be the case it is to be commended, but one disturbing feature is that far too small a part of the total wealth produced is available for ploughing back into industry, so that even comparatively powerful firms constantly find themselves suffering from obsolete methods of production and from outdated equipment. Really large undertakings are out of reach of individual entrepreneurs or of companies in the private sector of the economy, and despite the wealth generated by private enterprise it is only the state, the largest beneficiary of this wealth, which is able to finance the largest projects.*

* It is true that a fair degree of distribution of the wealth produced is to be expected also as a result of the joint-stock technique of financing enterprises. It may be argued,

A much more sinister feature of the present situation, however, is that in most developed countries a very great deal of the wealth scooped off by the state goes towards the maintenance of a vast parasitic bureaucracy. It is also true that this bureaucracy is assuming more and more legal authority to make many of the decisions upon which the conduct of business depends, and as a result many serious errors of judgement have been made. Unfortunately, the brunt of these mistakes must be borne by the community at large, and particularly by the business community, and not by the bureaucrats responsible for them; indeed, these officials are not infrequently free to continue in their secure pensionable posts and to go on to make further mistakes in the future while industrialists and workers who suffer from the errors must pick themselves up as best they may and try to rehabilitate themselves. Such a situation can impose a dead weight of inertia upon the society of a developed country and it may produce a general *malaise* which can slowly poison the whole body. This feature and its results have clearly been described by Trevor-Roper:* "Any society, so long as it is, or feels itself to be, a working society, tends to invest in itself: a military society tends to become more military, a bureaucratic society more bureaucratic, a commercial society more commercial, as the status and profits of war or office or commerce are enhanced by success, and institutions are framed to forward it. Therefore, when such a society is hit by a general crisis, it finds itself partly paralysed by the structural weight of increased social investment. The dominant military or official or commercial classes cannot easily change their orientation; and their social dominance, and the institutions through which it is exercised, prevent other classes from securing power or changing policy. If policy is to be changed to meet new circumstances we are more likely to find such a change, in the first instance, either in a complex elastic society—what today we would call a liberal society—in which different interests have separate, competing institutions, or in a less mature society: a society whose institutions have not been hardened and whose vested interests have not been deepened by past commitment." It can hardly be denied that our technological society did in fact develop during a liberal period, and while this made possible considerable abuses of power it also opened the way for a burgeoning of economic development unparalleled in the whole history of the world. It set the stage for the emergence of a society full of vitality, a society which laid the foundations of the wealth which we enjoy today (p. 309 above).

---

however, that the widespread development of this technique is an inevitable result of the prevailing situation under examination.

* In *The Rise of Christian Europe*, Thames and Hudson, 1965, p. 184.

There is little doubt that if the developed territories are to continue to grow in a healthy and vigorous manner a new source of vitality must be found; ways must be sought of throwing up inspiring leadership, of stimulating a "will to work" among employees generally, and of ensuring that individual members of the society are brought to feel that they form a real part of the community. This is not the case at present, and this is largely a corollary of the fact referred to above that with modern methods of manufacture the production of great wealth requires the services of only a part of the work force of the community. Although for convenience in this book we follow standard practice and refer to "developing" and "developed" territories, we must not forget that large numbers of inhabitants in the "developed" countries live under conditions of near penury. It is said, to take one possible case, that a quarter of all U.S. citizens live very close to the starvation line. In the agricultural sphere, for example, field workers in the San Joaquim Valley, California, one of the richest agricultural regions in the world, are among the most poorly paid workers in the whole community, while the plight of farm-workers in the hill country of the southern Appalachians was referred to earlier (p. 228 above). And the difficulty is by no means confined to the countryside. A case in point was mentioned in Chapter X, where it was pointed out that in recent years there has been a redeployment of the automobile industry of the U.S.A. in accordance with the changing demands of the economic situation (this has happened, for instance, as new assembly plants have been established—p. 239). This can be smoothly stated, but the lot of those workers affected by the adjustments is often far from smooth. The closing of the Studebaker factory in South Bend, Indiana, for instance, has thrown thousands of former employees out of work; in this case great efforts are being made to re-educate the dispossessed workers and to equip them with other skills (such efforts are becoming increasingly common in the U.S.A.), but the taking up of a new way of life is not easy for men of advanced years. Chicago used to boast the largest stock-yards in the world, and these yards used to employ thousands of workers, many of them Negroes.* Now these stock-yards are closed and a major city is faced with a colossal unemployment problem. This kind of situation is widely paralleled in the modern industrial world as the cases of New England (pp. 130–134) and Lancashire (p. 215) testify.

But the problem is not simply one of redeployment of labour; it also arises from the fact that with modern efficiency techniques, and as a result of automation greater industrial output can today be achieved with fewer workers than ever before. It is stated that in every year since 1957

* It is said that more Negroes live in Chicago than in any other city in the world.

industrial production in the U.S.A. has increased by 3%, but over the same period the number of jobs available in industry has *decreased* by 775,000. More and more work is now being undertaken by computers (the production of which has revolutionised the industrial structure of New England, p. 132 above), so that fewer and fewer workers are needed. It has been said that sixteen workers are even now able to produce all the electric light bulbs needed in the whole of the U.S.A.! Workers are becoming increasingly "button-pushers," and as the song pertinently enquires, when there are even more advanced machines to press the buttons, "What will happen to you and me?" The United States' Under-Secretary of Labour, Mr John F. Henning, has calculated that by 1970 the U.S.A. will need to provide for its workers $36\frac{1}{2}$ million *new* jobs simply to maintain existing levels of employment. This figure takes into account 24 million present jobs which will be eliminated by automation and an increase of $12\frac{1}{2}$ million in the total labour force which will accrue as a result of population increase.

This, then, is the critical situation which faces the developed countries at the present time; the dead hand of bureaucracy threatens to smother them while too many of their citizens feel as though they are outcasts through no fault of their own. These folk have lost their roots in their own communities and can therefore feel no enthusiasm for the societies in which they live.* At the same time it often seems that progressive leadership is penalised by excessive taxation, and discouraged as more and more initiative is perforce surrendered to the state. One thing is certain; unless ways can be found, and found quickly, of solving these problems the "maturely developed" territories could move into a stage of "senility," with dire consequences for themselves and for the rest of the world. From the economic point of view, the most fundamental necessity is that of evolving a system whereby the wealth produced by a minority can be shared among all members of society—and with increased numbers of labour-saving devices being invented every year this problem is becoming extremely urgent—and at the same time allowing considerable rewards to economic leaders. A partial answer to the problem of sharing will naturally develop as all kinds of tertiary activities expand—for example, people will travel more, they will probably be generally more highly educated, and they will read more literature of various forms—but a full solution to the problem cannot be achieved along these lines. A completely new attitude to work and its rewards may be necessary, and a society which has traditionally enslaved itself to work and its rewards may

* This situation helps in large measure to account for the considerable amount of emigration today from the United Kingdom.

find it necessary to subsidise many of its members who do not work.* It has been suggested that this could be done easily enough by fixing a minimum personal income; incomes which rise above that level could be taxed and incomes which fall below it could be augmented by tax rebates. But this problem is not simply one of economic adjustment; moral, social and religious issues are also involved. Is it morally right to subsidise idleness?† Is it feasible, in an age which is bedevilled by the problem of delinquency (and it is becoming increasingly clear that freedom from economic necessity by no means eradicates delinquency; it may even encourage it, especially among the young and active members of society) to pay people to do no work? Is it possible to substitute comparatively speedily in the human mind some principle other than the belief that work is the most ennobling of all activities, that economic effort is productive of good character, and that material rewards should be related to crude human effort, however imperfectly? And, if it is possible to do this, is it desirable? And what of those members of society who are selected for actual productive effort? How are they to be selected, and is it reasonable from their point of view that they should work to keep others in idleness?

A question of comparable importance is that of evolving a social pattern, in which those with powers of leadership and those with unusual capabilities will be encouraged to develop their talents without frustrations from an entrenched bureaucracy. No society can for long outstrip its leaders, and it is of the utmost importance that those in positions of authority should be men of the highest calibre who are free to lead and who can expect rewards commensurate with their contributions to society.

These are questions not easily answered, but they give some indication of the magnitude of the further economic and social revolution which is vital if the developed countries are to make continued progress. It is surely not beyond the wit of man to devise answers to the problem of surplus work potential, in a world of which so many parts are desperately poor and in need of all the help they can get. It could very well be that the developed countries will find that their own salvation lies along the path of helping others. After all, these developed societies still retain a great deal of liveliness and vigour—a vigour which Professor Geyl has described

* The term "enslaved itself" may seem unduly strong but it can probably be justified when one considers the vast amount of human suffering and misery which has been borne, often unnecessarily by workers in the past—and which is by no means unknown even today.

† It may be argued that idleness is subsidised even today by unemployment benefits of various kinds, but a different principle is here involved. Unemployment benefits are today viewed as temporary expedients tiding workers over periods of unemployment; what is here at issue is a continuing subsidy paid to members of society who may well never expect to work.

as the "vitality of western civilisation."* It remains for this vitality to show itself equal to the tasks ahead.

SUGGESTIONS FOR FURTHER READING

Crawford, Sir J., "Problems of International Trade in Primary Products," *Progress*, vol. 50, no. 297, 1/1964.
Debenham, F., "The Water Resources of Central Africa," *Geographical Journal*, September 1948.
Debenham, F., "Livingstone's Africa," *Geography*, May 1951.
Fryer, D. W., *World Economic Development*, New York, 1965.
Jarrett, H. R., *Africa*, second edition, London, 1966.
Nystrom, J. W., and Malof, P. M., *The Common Market: European Community in Action*, Princeton, N.J., 1962.
Perham, M., *African Outline*, London, 1966.
Reddaway, W. B., "External Capital and Self-help in Developing Countries," *Progress*, vol. 51, no. 286, 4/1965–66.
Sayers, R. S., *The Vicissitudes of an Export Economy: Britain since 1880*, Sydney, 1965.
Trevelyan, G. M., *Illustrated English Social History*, London (Pelican Book,) 1964, vol. 4.

* This vitality is well illustrated by the remarkable achievement of Britain in the post-war period. By the early 1960s the United Kingdom increased the volume of her exports to nearly two and a half times the pre-war figure, while in 1966 she was exporting nearly twice as much per head of population as the U.S.A. and nearly three times as much per head as Japan. Sayers doubts "whether any great industrial nation has ever in peace-time experienced such a tremendous turnover of activity." *See* also footnote, p. 15 above.

Index

ABADAN, 259, 261
Abbey Margam, 100, 185, 187, 243, 292
Aberdare, 288
Abu Dhabi, 259, 260
Adelaide, 106, 108
Africa, 27, 48, 49, 53, 59, 72, 74, 221, 310, 326
Ahmedabad, 221
aircraft industry, 99
Akosombo dam, 61
Alabama, 212
Albertville, 38
Alexander, J. W., 143
Algeria, 31, 260, 292
Aligarh, 153-5
Allegheny Plateau, 56
alumina, 24, 64, 66
aluminium, 24, 61, 64, 66, 80
American Optical Co., 111
Angola, 38
anthracite 52, 54
anti-pollution laws, 74
Appalachians, 208, 212, 227, 337
Arctic Circle, 57
Ardrossan, 193, 204
Argentina, 7, 48, 306
Arkwright, 10, 223, 224
Asia, 53, 64, 326, 327
Aswan, 61, 63, 103, 108, 331
Atbara R., 64
Atlantic Ocean, 7, 14, 41, 185
Australia, 7, 11, 48, 53, 79, 80, 95, 97, 103, 106, 110, 127, 135, 172, 193-204, 239, 254
Austria, 327
Avilés, 139
Avonmouth, 267-9
Avon R., 280

Babbit, 35

backyard furnaces, 189-92
Balsas R., 319
Baltimore, 96, 188
Banbury, 80, 244
Bankstown, 111
Bantry Bay, 252
Barcelona, 139
Barmen, 223
Barnaul, 220
Baroda, 221
Barry, 289
Baton Rouge, 264
bauxite, 24
Beaver, S. H., 214, 289
Beira, 38, 39, 40
Belgium, 144, 223
Bell Is., 129, 130
Bessemer, H., 9, 174, 176, 183, 185, 288
Bethlehem, 186
Bilbao, 137, 138
Billingham on Tees, 294
Birmingham (Alabama), 87, 182
Birmingham (England), 5, 67, 79, 80, 89, 112-13, 242, 243
bituminous coal, 51
Black Country, 183
Blackstone R., 224
Blaenavon, 289
Blisworth, 289
Bolivia, 291
Bombay, 220, 221
Boston (New England), 132, 224
Boulton, M., 5, 12
Bradford, 112
Bramah, 12
Brazil, 172, 310
Bridgend, 293
Brisbane, 98, 158
Bristol, 13
Broken Bay, 134

Broken Hill (N.S.W.), 40, 193, 194
Buchanan, K., 191-2
Buffalo, 57, 86, 240, 241
Bukama, 38
Bulawayo, 59
Burma, 70, 247

Cadiz, 139
Cairo, 108
Calgary, 87
California, 81, 315, 316, 337
Canada, 7, 40, 87, 105, 186, 189, 232, 239, 290
Canadian Shield, 34
Canton, 218
Cape Breton Is., 129
Cape of Good Hope, 249, 250
Cape Town, 38
capital, 19, 41, 101-2, 123, 138, 188, 207, 214, 219, 221, 225, 259, 303, 304, 305, 306, 307, 312, 329, 330
Cardiff, 289, 293
Cartagena, 139
Castle Donnington, 68
Catalan forge, 162
Cawnpore, 221
Central Lowlands (Scotland), 182
Charlotte Dundas, 7
chemical industry (*see also* petrochemical industry), 78, 126
Chesapeake Bay, 187
Cheshire, 126
Chiapas Highlands, 315
Chicago, 17, 18, 19, 29, 86, 99, 100, 187, 241, 337
Chihuahua, 322
Chile, 48, 186, 291
China, 44, 53, 64, 82, 156, 160, 189-92, 208, 215, 218-19, 221, 222, 232
Cleveland, 240, 241, 301
climate, 20, 42, 84, 97-100, 136, 137, 146
clothing industry, 78, 89, 217
Clydach, 290
Clyde R., 7, 214, 215
coal, 23, 48, 49, 51-6, 58, 59, 67, 69, 127, 128, 129, 134, 164, 170, 195, 198, 199, 288 *ff.*, 295-302, 319, 322
Coalbrookdale, 127

coal equivalent, 45
coal gas, 22
coal mining, 53-6, 292, 295-302
Coatzacoalcos, 33
Cockburn Sound, 204
Coimbatore, 221
Colombia, 53, 306
Colorado, 84, 100, 102, 103
Connecticut, 79
copper, 23, 27, 33-4, 40, 58, 234, 290, 291, 319
Copper Belt, 28, 38, 58, 59
Cordoba, 323
Corn Belt, 141, 156
Cort, H., 173
cotton, 9, 10, 11, 14, 15, 23, 136, 206 *ff.*, 218 *ff.*, 228, 321
Cotton Belt, 136, 227
Coventry, 80, 112, 113, 115, 242
Cowley, 243
Cracow, 284
Crawford, J., 332
Crompton, 10
Cwm, 292

Dagenham, 100, 137, 242, 243, 244, 245
Dahran, 75
Dakota, 112
Dallas, 81
Damascus, 8, 161
Damodar scheme, 64
Dangerfield, 187
Darby, A., 165
Dar es Salaam, 38, 40
Das Is., 259, 260
Davidson, B., 13, 14
Debenham, F., 331
Dee R., 126
Deere, John, 17
Delaware, R., 186
Delhi, 221
Denmark, 252, 326
Detroit, 236, 239, 240, 241, 322
Dolgelly, 112
Don R., 280
Dowlais, 288, 289
Drake, Colonel, 72, 247, 253, 254
Duluth, 186
Dundee, 21

INDEX

Dury, G. H., 80, 112, 115, 193
East Midlands, 23, 26, 29
Ebbw Vale, 108, 189, 243, 288, 289, 292
Edison, 57
Egypt, 63, 64, 103, 108, 249, 306
Elberfeld, 223
electricity, 57–69, 70, 75
El Infiernillo, 319
Elisabethville (see Lubumbashi)
entrepreneurs, 84, 312
Erie Canal, 86, 226, 240, 300, 301
Essen, 17, 111
Estall, R.C., 104, 111, 131, 183
Europe, 53, 66, 225, 235, 260, 261, 312, 326, 332
Europort, 252, 264

factory system, 10
Fairless Hills, 183, 186
Fall R., 224
false start, 127–30
Farangbaya, 37
Fawley, 259, 269
Fforest-fach, 293
Fifield, 193
Finnart, 263, 267
fjords, 65
Flint (U.S.A.), 240, 241
Florida, 100
Fontana, 187, 189
footloose industries, 25
Ford, H., 111, 112, 115, 236, 241, 243
Forth and Clyde Canal, 7
France, 9, 11, 12, 48, 79, 218, 223, 232
Frasch process, 32–3
Fraser R., 65
French Revolution, 12
Fryer, D.W., 152, 246
fuel, 19, 44–76, 165, 229, 233, 262, 263

Gary, 18, 100, 187
Gatooma, 59
Geelong, 193
Genoa, 267
Georgia, 239
Germany, 36, 41, 48, 53, 57, 79, 111, 129, 218, 219, 222–3, 232, 238, 267, 290

Geyl, P., 339
Ghana, 57, 61, 62, 98, 103, 262, 329
Ghardaia, 27
Gilchrist Thomas, 9, 176, 178
Gilewska, S., 283–285
Gold Coast (see also Ghana), 102
Gori, 220
Gorseinon, 290, 292
government activity, 20, 102–10, 188–9, 305
Grangemouth, 267
Great Lakes, 41, 42, 99, 143, 240, 300
Greece, 327
Green Belt, 106–7
greenfield sites, 100–1, 113, 114
Green Wedge, 106–7
Grimsby, 93, 124
Guadalajara, 321, 323
Guatemala, 315
Gulf of California, 315
Gulf of Mexico, 32, 90, 136, 258, 315, 318, 319
Gwelo, 59

haematite, 29, 34, 37, 168
Halifax, 112
Hamilton (Canada), 105, 130, 189
Hannover, 107
Hargreaves, 10
Harspranget, 57
Hassi Messaoud, 27
Hawkesbury R., 134
Hebrides, 210
Helwan, 108, 109
Hiner, O. S., 124
Hirwaun, 288, 293
Hobart, 193
Hockley Brook, 5
Honduras, 282
Hong Kong, 223
Houston (Texas), 186, 187
Hudson R., 280, 301
Huelva, 252
Hull, 87
Humber R., 69, 93, 124, 264
Hunter R., 134, 194, 196
Hwang-ho project, 64
Hwang-ho R., 219
Hyderabad, 161

hydrocarbons, 51, 262
hydro-electricity, 46, 60–7, 227, 318, 323

Illawarra (N.S.W.), 82, 134, 202
Illinois, 156
Immingham, 93, 264
India, 11, 44, 53, 64, 82, 160, 208, 215, 218, 220–1, 222, 260, 306, 313, 327, 328
Indiana, 156, 337
Industrial Revolution, 4–18, 19, 152, 289
inertia, 113–15
Iowa, 156
Ipswich, 21
iron, 8–9, 17, 23, 24, 161 ff.
Iron Knob, 193
iron manufacture, 8, 89, 93, 95, 127, 128, 129, 138, 161 ff., 193 ff., 288 ff., 313, 321
Iron Monarch, 193, 194, 196, 204
iron ore, 22, 26, 29, 30, 34–8, 125, 127, 128, 129, 138, 163, 171, 182–5, 193, 204, 241, 319, 322
Irwell R., 96, 280
Isle of Grain, 259
isodapanes, 117–18
Italy, 48, 79, 109–10, 327
Ivanovo, 219, 220

Jamaica, 13, 66
Japan, 48, 82, 106, 208, 210, 213, 218, 219, 220, 221–2, 223, 248, 252, 254, 327
Jebba, 64
Jinja, 103
joint supply, 122
Jones, E., 9
Junner (Dr), 35
jute, 21

Kaduna, 114
Kafue R., 59
Kainji, 64
Kapiri Mposhi, 39
Karachi, 221
Kariba scheme, 57–60, 61
Karlsruhe, 267
Kasese, 38, 231

Katanga, 37 38, 310
Kay, John, 9, 152, 213
Kemano, 65
Kenney Dam, 65
Kentucky, 54, 298
Kerr, D., 130
Keta Krachi, 61
Kigoma, 38, 40
Killingworth Colliery, 6
Kinshasa, 38
Kirunavaara, 29, 40
Kitimat scheme, 64–6, 67, 80
Knowles, L., 11, 12
Kpong, 61
Krupp, 111
Kuwait, 253
Kwinana, 193, 202, 204

labour, 19, 22, 40–1, 69, 77–85, 90, 114, 122, 136, 208, 214, 216, 219, 221, 225, 243, 260, 265, 303, 307, 310–12
Labrador, 29
Lake District, 283
  Erie, 42, 241, 301
  Huron, 186
  Michigan, 42, 100, 187
  Peninsula, 130
  Superior, 26, 30, 35, 104, 186, 241
  Tanganyika, 38, 40
  Volta, 61, 62
Lanarkshire, 183
Lancashire, 23, 82, 89, 93, 97, 114, 115, 125, 126, 208, 210, 212, 214, 215, 223, 243, 245, 281, 337
Lancing, 241
Lark R., 281
Lavera, 267
Lawrence, 134, 224
Leeton, 158
Leningrad, 220
Leopoldville (see Kinshasa)
Lewis, W. A., 83, 308
Liberia, 282
Libya, 260, 264
lignite, 48, 51, 52
Lille, 223
limonite, 23, 26, 29, 37, 168, 169
linkage, 89, 123–5, 132, 200, 223
Lisbon, 279

INDEX 345

Lithgow, 129, 196
Liverpool, 6, 93, 244
Liverpool and Birmingham Railway, 6
Liverpool and Manchester Railway, 6
Llandarcy, 267
Llanelly, 290, 292
load on top system, 282
Lobito, 37, 38
locational triangle, 118–19
location of industry, 19, 25
London, 75, 78, 80, 87, 243, 244, 249, 259, 264
long run, 115
Lorraine, 23, 86
Los Angeles, 187, 189
Loughor R., 290
Louisiana, 32, 264
Lourenço Marques, 38
Lowell, 134, 224
Lower California, 314, 315
Lubumbashi, 38
Lula R., 57
Lunsar, 35, 37, 42
uton, 242, 243, 244, 245
Lyme Bay, 264

Madras, 221
Madrid, 139
Maerdy, 292
magnetite, 29, 34, 35, 168
Maiduguri, 331
Malaysia, 291
Malawi, 40
Malta, 79
management, 20, 83–5, 114, 219, 221
Manchester (England), 6, 208, 213, 220, 283
Manchester (New England), 224
Marampa, 35, 37
markets, 19, 41, 77, 88–91, 186–7, 213, 217, 225, 261, 265, 267
Marseilles, 267
Martin, 9, 174
Marulan, 196
Maryland, 187
Massachusetts, 104, 111, 229
Masser, I. F., 310, 311
Matadi, 38
Maudsley, 12

Mediterranean Sea, 261
Melbourne, 106, 158, 193
Merrimack R., 134, 224
Mersa el Brega, 264
Mersey R., 93, 280
Merthyr Tydfil, 288
Mesabi ores, 29, 30, 41, 42
Mexico, 32, 33, 83, 306, 314–25
Mexico City, 315, 321, 322, 323
Mezzogiorno, 109–10
Michigan, 111, 112, 115, 239
Mid-West, 16, 87, 99, 133, 157, 226, 240, 241, 242
Miles, William, 13
Milford Haven, 80–1, 252, 263, 264, 267
Milwaukee, 18, 240, 241
Minatitlan, 33, 320
mineral oil, 27, 46, 48, 50, 70–5, 137, 246 ff., 296
Minnesota, 29, 34, 42, 105, 156
Mississippi R., 265, 300
Missouri (State), 156
Mittagong, 127, 128, 196
Mogul Empire, 11
Mohave Desert, 314
Moisley, H. A., 126
Mombasa, 38
Monclova, 322, 323
Monongahela R., 55, 300
Monterrey 321, 322, 323
Montreal, 87, 130
Morgan, 203
Moscow, 220
Moss Vale, 134
Mount Isa, 40
Mountjoy, A.B., 108, 304, 312
Mpulungu, 40
multiplier, 308
multi-purpose project, 64
Murray R., 203
Murrumbidgee Irrigation Area, 158–9

Nagoya, 221
Nagpur, 221
Nancy, 86
Nantargw, 292
Nantyglo, 288
Nasmyth, 12
natural gas, 48, 71, 73, 74, 137, 273

346 INDEX

Naugatuck Valley, 79
Neath R., 290
Nechako R., 65, 66
Neilson, 91
Netherlands, 72, 223
Newark, 81
New Bedford, 224
Newcastle (N.S.W.), 80, 82, 128, 135, 170, 193, 194 ff.
Newcomen, 5, 11, 33
New England, 15, 16, 23, 84, 89, 90, 104, 111, 130–4, 136, 137, 201, 208, 212, 214, 216, 224–9, 236, 240, 242, 334, 337, 338
Newfoundland, 129
New Jersey, 239, 301
New Orleans, 16
Newport (Mon.), 289
New South Wales, 80, 82, 111, 127–9, 134–5, 158–9, 170, 187
New York, 6, 57, 75, 78, 86, 224, 278, 280
Niagara Falls, 57
Nielson, 9, 165
Nigeria, 64, 114, 253, 291, 331
North America, 47, 49, 53, 74, 83, 193, 258, 312, 326
Northern Rhodesia (*see also* Zambia), 27, 58
Northern Territory, 103, 104, 110
North Sea, 27, 261, 273
Nova Scotia, 129, 137, 186
Nowa Huta, 108
nuclear power, 46–7

O'Connor, A. M., 231
Ohio R., 300
Oil Creek, 253, 254
oil pollution, 281–2
oil refining, 78, 256–73, 313
oil tankers, 72–3, 248–51, 263–4, 267
Oklahoma, 187
Ontario, 105, 129
opencast mining, 26
Orizaba, 321
Osaka, 221
Oviedo, 137
Owen Falls, 57, 103
Oxford, 111, 112, 115, 242, 243, 244, 245

Pacific Ocean, 17, 259, 315, 318
Pakistan, 44, 260
pampa, 7
Panama, 282
Panama Canal, 87
Paricutin, 315
Paris, 78
Passaic R., 224
Patapsco R., 96, 187
Paterson, 224
Pawtucket, 224
Pebworth, 153
pelletising, 35, 171, 203
Pennines, 93, 212
Pennsylvania, 29, 54, 55, 70, 186, 241, 247, 251, 253, 257, 258, 297
Pepel, 36
Persian Gulf, 249, 259
petrochemical industry, 267–70, 324
Philadelphia, 86
Philbrick, A. K., 17
phosphate, 31
Pittsburg, 42, 53, 81, 114, 300
Pittsburg Plus system, 87
Pittsburg Seam, 54, 56
Ploeşti, 247
Poland, 48
pollution, 96, 278–82, 284
Pontiac, 241
Pontypool, 244
Pontypridd, 292
Popocatepetl, 315
Port Augusta, 204
 Francqui, 38
 Jackson, 134
 Kembla, 79, 128, 134, 135, 158, 170, 187, 193, 196 ff.
 Pirie, 193, 194
 Stephens, 134
 Talbot, 100, 185
Portuguese East Africa, 38
Pounds, N., 141, 142, 161, 164, 173, 189
power, 4–5, 19, 44–76
prairies, 16, 17, 18, 129, 143
Preston, 126
Prince George, 65
Providence, 224
Puebla, 321

INDEX

Quebec, 29, 129
Queensland, 40, 98
Que Que, 59
Quintana, E. F., 138, 140

Rand, 40, 95, 310
Rapid Bay, 193, 204
raw materials, 19, 21–43, 135, 138, 154, 168–70, 182–6, 241, 332
Rawstron, E. M., 69
Reading, 112
Reddaway, W. B., 47, 330
Rheinau, 67
Rhineland, 163
Rhine R., 67, 222, 261, 264, 267
Rhode Is., 224
Rhodesia, 38, 39, 58, 59, 262, 310
Rhondda, 292
Ribble R., 115, 280
Rio Blanco, 321
Roan Antelope Mine, 28
Romania, 70, 247
Rossendale, 208, 213
Rostow, W. W., 304, 306
Rotterdam, 264, 267
Rugby, 89
Ruhr, 111, 141, 142, 183, 222, 223, 261, 267

Sabinas Coalfield, 322
Sahara Desert, 27, 267
St Helens, 126
St Lawrence R., 41, 130
St Lawrence Seaway, 41, 86
St Mary R., 42
Salisbury (Rhodesia), 59, 310
Saltillo, 323
Salton Sea, 315
Salzgitter, 41, 108, 189
San Joaquim R., 337
Saragossa, 139
Saudi Arabia, 75, 259
Savannah, 7
Savery, 5
Sayers, R. S., 17
Scargill, D. I., 112
Schefferville, 29, 40, 41, 129
Scunthorpe, 185
Seven Isles, 42

Severn R., 269, 280
Seville, 139
Shanghai, 218, 219, 221
Sheffield, 17, 18, 79, 278
Shipbuilding, 98, 133, 203, 204
short run, 115
Shotton, 93, 243
siderite, 169
Siegerland, 162, 164
Siemens, 9, 174
Sierra Leone, 35–7
Sierra Madre, 315, 318
Silesia, 283–5
Silver Bay, 35
Slater, S., 223, 224
slave trade, 13, 14
Smith, Andrew, 33
Smith, W., 22, 25
Soho, 5, 12
Sonora Desert, 314
Soo Canals, 42
South Africa, 38, 40, 48, 53, 97, 259, 310
South Amboy, 301
South America, 7, 48, 53
Southampton Water, 263
South Bend, 337
Southbridge, 111
Southern Rhodesia (*see also* Rhodesia), 310–11
South Wales, 187, 245, 288–94
Spain, 83, 137–40, 151, 174, 292, 306, 312, 328
Sparrows Point, 96, 183, 186, 187
Special Areas, 291
Spencer Gulf, 204
spiegeleisen, 185
steel, 8, 9, 18, 87, 93, 94, 96, 100, 104, 105, 108, 109, 111, 113, 114, 125, 129, 130, 138, 139, 154, 160, 161, 162, 173–82, 183, 185, 186, 187, 188, 189, 190, 192, 193 *ff.*, 207–9, 234, 241, 243, 244, 251, 284, 289, 292, 321–2, 330
Stephenson, George, 6
Stockton and Darlington Railway, 6
Strasbourg, 267
Stratford-upon-Avon, 289
Styria, 162, 163

Suez Canal, 7, 221, 249, 250
sulphur, 32, 33, 129, 164, 288, 319, 320
Swansea, 289, 293
Sweden, 29, 40, 57, 99, 162, 164, 169, 292
Sydney (Nova Scotia), 129, 130, 137, 186
Sydney (N.S.W.), 111, 128, 134, 135, 158, 193, 197

taconite, 29–30, 34–5, 105
Tame R., 67, 279
Tampico, 319
Tanzania, 39, 40, 262
Tashkent, 220
Tawe R., 290
taxation, 29, 103–6, 136, 186, 225, 229, 305, 335
Tehuantepec, 33, 315, 320
Tees R., 183, 185
Tema, 61
Tennessee (State), 81
Tennessee Valley Authority (T.V.A.), 136, 318
Texas, 32, 33, 186, 187, 239
textile industry 9–11, 15, 82–3, 89, 114, 115, 125, 131, 133, 136, 137, 206–30, 321
Thames R., 100, 244, 279
thermal electricity, 60, 67–9, 78, 296
Tientsin, 218
Tijuana, 323
Titusville, 70, 247, 251, 256
Tokyo, 221
Toledo (Spain), 8, 161
Toledo (U.S.A.), 240
Toluca, 322
Tonkolili, 37, 38
Toronto, 130
Towcester, 289
Townsville, 98
transfer costs, 85–6
transport, 5–7, 23, 35–40, 77, 85–8, 117, 118, 158, 164, 239, 248–53, 259, 331
Treforest, 293
Trent R., 68, 69, 279, 280
Trevelyan, G. M., 152
Trevor-Roper, H., 336

triangular trading system, 13, 14, 15
Trostre, 292
Troy, 111, 112
Tsingtao, 218
Tunisia, 31
Turkey, 306
Tuxpan, 319
Tyne R., 280, 281

Uganda, 38, 57, 86, 103, 231
Ullswater, 283
Umm Shaif, 259
United Kingdom, 9, 48, 53, 70, 79, 108, 129, 135, 137, 210, 215, 217, 218, 223, 229, 232, 235 ff., 256, 260 ff., 267 ff., 278, 281, 306, 329, 333
Ural Mts., 108, 220
uranium, 46, 49
Uruapan, 315
U.S.A., 7, 9, 15, 16, 17, 29, 34, 35, 36, 41, 48, 66, 74, 75, 77, 79, 81, 82, 84, 85, 86, 87, 90, 92, 93, 96, 97, 98, 100, 104, 111, 114, 130, 131, 132, 133, 134, 135, 155–7, 160, 178, 185, 186, 187, 210, 213, 218, 219, 223–30, 232, 234, 235 ff., 252 ff., 291, 292, 295 ff., 306, 314, 316, 319, 320, 321, 322, 332, 333, 337, 338
U.S.S.R., 53, 218, 219–20, 232, 306
Utah, 84, 99, 168

Vancouver, 87
Velindre, 292
Venezuela, 72, 74, 186, 260, 262, 306
Veracruz, 320
Vistula R., 284
Volta River Project, 57, 61, 103

Wabana, 130
Wadi Halfa, 63
wages, 77–8, 81–2, 97, 208, 216, 222, 225, 229, 307
Wankie, 58
Waratah, 196
Warren, K., 17
waste products, 68, 96, 188, 257–8, 277–8, 283, 299
water power, 45, 46, 49, 212

water supply, 20, 67, 92–7, 187–8, 197, 203–4, 264–5
Watt, J., 5, 6, 11, 12, 13, 19
Weald, 163, 164
Weber, A., 22, 116, 117
West Africa, 11, 13, 103, 104, 152, 186, 215, 292
West Pennsylvania, 54, 56
Westphalia, 222
West Virginia, 54, 56, 212, 241
wheat, 7
Whiddy Is., 252
Whyalla, 80, 193, 202–4
Wigan, 126
Wilhemshaven, 267
Windsor (Canada), 130
Wise, M.J., 5
Wollongong, 135
Wolverhampton, 243

wood (as fuel), 45, 48, 49, 59, 163, 164
wool, 7, 11, 23, 131, 206
wootz steel, 161
Wuppertal, 223
Wupper Valley, 223

Yampi Sound, 193, 196
Yangtze Kiang, 219
Yawata, 189
Yeovil, 21
Yokohama, 221
Yorkshire, 23, 93
Youngstown 42, 114
Ystwyth R., 280
Yuan-Li Wu, 190, 191
Yucatan, 314, 315, 323

Zambia, 27, 28, 38, 39, 58, 59, 310

This b

~~16 DEC 19~~
~~MAR~~